Theory and Simulation Experimental Technology of Oil Generation From Nonmarine Source Rocks in China

Defan Guan · Zhongliang Ma · Yongqiang Zhao

Theory and Simulation Experimental Technology of Oil Generation From Nonmarine Source Rocks in China

石油工业出版社有限公司
PETROLEUM INDUSTRY PRESS

Springer

Defan Guan
Sinopec Petroleum Exploration
and Production Research Institute
Beijing, China

Zhongliang Ma
Sinopec Petroleum Exploration
and Production Research Institute
Beijing, China

Yongqiang Zhao
Sinopec Petroleum Exploration
and Production Research Institute
Beijing, China

ISBN 978-981-96-3144-5 ISBN 978-981-96-3145-2 (eBook)
https://doi.org/10.1007/978-981-96-3145-2

Jointly published with Petroleum Industry Press
The print edition is not for sale in China (Mainland). Customers from China (Mainland) please order the print book from: Petroleum Industry Press.

This Springer imprint is published by the registered company Springer Nature Singapore Pte Ltd.
The registered company address is: 152 Beach Road, #21-01/04 Gateway East, Singapore 189721, Singapore

Preface

Petroleum can be divided into two categories: natural petroleum and artificial petroleum. Natural petroleum is further classified into marine oil and nonmarine oil based on the sedimentary environment.

The concept of "artificial petroleum" originated in Europe, particularly in the United Kingdom. In 1694, an Englishman named Martin Eyre obtained a patent (UK Patent No. 330) for extracting bitumen, tar, and other petroleum products from oil shale through distillation. This marked the beginning of the production of "artificial petroleum". European organic chemists later developed the oil shale pyrolysis retort, which enabled mass production of "artificial petroleum". They discovered that the oil precursors in oil shale, known as kerogen, were insoluble in organic solvents. They also found that kerogen extracted from any kerogen-rich rock could be pyrolyzed to produce "artificial petroleum". Based on extensive production practices, European organic chemists established the concept of "artificial petroleum" on the principle of material synergistic reaction, which means that the reactants directly produce the final products without any intermediate products.

Espitalié et al. (1977) developed a Rock-Eval pyrolysis instrument that allowed for the extraction of kerogen from source rocks without prior treatment. This instrument heated a 100 mg sample of source rock to approximately 550 °C, causing all the kerogen present to be pyrolyzed and converted into petroleum. This breakthrough led to the generation of four main parameters (S1, S2, S3, Tmax) that could be used to evaluate the hydrocarbon potential of source rocks. The Frenchman B. P. Tissot and German D. H. Welte took notice of this Rock-Eval pyrolysis instrument and its experimental data, leading them to propose that both source rock and oil shale can generate oil. They suggested that when buried shallow, it should be called oil shale, and when buried deep, it should be referred to as source rock. In their 1978 book "Petroleum Formation and Occurrence—New Approaches to Oil and Gas Exploration", Tissot and Welte emphasized the theory of oil generation by kerogen thermal degradation, which involves the thermal degradation of source rocks to generate oil. The parameters obtained through thermal degradation experiments using the Rock-Eval pyrolysis instrument, namely S1, S2, S3, and Tmax, can be used to estimate the oil (gas) generation potential of source rocks. This laid the foundation for the emerging basin simulation technology, which allows for the quantitative evaluation of oil (natural gas) prospects.

However, the ideas proposed by Tissot and Welte equate the oil-generating depression with a retort pyrolysis furnace that continuously pyrolyzes oil shales over long geological timescales, transforming all organic matter into oil. This understanding is not only inconsistent with reality but also self-contradictory. Tissot and Welte acknowledged in their book that only a portion of the organic matter in source rocks can generate oil through long-term pyrolysis (S1), while the remaining portion remains as residual organic matter (S2) that cannot be pyrolyzed to generate oil. This indicates that the residual organic matter will never undergo pyrolysis to produce oil. Hence, it becomes questionable how practical it is to use S1 + S2 to evaluate the oil and gas generation potential of source rocks.

Furthermore, Tissot and Welte neglected a crucial concept known as "media" or "intermediates". In organic chemistry, reactants require intermediates to facilitate the completion of chemical reactions and the production of final products. Unfortunately, these two scholars failed to incorporate this aspect into their theoretical methodologies. Within subsurface geological conditions, if the oil (S1) produced from black mudstones does not undergo the necessary intermediates, it cannot be expelled and will remain dispersed within the source rocks. In the case of nonmarine source rocks in China, these intermediates typically consist of sandstones, as natural source rocks usually comprise a combination of mudstones and sandstones. This is why, in the extraction of tight oil, additional measures such as hydraulic fracturing and the injection of quartz sand or ceramsite into fractures are necessary. These measures serve to enhance the diversion capacity and promote efficient oil extraction.

The United States has been at the forefront of marine oil exploration and development. In fact, as early as 1859, it achieved a significant milestone by successfully drilling the first oil well. By 1923, annual marine oil production had reached 100 million tons, and it further rose to 455 million tons by 1970. Through years of practical experience in petroleum exploration and development, American petroleum geologists have come to realize that the theoretical research on oil generation in source rocks is a highly complex organic geochemical problem. It encompasses various disciplines such as petroleum geology, mineralogy, sedimentology, paleontology, organic chemistry, fluid mechanics, and more, along with related experimental and technical methodologies.

Some petroleum geologists in the United States even acknowledge that until the very last drop of oil is discovered, questions regarding the generation, migration, accumulation, and preservation of oil cannot be fully answered. Consequently, major oil companies in the United States focus extensively on finding more oil reserves and developing techniques to extract as much oil as possible. This explains why seismic surveying technology, drilling engineering, oil well fracturing methods, equipment, and other applied technologies in the U.S. consistently maintain an advanced global standing.

However, it is worth noting that petroleum geologists like J. M. Hunt have made significant contributions to fundamental theoretical research, particularly in relation to the origin and migration of marine oil. Hunt believes that the percentage of organic matter in source rocks that transforms into oil typically ranges from a few percent to 15%. Additionally, he suggests that the same percentage range of oil ultimately forms economically viable oil reserves. Experimental data supports the conclusion that in subsurface geological conditions, the process of pyrolysis, which converts organic matter into oil, occurs under the influence of catalytic cracking at low temperatures (50–200 °C).

It should be emphasized that J. M. Hunt was the first to propose the concept that "source rock should be a type of rock that forms under geological conditions, capable of generating oil and expelling oil in sufficient quantities to create commercial accumulations." This idea was put forth in his book "Petroleum Geochemistry and Geology" in 1979. Hunt's proposition revolutionized the assessment of source rocks by shifting the focus from solely relying on experimental data regarding rock thermal degradation and oil generation to a method based on the dynamic changes of source rocks in subsurface geological conditions, known as organic geochemical methods. This marked an important step towards accurately defining and evaluating the characteristics and processes involved in oil generation from source rocks.

China has a rich history in nonmarine oil exploration and development. It all started in 1907 when the first oil well was drilled in Yanchang County, Shaanxi Province. The significance of nonmarine oil generation in China was officially recognized by Pan Zhongxiang, a Chinese petroleum geologist, in 1941 at the meeting of the American Association of Petroleum Geologists (AAPG).

After the establishment of the People's Republic of China, there was a strong focus on large-scale exploration and development of nonmarine oil. This led to the discovery of major oilfields such as Karamayi, Daqing, Shengli, and others, with reserves exceeding 100 million tons. By 1978, China had become the largest producer of nonmarine oil globally, surpassing annual production of 100 million tons. These achievements in nonmarine oil exploration and development greatly boosted the confidence of Chinese petroleum geologists and inspired them to contribute to the theoretical research and practical production of nonmarine oil in China. However, due to historical reasons, China had been closed off for a long time, limiting the exposure of Chinese petroleum geologists to the latest advancements in source rock research.

Until the reform and opening of China in 1978, Chinese petroleum geologists relied heavily on the textbook "Petroleum Geology" by American petroleum geologist A. I. Levorsen for their understanding of source rocks. Their methods were primarily focused on evaluating a few experimental parameters in the organic matter deposition environment, such as ferrous iron and reduced sulfur. In this context, European organic chemists B. P. Tissot and D. H. Welte recognized an opportunity and came to China in 1978 to give lectures and exchange knowledge with senior Chinese petroleum geologists. They presented their monograph "Petroleum Formation and Occurrence—New Approaches in Oil and Gas Exploration," which had not gained much popularity in the oil geology community in Europe and America. The book introduced the concept that Rock-Eval pyrolysis instruments could provide quantitative evaluations of oil and gas prospects in basins based on the pyrolysis parameters of source rocks. It also proposed a new approach to oil and gas exploration using mathematical models and geochemical data.

Chinese petroleum geologists showed great interest in these ideas and accepted them without critical analysis. The theoretical and experimental methods presented in the book were widely adopted and applied in scientific research, teaching, and production. Unfortunately, this led to more than 40 years of research on nonmarine source rocks in China being influenced by the theory of "artificial petroleum" and its experimental methods. Many talented individuals in China conducted extensive research in the field of nonmarine petroleum using the theory and experimental techniques of "artificial petroleum." Numerous books were written analyzing source rocks from different geological ages in various oil-bearing basins in China using Rock-Eval pyrolysis instruments. However, these laboratory reports and experiments lacked theoretical significance or practical application value.

Since 2003, the authors and their research team have been dedicated to changing the research field of nonmarine source rocks in China and challenging the prevailing theory and experimental techniques of "artificial petroleum" that have caused confusion for a long time. With the support and assistance of the Science and Technology Development Department of China Petrochemical Corporation, they conducted in-depth research on the development characteristics of source rocks in typical oil-bearing basins in China. Through comparative analysis, the team developed an oil generation simulation experiment instrument that accurately simulates the subsurface conditions of source rocks. The focus of their research was to utilize this instrument and employ organic geochemical thinking methods and simulation experiment technology to study the oil generation theory and development of nonmarine source rocks in China. Their goal is to effectively guide petroleum exploration by addressing key issues in this field.

It is important to note that for many years, the research on oil generation in source rocks has been influenced by the theory of kerogen thermal degradation. However, it was discovered that under subsurface geological conditions, the oil-generating materials produced through pyrolysis are not the kerogen contained in the source rocks, but rather the soluble organic matter S1 present within them. Therefore, understanding the formation and evolution characteristics of this soluble organic matter S1 becomes crucial for the theoretical research on oil generation in source rocks.

The authors firmly believe that the innovation and development of the theory of oil generation in source rocks is the result of continuous exploration and summarization by experts, representing the collective wisdom of the field. They express gratitude to all petroleum geologists involved in this research, with special thanks offered to A. I. Levorsen, H. D. Hedberg, Pan Zhongxiang, J. M. Hunt, B. P. Tissot, D. H. Welte, R. E. Chapman, F. K. North, and other senior petroleum geologists. Furthermore, they have high expectations for the younger generation of petroleum geologists who are engaged in the theoretical research of source rocks.

Beijing, China
January 2022

Defan Guan

Contents

Part I
Introduction

Chapter 1
Two Ways of Defining and Evaluating Oil-Prone Source Rocks

Abstract This chapter delves into two key approaches: the kerogen thermal degradation theory of petroleum generation and the conceptual framework of organic geochemistry.

The kerogen thermal degradation theory traces its origins to European organic chemists who pioneered the extraction of "synthetic oil" through oil shale retorting. Grounded in the principles of chemistry, the Rock-Eval pyrolysis method was developed to assess the potential and yield of industrial "synthetic oil" generation from oil shale. However, applying these theoretical and experimental techniques to evaluate natural petroleum generation neglects the geological complexities of subsurface environments. In contrast, Hunt, an American scientist, advocated for a closer integration of organic chemistry with petroleum geology. He emphasized the use of organic geochemistry methodologies to address fundamental questions related to source rocks, such as petroleum migration and accumulation. Despite its promise, this approach has yet to establish a comprehensive research methodology or develop specialized simulation instruments for experimental studies.

1.1 Thermal Degradation of Kerogen—An "Artificial Petroleum" Generation Theory

During the eighteenth century European Industrial Revolution, organic chemists in Europe utilized a process known as dry distillation to extract oil and related products from oil shale. This extraction process was referred to as "artificial petroleum." Oil shale consists primarily of kerogen, which is insoluble in organic solvents, and a small amount of bitumen, which is soluble. To generate oil from kerogen, it needs to be heated to approximately 550 °C, while bitumen only requires heating to around 250 °C. European organic chemists developed a Rock-Eval pyrolysis instrument based on the principles of organic chemistry's synergistic reactions. This instrument allowed them to define and evaluate the total amount of "artificial petroleum"

D. Guan et al., *Theory and Simulation Experimental Technology of Oil Generation From Nonmarine Source Rocks in China*, https://doi.org/10.1007/978-981-96-3145-2_1

produced through the pyrolysis of organic matter in oil shale (or other kerogen-rich rocks). By employing a specific temperature program, approximately 100 mg of the sample were gradually heated to 550 °C in an inert gas environment. Throughout the heating process, soluble organic matter was initially vaporized at a moderate temperature, and the quantity of these hydrocarbons was measured using a flame ionization detector (referred to as S1).

As the pyrolysis temperature increased, kerogen underwent pyrolysis to produce hydrocarbons (S2), as well as oxygen-containing volatiles such as carbon dioxide and water (S3).

The fourth parameter, known as Tmax or peak temperature, corresponds to the temperature at which the maximum hydrocarbon generation occurs during pyrolysis. To evaluate the rock, the data obtained from these four main parameters (S1, S2, S3, and Tmax) for each pyrolyzed rock sample are analyzed.

The quantity of S1 represents the portion of the original oil and gas generation potential that has been effectively converted into hydrocarbons, specifically bitumen A. S2 indicates the oil and gas generation potential of kerogen, which is the potential yet to be converted into hydrocarbons. Therefore, the sum of S1 and S2 represents the ultimate oil and gas generation potential of the rock, measured in kilograms per ton of rock (kg/t). S3 represents the oxygen-containing volatile compounds, carbon dioxide, and water contained in the rock. Tmax is primarily used to assess the thermal maturity stage of the rock. Additionally, the ratio of S2 to total organic carbon (TOC) is known as the hydrogen index (HI), while the ratio of S3 to TOC is referred to as the oxygen index (OI). These two indexes can help determine the type of kerogen present. In summary, by analyzing the aforementioned parameters, it is possible to estimate the oil and gas generation potential of rocks. The evaluation of oil and gas generation potential mentioned here refers to the total amount of oil and gas produced by the pyrolysis of all organic matter contained in the rocks, represented by S1 + S2 (kg/t).

In 1977, Frenchman B. P. Tissot and German D. H. Welte proposed the theory of kerogen thermal degradation oil generation based on the acquisition and interpretation of experimental parameters. This marked the first application of chemical thinking and laid a theoretical foundation for the production of "artificial petroleum" in the oil shale industry.

In 1978, Tissot and Welte collaborated on the publication of "Petroleum Formation and Distribution—A New Approach to Oil and Gas Exploration." This book provided a comprehensive analysis of oil shale and source rock, highlighting their similarities. For instance, they stated that the kerogen found in oil shale is not significantly different from that in source rocks. They also noted that the pyrolysis process used to produce "artificial petroleum" from kerogen is similar to the generation of petroleum from deeply buried source rocks, as both processes occur at high temperatures and yield similar results. Furthermore, Tissot and Welte classified rocks containing rich organic matter of type I or type II kerogen as good source rocks if they have sufficient burial depth. In contrast, they considered oil shale as a type of source rock with specific characteristics. They emphasized that there is no substantial difference between the oil-generation potential of a source rock and the total amount of oil and gas generated during thermal cracking. In their view, a shallow immature source rock is fundamentally similar to oil shale. The theoretical perspective presented by Tissot and Welte equates source rocks with oil shale, with the only distinction being the depth of burial. This viewpoint allows for the transfer of pyrolysis methods, evaluation parameters, and calculation formulas used in oil shale research to the evaluation and application of source rocks in oil exploration (as shown in Table 1.1).

It is evident that the two scholars conceptualized the oil-generating depressions, where source rocks are located, as pyrolysis and retort furnaces for oil shale. Within these depressions, oil generation and expulsion processes continuously occur. By calculating the amount of oil generated by the source rock, one can determine the total oil potential resulting from the pyrolysis of all organic matter contained within the source rock. This total oil potential can be seen as a typical experimental organic chemical reaction, described as follows: the source rock (represented as rock t) is heated to 550 °C, resulting in the formation of S1 and S2 (expressed in kg). Thus, based on the total oil generation represented by S1 + S2, it becomes possible to evaluate the oil potential of the sedimentary basin or the oil-generating depression within the basin. The book "Petroleum Formation and Occurrence—New Approaches to Oil and Gas Exploration" discusses various formulations such as the "mathematical model—a quantitative approach to oil and gas prospective evaluation," the "application of mathematical models in petroleum exploration," and "determining the most favorable oil accumulation zone in the basin: a geochemical approach to prospect assessment." These formulations have sparked considerable interest among petroleum geologists.

However, there can be misunderstandings regarding certain concepts, particularly concerning the value of S2. Both scholars argue that S2 represents "residual organic matter that has not previously generated oil" within rocks. This residual organic matter can be further subjected to pyrolysis using Rock-Eval pyrolysis instruments or carbonization furnaces to produce "artificial petroleum." By completely pyrolyzing oil shale, the sum of S1 + S2 can be used to calculate the total amount of "artificial petroleum" generated. However, when considering source rocks, the residual organic matter has not undergone pyrolysis to generate oil under subsurface geological conditions. This implies that it will never undergo pyrolysis to generate oil. Therefore, the value of S2 is effectively zero, and the sum of S1 + S2 equals S1. In other words, only S1 can be used to represent the oil generation potential of source rocks. The pyrolysis experiment method in the context of internal organic chemistry aims to obtain the value of S2 through source rock pyrolysis. However, this value does not hold significance for hydrocarbon potential evaluations. Tissot and Welte, both esteemed petroleum geologists, have gradually shifted their focus away from petroleum geology and leaned more towards organic chemistry thinking. Over the years, they have become proficient in employing organic chemistry methodologies to study source rocks. However, it is evident from their monographs that they lack a proper understanding of the principles and applications of organic geochemistry. Consequently, they struggle to grasp how to utilize organic geochemistry thinking and experimental techniques to investigate the theory of oil generation from source rocks. In their works, they mistakenly interchange organic chemistry with organic

Table 1.1 Comparison of research methods on oil shale and source rock

Term	Oil shale	Source rock
Pyrolysis conditions	Final heating to 500–550 °C in the absence of air	Gradual heating to 550 °C under inert gas atmosphere
Selection of pyrolysis temperature range	0–350 °C: dehydration and drying stage, when the temperature reaches 200 °C, asphalt, water, and gas are generated; 350–500 °C: generating into a large amount of shale oil; 500–550 °C: final temperature of pyrolysis	0–350 °C: dehydration and drying stage, when the temperature reaches 200–350 °C, free hydrocarbons S1 are generated; 350–500 °C: crude oil, wet gas S2 and oxygen-containing volatiles S3 are generated; 500–550 °C: generating dry gas
Evaluation index of organic matter abundance	TOC content, oil generation potential (S1 + S2), bitumen "A", kerogen type, organic matter maturity (Ro), maximum pyrolysis peak temperature, H/C, O/C, hydrogen and oxygen indexes	TOC content, oil generation potential (S1 + S2), bitumen "A", kerogen type, organic matter maturity (Ro), maximum pyrolysis peak temperature, H/C, O/C, hydrogen and oxygen indexes
Quality evaluation standards for oil shale and source rocks	Oil shale quality evaluation is based on oil content (oil content refers to the mass fraction of shale oil in oil shale): 1 low grade, $3.5\% < w \leq 5\%$; 2 medium grade, $5\% < w \leq 10\%$; 3 high grade, $w > 10\%$	Source rock quality evaluation based on S1 + S2 (hydrocarbon/rock): 1 non-source rock, less than 2 kg/t; 2 medium source rock, 2–6 kg/t; 3 good source rock, more than 6 kg/t
Oil shale and source rock oil generation calculation formula	Qshale oil = ShρW In the formula, Qshale oil is the amount of generated shale oil, 10^4 t; S is the distribution area of oil shale, km^2; h is the thickness of oil shale; m, ρ is the density of oil shale, t/km^3; W is shale oil content, %	-8 Qoil = 10 S hρCrRcHr In the formula, Qpetroleum is the oil generation rate of source rocks, t; S is the distribution area of effective source rocks, km^2;h is the average thickness of effective source rocks; m, ρ are the density of source rocks, t/km^3; Cr is residual TOC content of source rock, %; Rc is organic carbon recovery coefficient; Hr is oil production rate of source rock at this evolution stage, kg/t

Note The data of oil shale is from Liu Zhaojun (2009); the calculation formula of oil generation from source rock is from Zhao Wenzhi (1999); other source rock data are from Tissot and Welte (1978)

geochemistry, confuse "artificial petroleum" with natural petroleum, mix up the theory of thermal degradation of kerogen with the theory of thermal degradation of source rocks, and fail to differentiate laboratory experiments of organic chemical pyrolysis from organic geochemical simulation experiments. Unfortunately, this confusion has led them to misinterpret pyrolysis experiments in the context of oil generation from source rocks. To elaborate further, the following aspects illustrate the specific areas where these misunderstandings arise.

13) The essence of the theory of kerogen thermal pyrolysis for oil generation lies in the utilization of organic chemical reactions to demonstrate that both soluble and insoluble organic matter can produce oil (and natural gas) through high-temperature pyrolysis. Therefore, the theory of thermal degradation of kerogen, which investigates oil generation, belongs to the realm of organic chemistry. It forms part of the larger discipline system of chemistry within the six natural sciences encompassing mathematics, physics, chemistry, astronomy, earth, and biology. The various conditions necessary for organic chemical reactions, such as the substances involved, the specific reaction conditions, and the corresponding containers, are all human-designed, manufactured, and operated. This is why the oil generated by Rock-Eval pyrolysis apparatus or oil shale carbonization furnace is referred to as "artificial petroleum".

In contrast, the theory of thermal degradation of source rocks for oil generation examines the generation of oil from organic matter under subsurface geological conditions. This research area falls under the domain of organic geochemistry, which belongs to the earth science branch of the six natural sciences—mathematics, physics, chemistry, space, earth, and biology. Studying subsurface geological conditions necessitates knowledge from various fields, including petroleum geology, petrology, sedimentology, paleontology, and organic chemistry. It represents a complex research topic involving related simulation experiments. Consequently, it is fundamentally flawed to extrapolate the evaluation of natural petroleum's oil source using data obtained solely from the experimental method of "artificial petroleum". Such an approach deviates from reality and employs an erroneous line of thinking.

23) The theory of thermal degradation of kerogen for oil generation relies on the S1 + S2 value obtained from Rock-Eval pyrolysis to assess the potential for oil (gas) generation in rocks. However, this method is only applicable to oil shale because the S1 + S2 found in oil shale is produced through carbonization in a furnace, resulting in the creation of "artificial petroleum" through pyrolysis. This approach is not suitable for source rocks. In source rocks, the S2 value represents the organic carbon that remains unpyrolyzed and has not generated oil. It is impossible for oil to be generated indefinitely under subsurface geological conditions. Therefore, the S1 + S2 value in source rocks essentially represents only the S1 value, which represents the actual oil (gas) generation potential.

Furthermore, if the oil (gas) generation potential value obtained from pyrolyzing source rocks is the S2 value from Rock-Eval pyrolysis, it is purely a hypothetical value. This means that even if all the S2 in the source rock were pyrolyzed to generate oil, it does not actually exist in the subsurface. Using S1 + S2 to calculate the oil (gas) generation potential of source rocks will inevitably result in an overestimation of the amount of oil. Consequently, when evaluating oil resources, petroleum geologists are compelled to use artificially selected parameters like the "oil expulsion coefficient × accumulation coefficient" to estimate the subsurface amount of oil resources. This leads to the paradoxical situation of calculating the actual amount of oil resources in a basin using the amount of oil generated by source rocks that do not exist under subsurface geological conditions. Unfortunately, these artificially calculated oil resource estimates have become crucial parameters for oil companies and even national development plans.

33) The theory of kerogen thermal degradation and oil generation is based on the principle of synergistic reactions among organic chemical substances. In this theory, the reactants directly undergo reactions and generate final products without going through any intermediates. Therefore, the total amount of oil (gas) generated by pyrolyzing all the organic matter in various rocks can be quantified using the sum of S1 + S2 based on laboratory Rock-Eval pyrolysis. However, under subsurface geological conditions, the pyrolysis and oil generation processes of source rocks must involve intermediates.

The organic matter within source rocks is dispersed, and the oil produced by pyrolysis is also sporadically distributed. Without two-step or multi-step reactions involving intermediates, these scattered oils cannot be expelled (primary migration) from source rocks composed of mudstone or shale; instead, they remain trapped within them. The intermediate necessary for this process is sandstone (or a carbonate rock with better porosity and permeability). Its function is similar to the quartz sand or ceramic materials injected during the exploitation of tight oil reservoirs. Fracturing processes, often employed in the production of tight oil reservoirs with low porosity and permeability, involve injecting high-pressure fluids to create fractures. Additionally, large quantities of quartz sand or ceramics are injected into these fractures. The presence of sand grains increases conductivity, allowing the oil, which would otherwise have difficulty flowing through the tight reservoir, to be transported out of the shale via the sand grains acting as intermediaries.

In summary, it is evident that the theory of thermal degradation of kerogen for oil generation is solely a theory applicable to "artificial petroleum". Utilizing this theory and the experimental techniques of "artificial petroleum" to evaluate natural petroleum resources is clearly an erroneous approach that deviates from reality.

1.2 Introducing an Organic Geochemistry Thought Framework

After obtaining his chemistry doctorate from Pennsylvania State University in 1946, J. M. Hunt began working in petroleum exploration and survey at the American Petroleum Corporation. Over the years, he collaborated with renowned petroleum geologist Prof. A. I. Levorsen and engaged in discussions about various challenges encountered in oil exploration. J. M. Hunt skillfully merged organic chemistry with petroleum geology, effectively applying organic geochemical thinking methods to analyze source rocks and explore fundamental theoretical aspects such as oil migration and accumulation. This transformation enabled him to transition from an organic chemist to a proficient petroleum geologist utilizing the organic geochemical thinking approach for studying source rocks.

In 1979, J. M. Hunt authored and published the book "Petroleum Geochemistry and Geology," wherein he introduced the concept of organic geochemical thinking. The key points put forth in the book are as follows:

(1) Source rocks are defined as rocks capable of generating oil and facilitating oil migration and accumulation to form economically viable petroleum reserves under specific geological conditions. These rocks typically contain organic matter ranging from 0.5 to 5%. Argillaceous source rocks generate oil that can undergo primary migration, resulting in sufficient oil production required for the formation of commercial reservoirs. Evaluating the quality of source rocks involves extracting them with organic solvents like benzene, methanol, and acetone. Source rocks are then classified and assessed based on the amount of dissolved oil.

(2) Statistical data suggests that in oil-bearing basins, the conversion of organic matter in oil-source rock formations into oil ranges from a few percent to fifteen percent. The amount of oil generated, eventually leading to the formation of commercial accumulations, falls within this range. Table 1.2 provides the conversion percentage of organic matter to hydrocarbons. In sedimentary rocks overall, approximately 3% of organic matter can be transformed into C^+ hydrocarbons. The 15 ratio of hydrocarbons to organic matter in modern sediments is also relatively low.

However, C^+ hydrocarbon products from rock samples obtained in ancient sediments 15 from oil-producing regions can reach up to 10% of the total organic matter.

(1) The importance of source rocks is emphasized, as they consist of fine- grained sediments and organic matter within sedimentary basins. Throughout the geological evolution of these basins, both the diagenetic evolution of sediments and the thermal evolution of organic matter occur simultaneously. Oil generation from source rocks occurs through a catalytic cracking reaction process under subsurface geological conditions. Clay minerals not only play a catalytic role in the thermal evolution of organic matter but can also undergo mineral

Table 1.2 Distribution of C^{+} hydrocarbons and related organic matter in sediments (after Hunt 1986)

Sediments		Residual organic matter/%	Extractable organic matter/%	
			NSO bitumen	C^{+} hydrocarbons 15
All sedimentary rocks		94	3	3
Ancient carbonate rocks		74.5	13.8	11.7
Ancient shales		95.7	2.9	1.4
Contemporary sediments	Average	95.5	4.2	0.3
	Cariaco Trench (4 m)	94.6	5.2	0.2
	Cariaco Trench (85 m)	98.1	1.7	0.2
	Caspian Sea	93.6	6.1	0.3
	Orinoco Delta	94.5	5.0	0.5
	Mediterranean sea	94.8	4.9	0.3
Ancient sediments	Cherokee, Kansas	96	3	1
	Frontier, Wyoming	92	6.3	1.7
	Wilcox, Louisiana	92	6	2
	Winnipeta, Montana	95.9	1.4	2.7
	Dufresne, Alberta	89	7.3	3.7
	Fossilia, Wyoming	92	3.4	4.6
	Woodford, Oklahoma	89.2	5.8	5
	Green River, Utah	72.3	21.4	6.3
	Travers, Michigan	79.4	13.3	7.3
	Zechstein, Denmark	73.4	17.4	9.2
	Madison, Montana	68.6	18.9	12.6

Note Extractable fractions are soluble in mixed solvents. For example, with a mixed solvent of 70% benzene, 15% methanol, 15% acetone. There is no loss in the extracted organic matter by removing the solvent in its boiling point

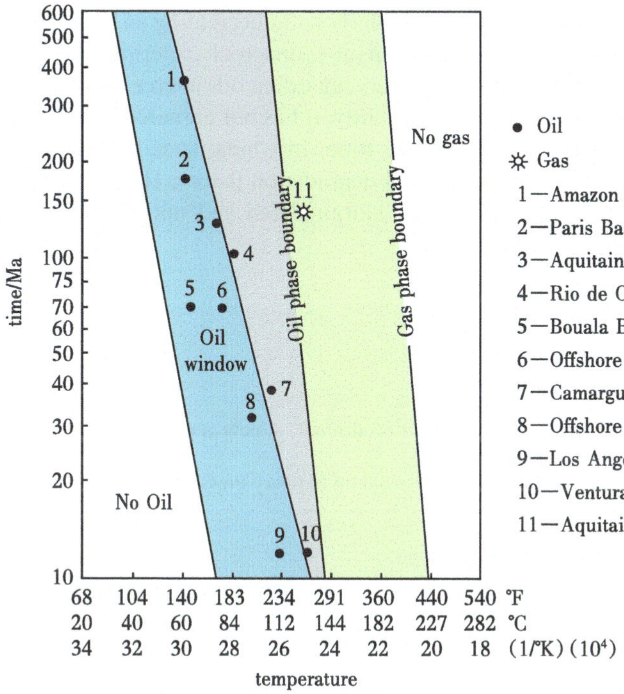

Fig. 1.1 The time–temperature relationship of petroleum generation (after Hunt 1986)

transformation to increase the porosity of shale. For instance, the transformation of montmorillonite into illite can release crystallized water, increasing shale porosity from 5 to 10%. The temperature range at which significant petroleum generation occurs is between 60 and 150 °C. Beyond this range, at temperatures of 200 and 250 °C, organic matter primarily generates methane, and above 250 °C, hydrocarbon generation largely ceases (Fig. 1.1).

(2) It is believed that conducting high-temperature pyrolysis experiments on source rocks using a Rock-Eval pyrolysis instrument significantly alters the original state of organic matter within the source rocks. These experiments merely illustrate the relationship between insoluble organic matter, such as kerogen, and high-temperature pyrolysis. They cannot be used to accurately represent the characteristics of oil and gas generation from source rocks, as such high-temperature conditions do not exist within subsurface geological environments. The correct approach is to conduct simulation experiments that replicate the highest temperature range (200–250 °C) found in subsurface geological conditions.

It is evident that Hunt's organic geochemical thinking aligns with reality and significantly differs from the organic chemistry thinking methods of Tissot and Welte. However, since Hunt has not developed practical research methods and has only remained at the theoretical thinking stage, particularly without creating an instrument for simulating subsurface geological conditions in oil-source rock experiments, it has been challenging to obtain quantitative parameters regarding oil- source rock generation through simulation experiments. Consequently, it has not garnered significant attention from petroleum geologists for a long time. In China, specifically, influenced by the preconception of kerogen thermal degradation theory, Hunt's organic geochemical thinking method has often been marginalized and not given proper attention and consideration.

References

Tissot BP, Welte DH (1982) Petroleum formation and occurrence. A new approach to oil and gas exploration (Translated by Hao et al. in Chinese)

Hunt (1986) Petroleum geochemistry and geology. Petroleum Industry Press, Beijing (in Chinese)

Chapter 2
Thoughts and Methods of Theoretical Research on Hydrocarbon Generation of Lacustrine Source Rocks in China

Abstract This chapter presents a systematic comparative analysis of the petroleum geological evolution characteristics in 34 lacustrine oil-bearing basins (including depressions and sags) across China. It focuses on key aspects of source rock research, such as the depositional environments of source rocks across various geological periods, the thermal evolution of organic matter, and the mechanisms of petroleum generation, primary migration, and secondary migration. The study consolidates conceptual approaches related to the petroleum generation theory for lacustrine source rocks in China and provides detailed discussions of the following theoretical frameworks: Basin formation, petroleum generation, and reservoir formation theory; Limited-space petroleum generation theory of source rocks; Primary migration and reservoir formation in fluvial-deltaic sedimentary systems; Low-temperature catalytic cracking theory for petroleum generation in source rocks. These frameworks are integral to advancing the understanding of the petroleum geological characteristics and evolutionary processes in China's lacustrine oil-bearing basins.

Sedimentary basins serve as the fundamental geological units where petroleum accumulates. Regardless of their size, morphological features, or genesis types, any basin that has undergone processes of petroleum generation, migration, and accumulation is referred to as a petroleum-bearing basin. Conversely, any basin containing petroleum should exhibit similar characteristics and undergo comparable petroleum geological evolution processes related to generation, migration, and accumulation of petroleum. By adopting this type of reasoning, a systematic comparative analysis has been conducted on the petroleum geological evolution characteristics of 34 nonmarine petroleum-bearing basins (or depressions) in China. This analysis specifically focuses on the developmental environment of source rocks during different geological epochs, conditions for the thermal evolution of organic matter, petroleum generation, primary and secondary migration characteristics, and other key aspects of source rock research. The proposed theoretical thinking methods include the theory of basin formation, oil generation, and reservoir formation, as well as the limited space oil generation theory for source rocks. The study also encompasses the theory

D. Guan et al., *Theory and Simulation Experimental Technology of Oil Generation From Nonmarine Source Rocks in China*, https://doi.org/10.1007/978-981-96-3145-2_2

of petroleum generation, primary migration, and accumulation in river-delta sedimentary systems, along with the theory of low-temperature catalytic cracking for source rock oil generation.

2.1 Theory of Basin Formation, Oil Generation and Reservoir Formation

Through the analysis of the petroleum geological characteristics and evolutionary history of 34 nonmarine oil-bearing basins (or depressions) in China, it is clearly seen that these oil-bearing basins (or depressions) have undergone three development stages: continuous subsidence, overall uplift, and overall shrinkage (Table 2.1).

2.1.1 Continuous Subsidence Stage of the Basin

The formation of a basin and the accumulation of sedimentary materials are marked by the continuous subsidence of the basin. Most basins have a deposition rate exceeding 200 m/Ma, with the deposition rate of the main source rocks being particularly notable. For instance, in the Dongying Depression of the Bohai Bay Basin, the deposition rate of the main source rocks in the upper part of the fourth member and lower part of the third member of the Shahejie Formation can reach up to 300 m/Ma. This high deposition rate of source rocks facilitates the preservation of organic matter effectively. During this depositional stage, the sedimentary thickness can exceed several thousand meters. Consequently, as the continuous subsidence of the basin reaches its end, the main source rocks in oil-generating depressions enter a large-scale oil generation stage.

Throughout the continuous subsidence stage of the basin, different subsidence rates in various areas lead to the formation of three distinct sedimentary structural units: depressions, slopes and uplifts. Depressions contain the thickest and most complete sediment sequences, making them the areas where organic matter is best preserved. Therefore, the depression area in an oil-bearing basin is the most favorable zone for source rock development. Slopes primarily surround depressions and can be categorized into gentle and steep slopes. Gentle slopes have relatively smooth topography, small inclines, and consist mainly of fluvial deposits, such as sandstones interbedded with mudstones. Steep slopes have significant angles and often exhibit mixed accumulations of alluvial fans and diluvial fans with varying particle sizes. Uplifts represent areas with minimal or no deposition during this stage, often serving as sediment source regions that undergo erosion.

The continuous subsidence stage of the basin involves the accumulation and compaction of diverse sedimentary materials, as well as the pyrolysis of organic matter to generate oil. Therefore, research on the oil generation theory of nonmarine

Table 2.1 Development characteristics of 34 nonmarine oil-bearing basins (or depressions) in China during the Mesozoic and Cenozoic eras

Name of basin	The overall development stage of the basin	Stage of continuous subsidence	Stage of overall uplifting	Stage of overall shrinkage
1. Songliao Basin	Cretaceous—Quaternary	Deposition period of Qingshankou Formation—Nenjiang Formation	The end of deposition of Nenjiang Formation	Deposition period of the Sifangtai Formation—Quaternary
2. Hailar Basin	Cretaceous—Quaternary	Deposition period of Tongbomiao Formation—Yi Min group	The end of deposition of the Yimin Formation	Deposition period of the Qingyuangang Formation—Quaternary
3. Kailu Basin	Jurassic—Quaternary	Late Jurassic	The end of deposition of the Fuxin Formation	Deposition period of the Yaojia Formation—Quaternary
4. Iran-Iton Basin	Paleogene—Quaternary	Paleogene	The end of deposition of the Qijia Formation	Deposition period of the Chaluhe Formation—Quaternary
5. Erlian Basin	Cretaceous—Quaternary	Early Cretaceous	The end of deposition of the Saihantara Group	Late Cretaceous—Quaternary
6. Liaohe Basin	Paleogene—Quaternary	Paleogene	The end of deposition of the Dongying Formation	Deposition period of Guantao Formation—Quaternary
7. Huanghua Basin	Paleogene—Quaternary	Paleogene	The end of deposition of the Dongying Formation	Deposition period of Guantao Formation—Quaternary

(continued)

Table 2.1 (continued)

Name of basin	The overall development stage of the basin	Stage of continuous subsidence	Stage of overall uplifting	Stage of overall shrinkage
8. Jizhong Basin	Paleogene—Quaternary	Paleogene	The end of deposition of the Dongying Formation	Deposition period of Guantao Formation—Quaternary
9. Jiyang Basin	Paleogene—Quaternary	Paleogene	The end of deposition of the Dongying Formation	Deposition period of Guantao Formation—Quaternary
10. Dongpu Basin	Paleogene—Quaternary	Paleogene	The end of deposition of the Dongying Formation	Deposition period of Guantao Formation—Quaternary
11 Biyang Basin	Paleogene—Quaternary	Paleogene	At the end of deposition of the Liaozhuang Formation	Deposition period of Shangsi Formation—Quaternary
12. Subei Basin	Paleogene—Quaternary	Paleogene	The end of deposition of the Sanduo Formation	Deposition period of the Yancheng Formation—Quaternary
13. Jianghan Basin	Paleogene—Quaternary	Paleogene	The end of deposition of the Jinghezhen Formation	Deposition period of Guanghuasi Formation—Quaternary
14. Sanshui Basin	Paleogene—Quaternary	Paleogene	The end of deposition of the Huayong Formation	Eocene—Quaternary
15. Baise Basin	Paleogene—Quaternary	Paleogene	The end of deposition of the the Jianduling Formation	Pliocene—Quaternary

(continued)

Table 2.1 (continued)

Name of basin	The overall development stage of the basin	Stage of continuous subsidence	Stage of overall uplifting	Stage of overall shrinkage
16. Fushan Sag	Paleogene—Quaternary	Paleogene	At the end of deposition of the Lanzhou Formation	Neogene-Quaternary
17. Pearl River Mouth Basin	Paleogene—Quaternary	Paleogene	At the end of deposition of the Zhuhai Formation	Neogene-Quaternary
18. Qiongdongnan Basin	Paleogene—Quaternary	Paleogene	The end of deposition of the Lingshui Formation	Neogene-Quaternary
19. Yinggehai Basin	Paleogene—Quaternary	Paleogene	The end of deposition of the Lingshui Formation	Neogene-Quaternary
20. Beibu Gulf Basin	Paleogene—Quaternary	Paleogene	The end of deposition of the Qianzhou Formation	Neogene-Quaternary
21. Jiuquan Basin	Cretaceous—Quaternary	Cretaceous	Late cretaceous	Oligocene-Quaternary
22. Tuha Basin	Jurassic—Quaternary	Jurassic	Late Jurassic	Cretaceous—Quaternary
23. Ordos Basin	Late Triassic—Early Cretaceous	Late Triassic—Middle Jurassic	Late Middle Jurassic	Late Jurassic—Quaternary
24. Junggar Basin	Late Paleozoic—Quaternary	Permian—Jurassic	Late Jurassic	Cretaceous—Quaternary
25. Qaidam Basin	Paleogene—Quaternary	Paleogene-Neogene	Late Neogene	Quaternary
26. Yanqi Basin	Triassic—Quaternary	Late Triassic—Jurassic	Late Jurassic	Cretaceous—Quaternary
27. Sichuan Basin	Late Triassic—Quaternary	Late Triassic—Middle Jurassic	Late Middle Jurassic	Late Jurassic—Quaternary
28. East China Sea Basin	Paleogene—Quaternary	Paleogene	The end of deposition of the Huagang Formation	Neogene-Quaternary

(continued)

Table 2.1 (continued)

Name of basin	The overall development stage of the basin	Stage of continuous subsidence	Stage of overall uplifting	Stage of overall shrinkage
29. Santang Lake Basin	Late Paleozoic—Quaternary	Late Permian—Middle Jurassic	The end of deposition of the Xishanyao Formation	Middle Jurassic—Quaternary
30. Baiyun Chagan Sag	Cretaceous—Quaternary	Early Cretaceous	The end of deposition of the Saihantala Formation	Late Cretaceous—Quaternary
31. Jinggu Basin	Paleogene—Quaternary	Paleogene	At the end of deposition of the Mengla Formation	Neogene-Quaternary
32. Lumpola Basin	Paleogene—Quaternary	Paleogene	The end of deposition of the Dingqinghu Formation	Neogene-Quaternary
33. Jidong Depression	Paleogene—Quaternary	Paleogene	The end of deposition of the Dongying Formation	Deposition period of Guantao Formation—Quaternary
34. Kuqa Depression	Triassic—Quaternary	Middle Triassic—Jurassic	Late Jurassic	Cretaceous—Quaternary

source rocks primarily focuses on analyzing the subsidence and deposition history of the basin during this stage of development. Additionally, it investigates the corresponding environments for organic matter formation and preservation, the lithological composition of mudstones and sandstones, temperature and pressure conditions, and the thermal decomposition of organic matter. Both internal and external factors influencing oil generation are examined in this context.

2.1.2 *The Overall Rising Stage of the Basin*

During this stage of development, various forms of energy accumulated during the continuous subsidence of the basin, such as high temperature and high pressure, are gradually released. This leads to the unloading and decompression of sediments. The main indicators of this process are the varying degrees of denudation observed in the basin's strata, with denudation rates typically ranging from tens to hundreds of meters per million years. Different sedimentary structural units within the basin experience different rates of uplift and exhibit varying amounts of denudation. Generally, the basin margin rises at a relatively fast rate, resulting in significant denudation. The slope, particularly the gentle slope, is dominated by sandstone due to the "sandstone rebound" effect. This effect leads to a higher rate of uplift and a larger amount of denudation. In areas where depressions and mudstones are more prevalent, the relative rate of uplift and denudation is smaller.

As a result of the varying denudation amounts among different sedimentary structural units during this stage, there are noticeable differences in sediments unloading and decompression. The oil present in the source rocks within the oil-generating depressions, which are characterized by high temperatures and pressures, migrates towards the slope area. It subsequently accumulates and gathers in favorable traps, forming oil accumulations.

2.1.3 *The Overall Shrinking Stage of the Basin*

After undergoing the stages of continuous subsidence and overall elevation, the basin transitions into a stage of overall shrinkage. During this stage, the basin gradually reaches a state of crustal balance with the surrounding areas. This is characterized by minor subsidence and uplift denudation within the basin. Through these small-scale movements, the process of oil formation and accumulation is further adjusted, refined and finalized. This stage of development extends into the Quaternary period.

The theoretical framework for understanding the basin's formation, oil generation, and reservoir formation aims to comprehend the complex geological processes involved in the generation, migration, accumulation, and preservation of oil within the basin. These processes are the result of various sediments within the basin undergoing a series of physical and chemical reactions. These reactions occur across the three stages of continuous subsidence, overall rise, and overall shrinkage. During the continuous subsidence stage, the diagenetic evolution characteristics of source rocks in oil-generating depressions, the thermal evolution characteristics of organic matter, and their interrelationship play crucial roles. Understanding and analyzing these factors are essential aspects of the theory of oil generation from source rocks. Therefore, a rational division of the three aforementioned development stages of sedimentary basins and an analysis of the characteristics during the continuous subsidence stage are prerequisites for studying the theory of oil generation from source rocks.

2.2 Theoretical Contemplation on Oil Generation in Confined Source Rocks

Based on sedimentological principles, when the continuous subsidence of the basin comes to an end, various sediments, including source rocks, have filled the entire space of the oil-generating depression. This provides a suitable environment for the thermal degradation and oil-generating reactions of dispersed organic matter within the source rocks. During the extensive oil generation stage, the only space capable of containing oil, gas, and water is the pore space, which is characterized by its porosity. As the overlying strata exert pressure, the porosity of the source rocks decreases, while the thermal degradation rate of organic matter accelerates with increasing burial depth and temperature.

As dispersed organic matter gradually transforms into oil and fills the pore space of the source rocks, the porosity of the source rocks no longer decreases, and the oil generation process slows down, resulting in a temporary equilibrium. Therefore, at the end of the continuous subsidence phase, the oil generated from the thermal degradation of dispersed organic matter in the source rocks is distributed in a state of being miscible with sporadic oil droplets and pore water. However, the sporadic oil droplets within mudstone or shale, which form the source rocks, face difficulties in accumulating into continuous oil flows and migrating out of the source rocks during the primary migration process.

Nevertheless, if several layers of sandstone (or other rock formations with similar physical properties) exist within the mudstone or shale, the characteristics of the source rocks change significantly. Firstly, sandstone has higher thermal conductivity compared to mudstone or shale, making the organic matter near the sandstone more prone to thermally degrade and oil generation. Secondly, the well- developed intergranular pores in sandstones not only contribute to the overall pore space but also result in lower internal pore fluid pressure compared to mudstones or shales. As a result of pressure differences, dispersed oil droplets within mudstone or shale can vertically form oil flows towards the sandstones through the bedding, facilitating the primary migration process of oil (Fig. 2.1).

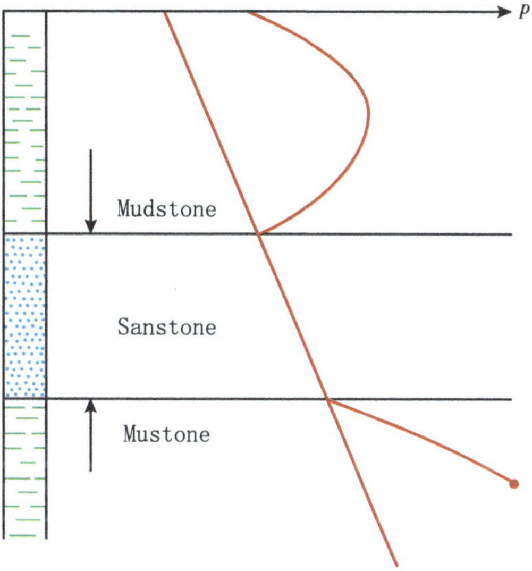

Fig. 2.1 Pressure-depth relationship diagram of sandstone sandwiched in abnormal pressure mudstone layer

Based on this concept, in 2009, the Wuxi Petroleum Geology Research Institute of Petroleum Exploration and Development Research Institute of the SINOPEC created a device for conducting simulation experiments on the generation and expulsion of hydrocarbons from formation pores under the influence of thermal and pressure factors. The research team then proceeded to conduct these simulation experiments. The source rock samples selected for the experiments were obtained from well Wang 24 in Henan Oilfield and well Pu 1–154 in Zhongyuan Oilfield. The experimental process and the resulting data are presented in Figs. 2.2, 2.3 and 2.4.

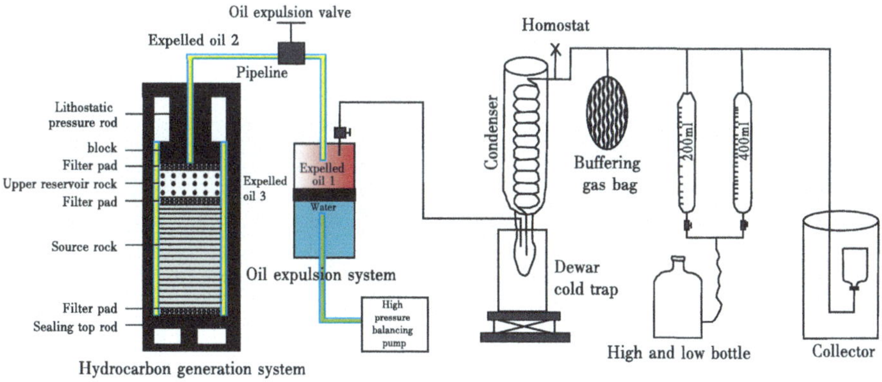

Fig. 2.2 Illustration of experiment setup for simulating hydrocarbon generation and expulsion in limited pore space

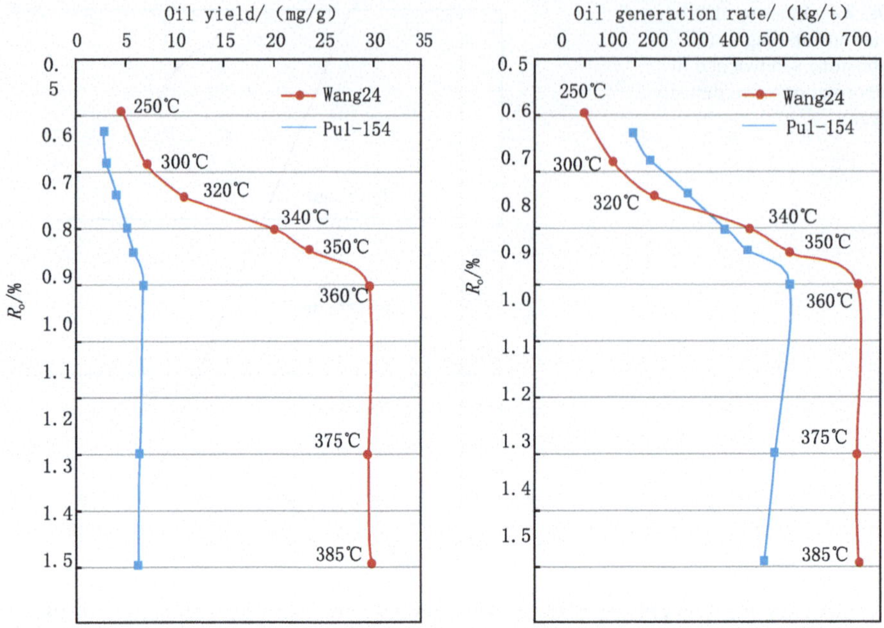

Fig. 2.3 Evolution curve of oil yield and oil generation rate in the simulation experiment of hydrocarbon generation from limited space at continuous subsidence stage

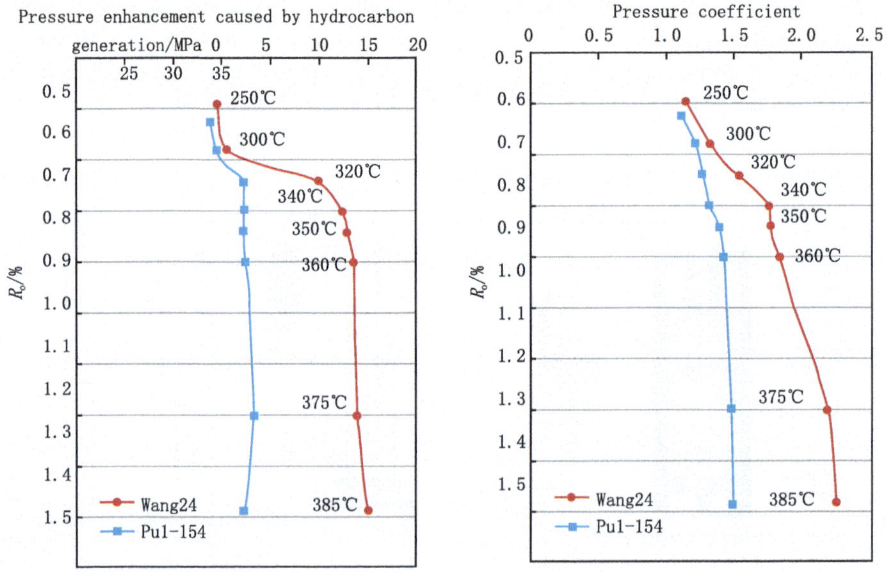

Fig. 2.4 Evolution curve of pressure enhancement caused by hydrocarbon generation in the simulation experiment of hydrocarbon generation from limited space at continuous subsidence stage

The oil generation simulation experimental instrument developed based on the theoretical concept of limited pore space in source rocks demonstrates notable differences compared to the Rock-Eval pyrolysis instrument in terms of both theoretical assumptions and experimental outcomes.

The oil generation process, as illustrated by Tissot, does not exhibit the characteristic "big belly curve" during the oil generation stage when Ro (vitrinite reflectance) ranges from 0.7 to 1.3%. Figures 2.3 and 2.4 clearly indicate that despite the significant difference in TOC content between samples from well Wang 24 and well Pu 1–154 (4.55% and 1.3% respectively), and the distinct kerogen types (type I and type II), both samples reach a point of stable oil generation when the simulated temperature reaches 360 °C and Ro reaches 0.9%. This stable phase persists even as the simulated temperature increases to 385 °C and Ro approaches 1.4%. Beyond an Ro value of 1.3%, the peak of oil generation is surpassed, and a transition to gaseous hydrocarbon transformation occurs. This experimental result aligns with the current conditions observed in the primary source rocks of eastern China, such as the Paleogene Shahejie Formation in the Bohai Bay Basin, which maintains an Ro value of approximately 0.9%.

The process of oil generation from source rocks does not exhibit two distinct stages of evolution from low maturity to maturity. In the case of samples from well Wang 24 and well Pu 1–154, as the temperature increases from 300 to 360 °C, there is only a modest increase in Ro, specifically 0.22% (from 0.68 to 0.90%). However, the oil yield and generation rate of both samples rapidly rise to their maximum. This suggests that the primary oil generation process within source rocks can be completed with a burial depth of just a few hundred meters, thereby negating the existence of a so-called low-mature evolution stage or the formation of low-mature oil.

The pressure enhancement and pressure coefficient of source rocks under the limited pore space are closely linked to the organic matter content and kerogen type within the source rocks. Samples from well Wang 24, characterized by high organic matter content and excellent kerogen type, exhibit significant oil generation that rapidly fills the pore space, resulting in significant pressurization and abrupt changes in pore fluid pressure (Fig. 2.4). Conversely, samples from well Pu 1–154, with lower organic matter content and inferior kerogen types, do not completely fill the pore space under the same temperature and pressure conditions, leading to minimal pressure enhancement and pressure coefficients during the oil generation period.

The oil generation simulation experimental instrument developed in 2009 solely simulates the sedimentary characteristics and pressure conditions of source rocks filled with oil generation depressions (achieved by filling high-temperature and high-pressure reactors with rock samples). The simulation temperature is still set at 250 °C as the starting point primarily for the sake of saving simulation time. In contrast, the oil generation simulation experimental instrument developed in 2019 achieves a comprehensive simulation that accounts for the actual subsurface conditions of source rocks.

2.3 Theoretical Thinking of Oil Generation, Primary Migration and Accumulation in Fluvial-Delta Sedimentary Systems

In China, most of the continental sedimentary basins are characterized by inland saline or brackish lake environments with strong reducing conditions. During the continuous subsidence stage of these basins, source rocks primarily develop in the depressions where oil generation occurs. The geomorphic conditions vary significantly between steep slopes and gentle slopes within these depressions. Steep slopes have relatively steep topographical gradients, often leading to the abrupt accumulation of alluvial fans or diluvial fans. Sediment particles in these areas are heterogeneous in size and exhibit poor roundness. Towards the center of the lake basin, these sediments gradually disappear. Hence, these sediments do not meet the necessary conditions for source rock formation. In contrast, gentle slopes are flat and tend to form river delta deposits. Within these deposits, fluvial facies can be categorized into single-channel (straight or meandering rivers) or multi-channel (braided or networked rivers). Most of these fluvial deposits occur along the long axis of the lake basin. Clastic particles in these deposits are well-sorted and rounded due to long-distance transportation. When rivers flow into the lake basin, they create a delta sand body that extends towards the center of the lake at the shallow slope of the estuary.

The fluvial-deltaic sedimentary system consists of the fluvial facies deposition and the delta body. Generally, this system includes three sedimentary zones: ① The delta plain zone, located onshore between the fluvial facies and the lake shore, primarily consists of fine sand, silt, and muddy deposits in the distributary channel. ② The delta frontal zone, situated near the river mouth where it flows into the lake, is a coastal-shallow lacustrine area mainly composed of sheet sand and coastal-shallow lacustrine mudstone deposits. ③ The pre-delta mud zone is the sedimentary region that transitions from semi-deep lacustrine deposits to deep lacustrine facies. It is characterized by dark mudstone interbedded with thin silt.

The fluvial-deltaic sedimentary system undergoes changes because of lake basin uplift, subsidence, and climate fluctuations during different geological periods. These changes can manifest as either advancement or retreat towards the onshore direction of the riverbank. On vertical sedimentary profiles, this is reflected by alternations of mudstones and sandstones. Towards the lakeshore direction, sandstones are primarily interbedded with mudstones. For example, in the Songliao Basin, the Cretaceous Qingshankou Formation and the first member of the Nenjiang Formation serve as the main source rocks. They are predominantly composed of mudstone interbedded with sandstone. The main reservoir rock, the Yaojia Formation, consists mainly of sandstone interbedded with mudstone (Fig. 2.5).

The development of the fluvial-delta sedimentary system on the gentle slope of a lake basin during its continuous subsidence stage is essentially an oil-bearing system that encompasses oil generation, migration, accumulation, and preservation. A prime example of such a system is found in the Songliao Basin during the Lower

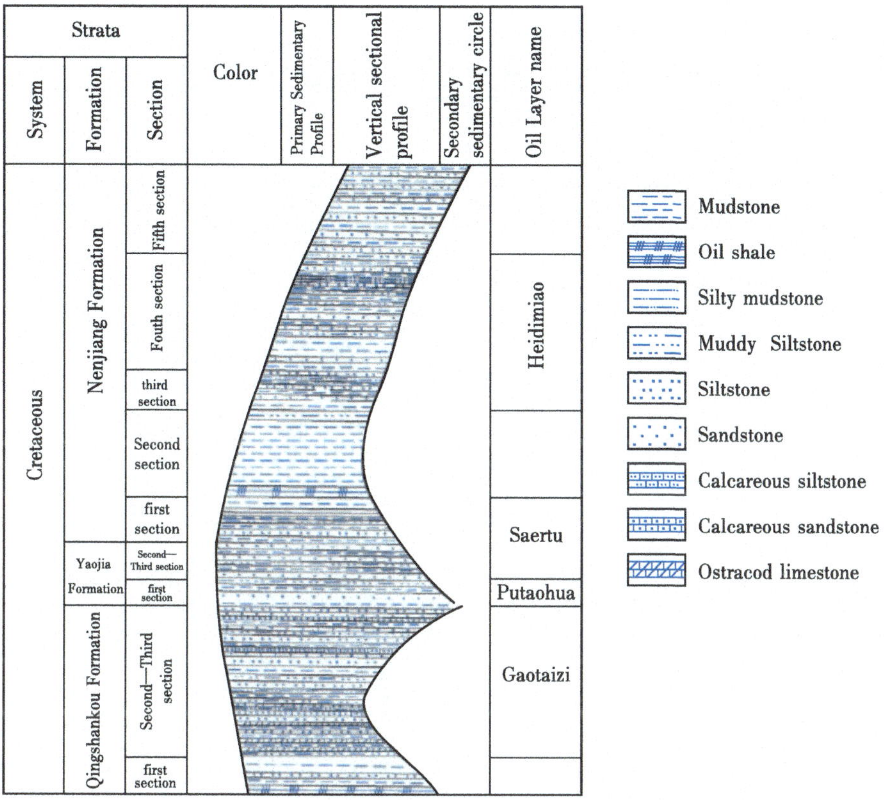

Fig. 2.5 Comprehensive stratigraphy column of the Cretaceous oil-bearing combination in Songliao Basin (after Wanli 1985)

Cretaceous period. As the water level of the lake transgression and regression, the delta underwent periods of retreat and advancement. During the deposition process, mudstone interbedded with sandstone in the pre-delta mud zone and sheet sand in the delta front zone were layered together. This resulted in the deposition of sandstone and mudstone as vertical interlayers, providing favorable conditions for both primary and subsequent secondary migration of oil from the source rocks in the Qingshankou Formation. This sedimentary system played a direct role in the formation of significant conventional and shale oil fields in Daqing (Fig. 2.6).

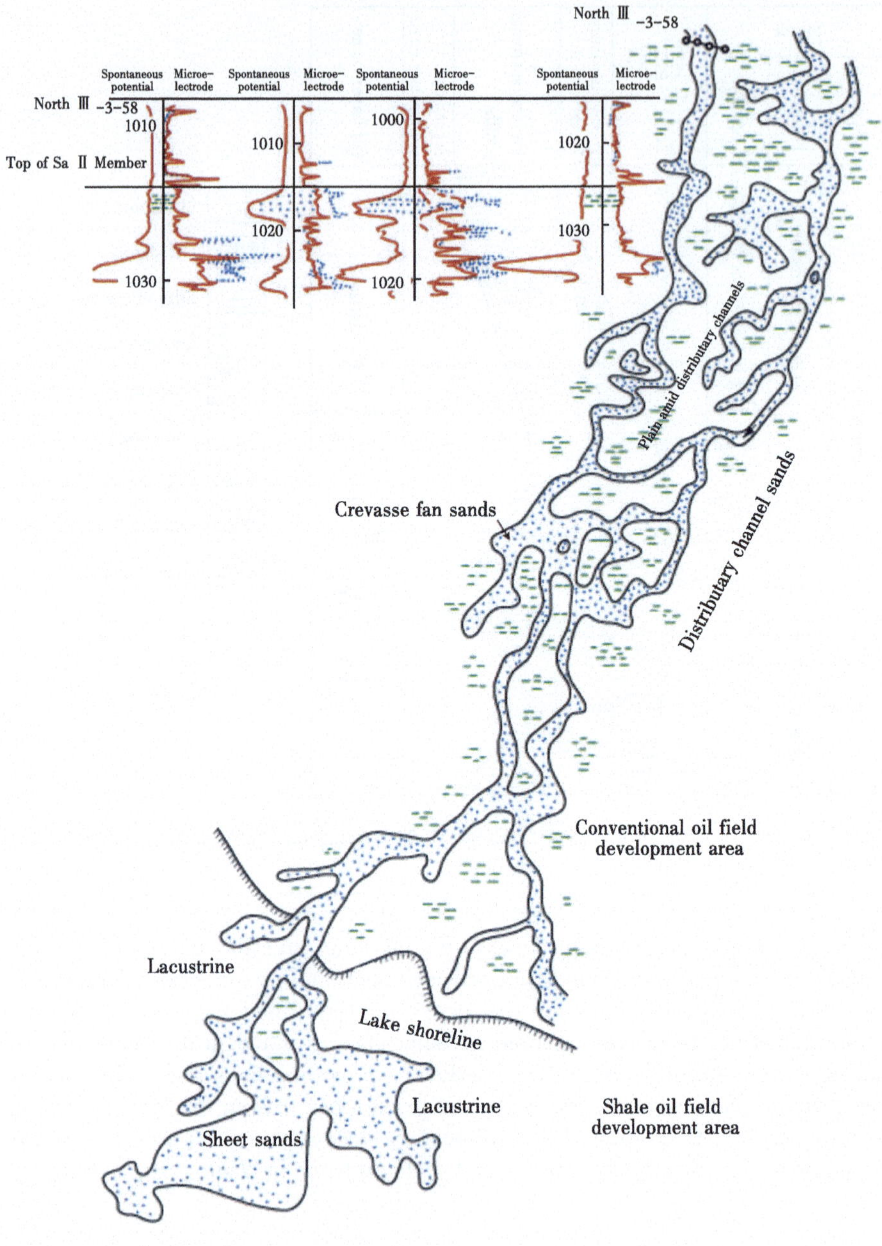

Fig. 2.6 Sandstone distribution of the Lower Cretaceous fluvial-deltaic sedimentary system in the Songliao Basin (after Hengjian and Wenyin 1981)

2.4 Theoretical Thinking of Oil Generation by Low-Temperature Catalytic Cracking of Source Rocks

China's nonmarine source rocks have two distinct features:

(1) The majority of nonmarine source rocks in China have oil-generating temperatures below 100 °C. For instance, in the upper Triassic source rocks of the Ordos Basin, where the geothermal gradient is only 3.1 °C per 100 m, these source rocks enter the stage of significant oil generation at a temperature as low as 58 °C. Except for Neogene source rocks in the Qaidam Basin, which enter the extensive oil generation stage at 126 °C, other nonmarine source rocks undergo substantial oil generation at temperatures below 100 °C (Fig. 2.7).

(2) China's nonmarine source rocks and crude oils exhibit higher concentrations of metals, rare metals, and rare radioactive metal elements. For instance, crude oils from oil-bearing basins such as Songliao, Bohai Bay, and Junggar contain Fe, Mn, Zn, Cu, and other metal elements at levels that are 1–2 orders of magnitude higher than the average sedimentary layers in China (Table 2.2).

Similarly, Daqing crude oils and Xinjiang crude oils have higher concentrations of metals and rare metal elements compared to the average values of China's continental crust (Table 2.3). This phenomenon suggests that these metals and rare metal elements might have originated from the deep mantle and subsequently entered the

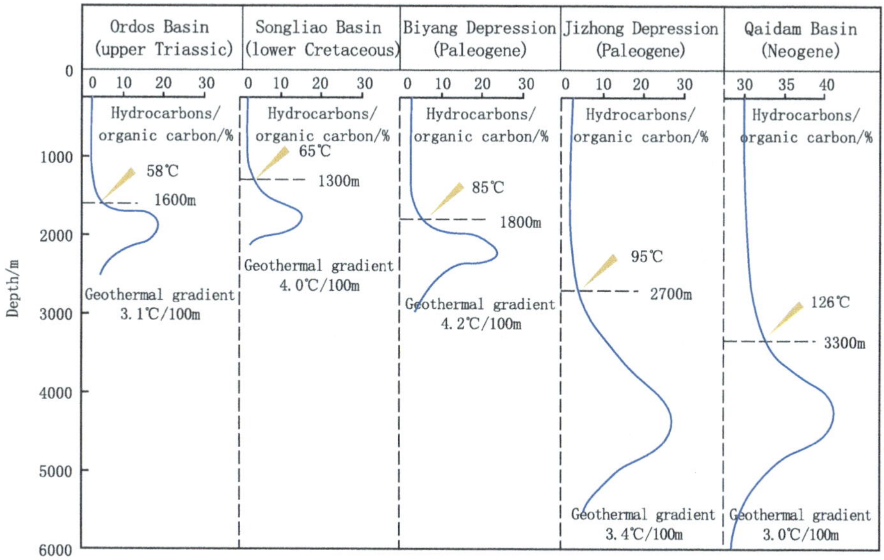

Fig. 2.7 The natural evolution curve of organic matter in Mesozoic-Cenozoic source rocks in China (after Jianyi and Difan 1991)

Table 2.2 Comparison of metal element content in crude oil of each oilfield and average content of sedimentary layers in China (after Zhanqian 2001)

Element	Chinese sedimentary formation	Daqing Oilfield	Liaohe Oilfield	Dagang Oilfield	Xinjiang Oilfield	Changqing Oilfield

oil-generating depression near the basin through deep faults. Furthermore, experimental evidence has confirmed that these metals and rare metal elements play a catalytic role in the generation of oil and gas from organic matter.

For instance, in a 1992 ultrasonic hydrolysis experiment conducted by the Daqing Petroleum Institute, a comparison was made between hydrocarbon generation with the addition of Pd (palladium) for hydrogen adsorption and hydrocarbon generation without Pd. The results showed that the addition of Pd increased gaseous hydrocarbon yields by 55–496% (Tables 2.4 and 2.5). Another example comes from a study conducted by the College of Oceanography at Oregon State University in the United States. They examined hydrothermal oil in immature diatomite sediments in the Guayaimas Basin of the Gulf of California. The study revealed that during the conversion of organic matter into oil and natural gas, hydrothermal fluid from crustal magma transformed immature organic matter into oil. Throughout this process, the temperature of the hydrothermal solution decreased from 350 to 315 °C, the pH value of the solution increased from 4 to 6, and the concentrations of metal elements Fe, Mn, Zn, and Cu in the hydrothermal solution decreased by 2–20%. These findings indicate that the thermal energy and metal elements present in the magma hydrothermal fluid expedited the thermal pyrolysis of organic matter, resulting in oil generation.

Based on the analysis of the aforementioned data, it can be inferred that nonmarine source rocks in China may serve as a model for oil generation through low-temperature catalytic cracking of organic matter. This concept is particularly evident in the oil generation process observed in the Upper Triassic Yanchang Formation source rocks in the Ordos Basin. The Ordos Basin, along with other oil-bearing basins such as Qaidam and Tarim, located in western China, are commonly referred to as "cold basins". These basins exhibit a geothermal gradient of approximately 3.0 °C/100 m. With such a low geothermal gradient, the primary source rocks enter a stage

Table 2.3 Comparison of petroleum symbiotic elements, Chinese continental crust elements and biological elements (after Zhanqian 2001)

Element	Crust under China	Daqing	Liaohe	Dagang	Xinjiang	Element property	Biomass
						Nonmetals (亲石元素)	
						Transition metals (亲铜元素)	
						Lanthanides (镧系元素)	
						Not visible (未见)	

Table 2.4 Gaseous hydrocarbon yields in the ultrasonic hydrolysis experiment (after Zhanqian 2001)

Sample	Weigh/g	Palladium hydrogen absorption capacity/mg	Yield of gaseous hydrocarbon/mL	Production increase/%
Jianghan mudstone after extraction				
Zhalainuoer coal				
Sporophyte coal				
Resin				

Table 2.5 Liquid hydrocarbon yields in ultrasonic hydrolysis experiments (after Zhanqian 2001)

Sample	Weigh/g	Palladium hydrogen absorption capacity/mg	Yield of total hydrocarbon/mL	Production increase/%
Resin				
Zhalainuoer coal				
Sporophyte coal				
Contemporary sediments from Er Ocean				
Jianghan mudstone after extraction				

of significant oil generation at a mere 58 °C. The reason behind this phenomenon is likely associated with the abundant presence of uranium elements within the source rocks.

According to Zhang Wenzheng et al.'s data, the average uranium content in the source rocks of the Chang 7 member of the Upper Triassic Yanchang Formation in the Ordos Basin is 51 μg/g (mass fraction), which is approximately ten times higher than the average uranium content in the continental crust of China (5.6 μg/g). The overall uranium concentration in the source rocks of the Upper Triassic Yanchang Formation in the Ordos Basin reaches approximately 0.8×10^8 t. Furthermore, Zhao Junlong et al.'s data indicates the discovery of over 10,000 various types of uranium mineralization and abnormal points and belts in the Ordos Basin, including one large-scale uranium deposit, four medium-sized uranium deposits, two small-scale uranium deposits, and dozens of ore points. These uranium resources are primarily concentrated around the basin's edges (Fig. 2.8). The central region of the basin is rich in sedimentary minerals such as oil, natural gas, and coal from the Paleozoic and Mesozoic eras. This distribution pattern suggests that a substantial amount of thermal energy was released through uranium nuclear fission, thus playing a crucial role in the formation of various sedimentary minerals.

Therefore, from a geological perspective, the previous notion that organic-rich dark mudstone alone serves as the source rock is limited. Source rocks should not only consist of abundant organic-rich mudstones but should also have close contact with permeable strata. These permeable strata can be in the form of sandstone layers with good porosity and permeability or carbonate rocks (such as ostracod shell limestone, algal limestone, dolomite, etc.) with physical properties comparable to sandstones. In other words, under geological conditions source rocks must possess not only the capacity to generate oil through thermal degradation but, more importantly, the ability to expel the generated oil and accumulate it into commercially valuable reserves.

The theory of primary migration of source rocks was first studied by petroleum geologists in the Shengli Oilfield in China. In 1990, Zhang Dunxiang et al. conducted a study on oil displacement and migration in mudstone interlayers and developed

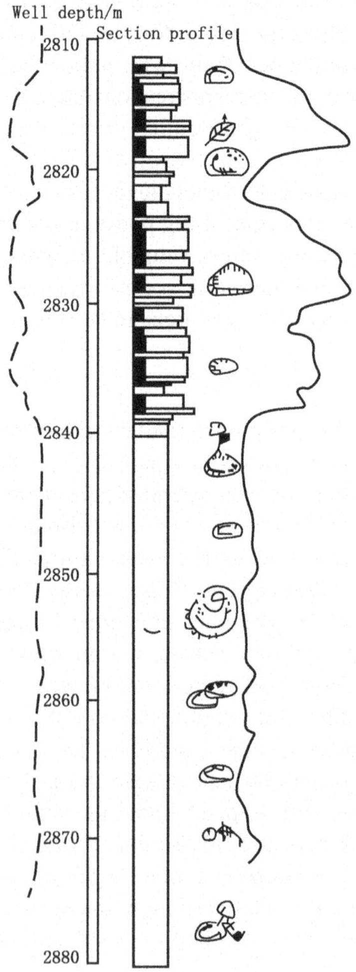

Fig. 3.1 Partial of vertical profile of well Liang-28

quantitative methods for analyzing oil in the Liangjialou Oilfield and the third member of the Shahejie Formation in the Dongying Depression. They collected a core sample from a well in the middle of the third member of the Shahejie Formation, which consisted of a 27.2 m oil-bearing layer of turbidite pebble-bearing sandstone, an overlying layer of lacustrine mudstone and dolomite measuring 0.8, and a 41.1 m layer of underlying lacustrine mudstone (Figs. 3.1). They obtained one block of oil sand and one crude oil sample from the oil layer, as well as 18 blocks of densely sampled mudstone at intervals of 1–3 m for organic geochemical analysis (Figs. 3.2)

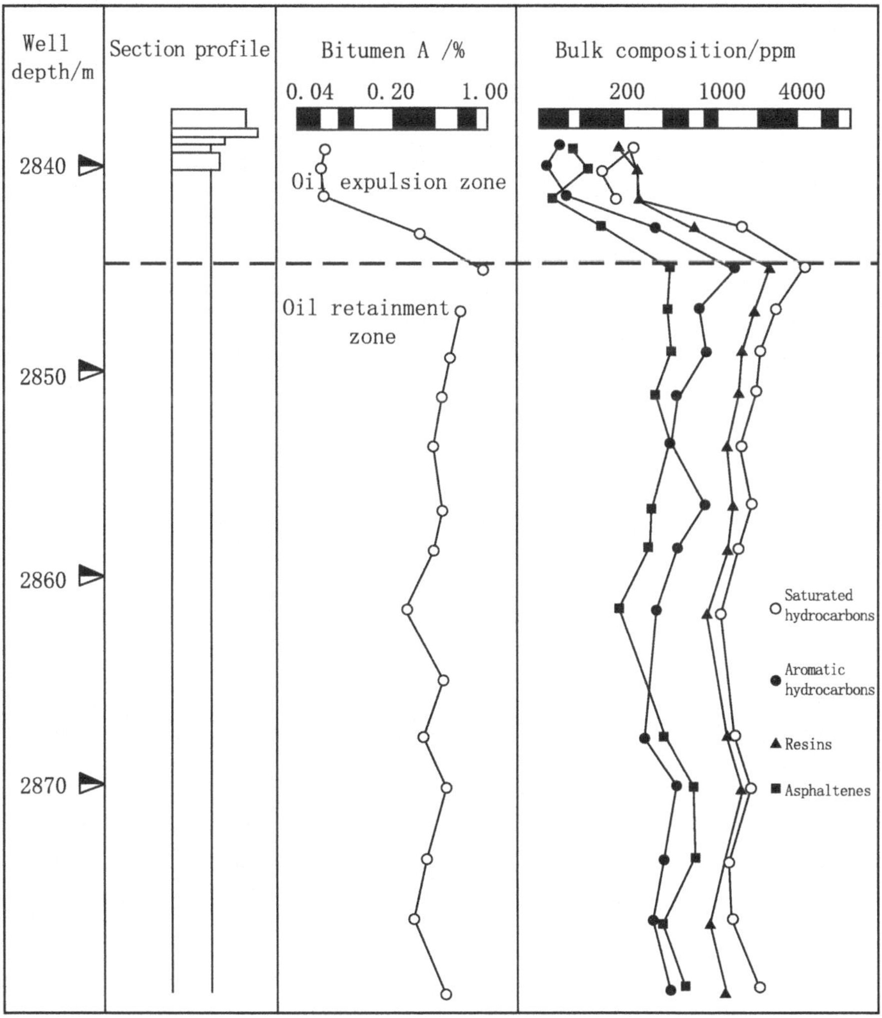

Fig. 3.2 Distribution of expelled and retained oil in well Liang-28 (after Dunxiang Zhang et al. 1990)

The research findings revealed that the primary migration of oil and gas in clastic rock sections did not result in the expulsion of oil from all mudstones but occurred within a specific thickness range. During the migration process, smaller, shorter-chain, simpler-structured, and low-polarity molecules were preferentially expelled, while larger, longer-chain, complex-structured, and more polar molecules were retained.

This indicates that under geological conditions, oil can only be expelled from the parts of source rocks that are close to the top and bottom interfaces or in proximity to permeable formations. Therefore, source rocks must consist of both mudstones and sandstones. The interbedding of sandstone within the mudstone enables primary migration of oil. Oil generated in the mudstone is initially stored as dispersed oil droplets in the mudstone pores. For these dispersed oil droplets to form commercially valuable oil reservoirs, they must migrate to the sandstone pores that are in contact with the upper and lower mudstone surfaces, facilitating primary migration in the source rocks.

Understanding the most favorable combinations of mudstone and sandstone interlayers for primary oil migration is crucial. This chapter aims to clarify the effective driving forces of primary migration of oil from source rocks through a mechanical analysis of the process. It also aims to establish a mathematical model of oil primary migration and propose a quantitative method for determining the thickness of the effective oil drainage zone in source rocks. By utilizing this quantification method, the efficiency of oil drainage zones in different sand-shale combinations can be compared and analyzed. Additionally, it will help identify the main controlling factors of primary oil migration from source rocks, providing insights into the oil-generation and accumulation mechanisms of source rocks and allowing for an objective evaluation of their oil-reserve capacity under subsurface geological conditions.

3.1 Calculation Model of Primary Migration

The primary process where driving forces overcome resistance. When the driving forces of migration migration of dispersed oil droplets in source rocks is essentially a are strong enough to surpass the resistance, the oil generated in the mudstone starts to migrate towards the sandstone interlayer. However, once the sandstone pores become filled with oil, the migration process is hindered. As a result, the maximum thickness of the effective oil drainage zone in the mudstone of the source rock depends on two factors: the driving forces and resistance within the mudstone itself, as well as the accommodation space provided by the sandstone interlayer.

3.1.1 *Driving Forces and Resistance of Primary Migration*

During the primary migration of oil in source rocks, several forces come into play, including formation pore fluid pressure, self-gravity, and capillary force (Fig. 3.3). By examining the magnitude and direction of these forces, we can gain insights into the specific roles they play in the expulsion process of oil from source rocks.

3.1.1.1 Pore Fluid Pressure

In source rocks containing interbedded sandstones and mudstones, the lithostatic pressure distribution in the sandstone layer corresponds to the hydrostatic pressure distribution. As a result, the pore fluid pressure on the sandstone-shale interface is equivalent to the normal hydrostatic pressure. In contrast, the lower part of the shale in mudstones typically experiences abnormally high pressure due to factors such as under-compaction, clay mineral conversion, hydrocarbon generation pressurization, and hydrothermal pressurization. Consequently, the pore fluid pressure in the mudstones exceeds the hydrostatic pressure at the same depth. This generates a significant difference in pore fluid pressure between the interior of the mudstone and the interface with the sandstone (Fig. 3.4). The pressure difference acts as an effective driving force for the primary migration of oil, with the direction of this pressure difference being from the interior of the mudstone towards the upper and lower contacting sandstone layers.

Once the direction of the pressure difference in the source rock is established, it becomes necessary to determine the magnitude of this force. In mudstone formations with overpressure, the abnormal pressure exerted on the pore fluid can be divided into two components: the normal hydrostatic pressure and the residual pressure exceeding the hydrostatic pressure (Fig. 3.5). This can be expressed through equation (3.1), where the first term represents the hydrostatic pressure and the second term represents

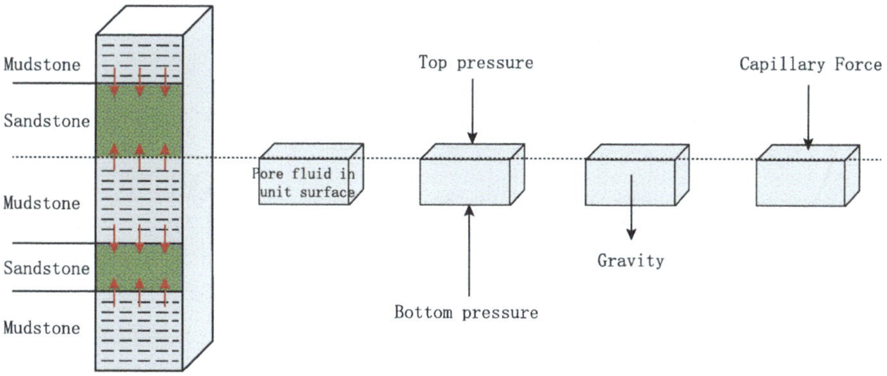

Fig. 3.3 Schematic diagram of pore fluid stress analysis

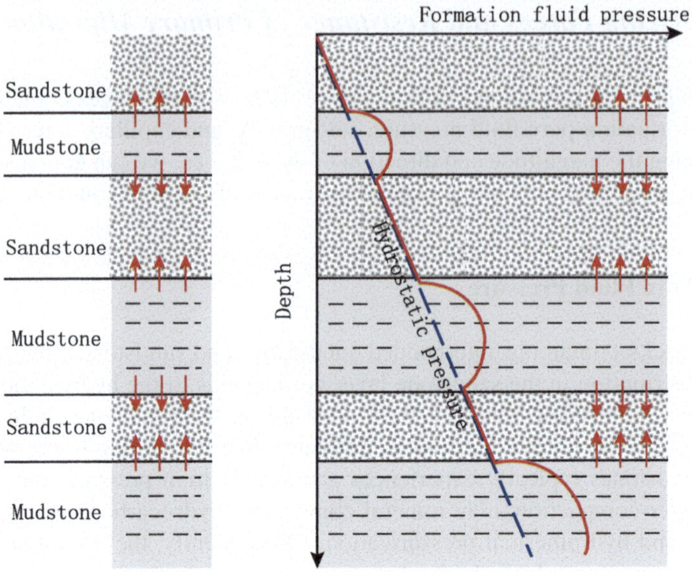

Fig. 3.4 Schematic diagram of pore fluid pressure distribution in source rocks

the residual fluid pressure.

$$p = \rho_w g Z + p_r \tag{3.1}$$

In the formula,
p—formation pore fluid pressure, Pa;
ρ_w—the density of water, kg/m^3;
g—Gravity acceleration, 9. 8N/kg;
Z—burial depth of pore fluid, m;
p_{r}—residual pressure of formation pore fluid, Pa;

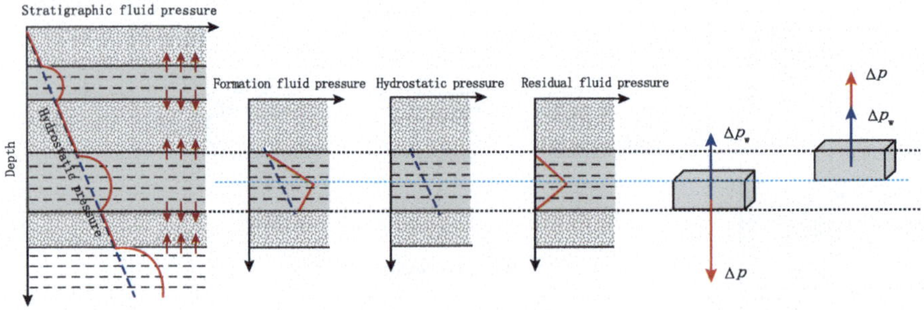

Fig. 3.5 Decomposition diagram of pore fluid pressure in mudstone

The hydrostatic pressure increases proportionally with the burial depth. When the pore fluid column, with a height of Δh per unit area, migrates upward or downward, the hydrostatic pressure difference is equivalent to the weight of that pore fluid column. The upward force exerted by the weight of the water column is balanced by the gravity of the pore fluid itself, and has negligible impact on the oil migration process, thus it can be disregarded.

The distribution of residual fluid pressure is zero at the upper and lower sandstone-mudstone interfaces, reaching its maximum in the middle of the mudstone layer. From the middle of the mudstone layer towards the upper and lower sandstone layers, the residual fluid pressure gradually decreases. Assuming that the residual pressure in the middle of the mudstone layer is (k-1) times the hydrostatic pressure at the same depth, the distribution of residual pressure increases linearly with increasing burial depth above the central mudstone layer and decreases linearly with increasing burial depth below the central mudstone layer. Therefore, for the pore fluid column with a height of Δh per unit area, the pressure difference generated by the residual pressure can be calculated using a linear equation and the initial boundary condition.

$$\Delta P = \frac{(k-1)\rho_w g Z_m}{H} . \Delta h \tag{3.2}$$

In the formula,

Δp—residual pressure difference acting on the pore fluid with height Δh per unit area, Pa;

k—formation pressure coefficient in the middle of mudstone;

ρ_w—the density of water, kg/m^3;

g—Gravity acceleration, 9.8 N/kg;

Z_m—burial depth in the middle of the mudstone, m;

H—half thickness of mudstone layer, m;

Δh—pore fluid height, m.

3.1.1.2 Gravity

Gravity is the weight of the pore fluid itself. For a pore fluid with a height of Δh per unit area, the gravity of it is as follows:

$$G = \rho_O g \Delta h \tag{3.3}$$

In the formula,

G—the gravity of pore fluid per unit area, N/m^2;

ρ_o—density of oil, kg/m^3;

g—Gravity acceleration, 9. 8 N/kg;

Δh—pore fluid height, m.

Gravity always acts vertically downward. As mentioned earlier, during oil migration, the direction of gravity and the hydrostatic pressure difference are opposite to each other. The combined effect of these forces results in a net buoyancy force caused by the density difference between oil and water. Due to the higher density of water compared to oil, the net buoyancy force is always directed upward. For oil droplets present in the pores of the upper mudstone, the net buoyancy force serves as the driving force for their migration. It propels them upward during the migration process. In contrast, for oil droplets in the pores of the lower mudstone, the net buoyancy force acts as resistance to their migration. It opposes their upward movement.

$$F = (\rho_w - \rho_o) g \Delta h \tag{3.4}$$

In the formula,

F—the net buoyancy force on the pore fluid with a height of Δh per unit area, N/m^2;

ρ_W—the density of water, kg/m^3;

ρ—density of oil, kg/m^3;

g—Gravity acceleration, 9.8 N/kg;

Δh—pore fluid height, m.

During the primary oil migration, as oil droplets gather and form continuous streams, the height of the oil column gradually increases, amplifying the effect of net buoyancy. Simultaneously, the pressure difference caused by residual pressure also increases. However, the net buoyancy force is typically less than 1% of the residual pressure difference, as indicated in Table 3.1 and formula (3.5).

Therefore, in the calculation process, the net buoyancy can be safely ignored.

$$\frac{F}{\Delta p} = \frac{(\rho_w - \rho_o)H}{(k-1)\rho_w Z_M} \tag{3.5}$$

In the formula,

F—the net buoyancy force on the pore fluid with a height of Δh per unit area, N/m^2;

Δp—residual pressure difference of pore fluid with height Dh per unit area, Pa;

ρ_w—the density of water, kg/m^3;

ρ_o—density of oil, kg/m^3;

Z_m—burial depth in the middle of the mudstone layer, m;

H—half thickness of the mudstone layer, m;

Table 3.1 Calculation results of the ratio of net buoyancy to residual pressure difference

Calculating parameters	Unit	Value	Calculating result
ρ_w	g/cm^3	1	
ρ_o	g/cm^3	0.8	
H	m	15	
K	No unit	1.2	
Z_m	m	1500	

k—formation pressure coefficient in the middle of mudstone.

3.1.1.3 Capillary Force

The capillary pressure resulting from the migration of non-wetting oil in porous media is primarily characterized by capillary pressure differences. The magnitude and direction of these differences are mainly influenced by the pore structure of the medium. Therefore, to analyze the distribution of capillary forces, it is necessary to first understand the pore structure of mudstone.

During mudstone compaction, drainage is relatively straightforward in the contact zone between mudstone and sandstone. Here, the pore volume decreases significantly. Conversely, water in the central region of the mudstone is more difficult to drain, resulting in under-compacted pores. Compared to the mudstone near the sand-mudstone contact zone, the central mudstone has larger pores and pore throats, while the mudstone near the contact zone has relatively smaller ones. When dispersed oil droplets or continuous oil columns migrate within the mudstone from the larger pore throats in the middle towards the sandstone layers, they need to pass through the relatively smaller pore throats near the sandstone-mudstone interface. Generally, the process of oil migration from the higher-pressure central area to the lower-pressure areas in the mudstone can be understood as the movement of an oil droplet (or oil column) from larger pore throats to smaller ones.

Since capillary pressure difference always acts from the end of the continuous oil phase with a smaller curvature radius to the end with a larger curvature radius, the capillary pressure difference generated during oil migration within the reservoir also opposes oil migration. Specifically, it points from the sandstone-mudstone interface towards the center of the mudstone. The magnitude of the capillary pressure difference is as follows:

$$p_c = 2\sigma\left(\frac{1}{r1} - \frac{1}{r2}\right) \tag{3.6}$$

In the formula,

p_c—capillary pressure difference, Pa;

σ—interfacial tension, N/m;

r_1—the average pore-throat radius of the mudstone near the sandstone-shale contact zone, m;

r_2—average pore-throat radius near the middle of mudstone, m.

When the continuous oil phase flows through pores and throats of varying thickness, the two sides of each curved throat experience equal deformation of the two-phase interface. Consequently, the capillary pressure difference on both sides of all the throats in the middle section is zero. The final capillary pressure difference is determined by the curvature of the oil-water interface at the front and rear ends of the continuous oil phase, and it is unaffected by the shape of the pore throat in the middle section (Zhixin 2000).

To calculate the capillary pressure difference, the pore structure of the mudstone is classified into two types: the sharply reduced pore structure near the sandstone-shale contact zone and the pore structure in the relatively under-compacted middle area of the mudstone. By classifying the pore structure in this way, it is only necessary to calculate the average pore-throat radius near the sandstone-shale contact zone and the average pore-throat radius in the middle of the mudstone when determining the capillary pressure difference. This classification allows for a quantitative characterization of the migration effect.

In summary, the primary migration process of oil in source rocks involves overcoming capillary resistance with the remaining pressure difference as the driving force. When the residual pressure difference is insufficient to overcome the capillary pressure difference, the oil generated in the source rock finds it difficult to pass through the narrow throats and instead remains trapped in the mudstone pores. However, with the accumulation of oil generated, these dispersed oil droplets gradually coalesce. As the flow continues, the residual pressure difference in the source rocks increases. Once the residual pressure difference exceeds the capillary pressure difference, the oil generated in the mudstone can break through the narrow throat blockage and enter the adjacent sandstone layer until it fills the entire sandstone pores. Therefore, the capillary pressure difference in the source rock determines the thickness of the oil stagnation zone in the mudstone. By subtracting the thickness of the oil retention zone at the upper and lower parts from the thickness of the mudstone, we can obtain the effective oil drainage zone thickness of the source rock.

For a given unit area of pore fluid, assuming the residual pressure difference is equal to the capillary pressure difference (Eq. 3.7), the thickness of the oil stagnation zone in the mudstone can be calculated using Eq. 3.8, and the thickness of the effective oil drainage zone can be calculated using Eq. 3.9.

$$\Delta p - p_c = (k-1)\rho_w g \frac{Z_m}{H}\Delta h - 2\sigma\left(\frac{1}{r_1} - \frac{1}{r_2}\right) = 0 \tag{3.7}$$

$$\Delta h = \frac{2\sigma H\left(\frac{1}{r_1} - \frac{1}{r_2}\right)}{(K-1)\rho_w g Z_m} \tag{3.8}$$

$$h = H - \frac{2\sigma H\left(\frac{1}{r_1} - \frac{1}{r_2}\right)}{(K-1)\rho_w g Z_m} \tag{3.9}$$

In source rocks that consist of alternating sandstone and mudstone layers, the mudstone layer is sandwiched between upper and lower sandstone layers that possess good porosity and permeability. Therefore, when considering the thickness of a single mudstone layer, it is divided into two parts: the upper mudstone (with a thickness of H), which expels oil upward, and the lower mudstone (also with a thickness of H), which expels oil downward. Using the appropriate equations, the thickness of the oil retention zone and the effective oil drainage zone can becalculated separately. If we neglect the net buoyancy resulting from the density difference between oil and water, we can apply the same equations (3.8 and 3.9) to calculate the thickness of the oil stagnation zone and the effective oil drainage zone in both directions.

3.1.2 Limited Storage Space of Source Rocks

Nonmarine oil-bearing basins often consist of interbedded sandstone and mudstone sequences, where mudstone or shale layers are interspersed with sandstone layers. The presence of sandstone layers plays a crucial role in controlling the primary migration and accumulation of oil. Firstly, sandstone layers provide an effective storage space for the generated oil, facilitating continuous oil generation reactions. Additionally, the pressure difference between the sandstone and mudstone layers promotes the accumulation of dispersed oil droplets within the mudstone. Under the influence of this pressure difference, the oil can overcome capillary resistance and achieve primary migration. However, once the oil from the mudstone fills the pores of the sandstone, the migration process becomes impeded. Therefore, apart from the driving force for primary migration, another critical factor determining the thickness of the effective oil drainage zone in the source rock is the limited storage capacity of the sandstone layer. When the pores of the sandstone become filled with oil, the combined volume of oil that can be expelled from the overlying and underlying mudstone layers is equivalent to the pore volume of the sandstone:

$$Ah_s\phi_s = Ah_{sh1}\phi_{sh1} + Ah_{sh2}\phi_{sh2} \tag{3.10}$$

If we assume that the thickness of the upper and lower mudstone oil drainage zones is directly proportional to their respective oil drainage powers (with a ratio of k', we can calculate the thickness of the oil drainage zone based on the limited storage space within the sandstone interlayer. The following equations are used for these calculations. Equation (3.11) is utilized to determine the thickness of the effective oil drainage zone in the mudstone layer above the sandstone, while equation (3.12) is employed to calculate the thickness of the effective oil drainage

$$h_{sh1} = \frac{h_s\phi_s k'}{\phi_{sh1}k' + \phi_{sh2}} \tag{3.11}$$

$$h_{sh2} = \frac{h_s\phi_s}{\phi_{sh1}k' + \phi_{sh2}} \tag{3.12}$$

zone in the mudstone layer below the sandstone. In the formula,

A—sandstone-mudstone contact area, m^2;

h_s—thickness of sandstone, m;

$\emptyset_s$—sandstone porosity;

Table 3.2 Calculation parameters of primary migration model

ρ_w	g	H	Z_m	h_s	ϕ_s	ϕ_{sh}
Water density	Gravity acceleration	½ mudstone thickness	Burial depth of mudstone center	Sandstone interlayer thickness	Sandstone interlayer porosity	Mudstone porosity

Table 3.3 Value range of empirical parameters of primary migration model

Oil-water interfacial tension σ	Pore throat radius r	Formation pressure coefficient k
15–30mN/m	0.01–0.2 μm	1.1–1.9

h_{sh1}—thickness of effective oil drainage zone of mudstone overlying sandstone, m;

f_{sh1}—porosity of mudstone overlying sandstone;

h_{sh2}—thickness of effective oil drainage zone of mudstone underlying sandstone, m;

f_{sh2}—porosity of mudstone underlying sandstone.

3.1.3 Thickness of Effective Oil Drainage Zone

In source rocks with high organic content, for the generated oil to be continuously expelled from the mudstone, it must possess sufficient oil expulsion capability and there must be ample storage space available. Both conditions are essential and must be met simultaneously. When performing calculations, it is necessary to determine the maximum thickness of the effective oil expulsion zone based on the oil expulsion capability and storage space, using Eqs. (3.9) and (3.11) or (Eq. 3.12). Subsequently, the smaller value obtained from these equations is considered as the effective thickness of the oil drainage zone in the source rock.

To calculate the thickness of the effective oil drainage zone in source rocks using the aforementioned model, the parameters involved can be obtained through statistical or experimental methods (refer to Table 3.2), or by referencing empirical parameters derived from source rocks in adjacent areas during the same time period (refer to Tables 3.3 and 3.4).

3.1.4 Application Calculation Examples

Liangjialou Oilfield is situated on the northern slope of an ancient nose-shaped uplift within the southern slope of Dongying Depression. It is adjacent to Boxing Sag in the west, a central uplift in the north, Lijin Sag in the northwest, and Niuzhuang Sag

Table 3.4 Formation pressure coefficient of main nonmarine oil-bearing basins in China

Basin	Formation			Formation pressure coefficient
Bohai Bay Basin	Cenozoic	Paleogene	Shahejie Formation	1.25–1.8
Songliao Basin	Mesozoic	Cretaceous	Qingshankou Formation	1.13–1.65
Sichuan Basin		Jurassic	Ziliujing Formation	1.3–1.8
Ordos Basin		Triassic	Yanchang Formation	1.12–1.56
Junggar Basin	Paleozoic	Devonian	Lucaogou Formation	1.2–1.67
		Carboniferous	Carboniferous	1.37–1.9

in the northeast. The oilfield covers an approximate area of 250 km^2. During the sedimentation period of the third member of the Shahejie Formation in Dongying Depression, the lake experienced significant expansion due to intensified crustal rifting and warm, humid climatic conditions. As a result, the main source rocks of the lower lake environment during the sedimentation period of and middle of the third member of the Shahejie Formation were developed. Although the study area was predominantly a calm deep the middle sub-member of the third member of the Shahejie Formation, geological events such as high-density currents rushing through shallow lakes during the rainy season led to turbidites in the deep lake area. These turbidity currents broke into the center of the lacustrine basin, creating a lithofacies structure where deep lacustrine mudstones sandwiched turbidite fans, with the turbidites inserting into the deep lacustrine mudstones.

This study focuses on a specific example of source rock analysis within the Liangjialou Oilfield. The source rock at a depth of 2810–2880 m in the middle sub-member of the third member of the Shahejie Formation in well Liang 28 was analyzed. The effective thickness of the oil drainage zone in the well was examined using the source rock analysis method and primary migration calculation model proposed in this paper. Firstly, based on the lithology data from well Liang 28, a combination of two layers of mudstone interbedded with a layer of sandstone within the interval of 2821–2859.45 m was identified as a single source rock combination on the section (Fig. 3.6). Subsequently, various calculation parameters were determined for further analysis (refer to Table 3.5).

According to the formula (3.9), the thickness of the oil discharge zone determined by the effective oil discharge power is calculated. To simplify the calculation process, the (1/r1-1/r2) in the calculation formula of the capillary pressure difference is simplified to 1/r1. The reason is that when r1 and r2 differ by 2 orders of magnitude, which means 1/r2 is about 1% of 1/r1, 1/r2 is negligible.

$$h = H - \frac{2\sigma H}{r(k-1)\rho_w g Z_m} = 6.18m$$

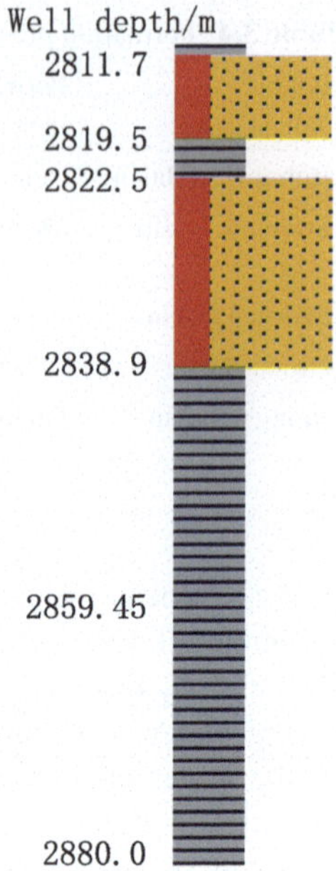

Fig. 3.6 Illustration of the vertical lithologic profile of well Liang 28

The thickness of the underlying mudstone oil drainage zone determined by the limited storage space is calculated according to equation (3.12). To simplify the calculation process, the approximate value of k′ is 1, and the average porosity of

$$h = \frac{h_s \phi_s}{2\phi_{sh}} = 15.58m$$

the upper and lower mudstone layers is taken as the mudstone porosity.

The effective thickness of the oil drainage zone in the source rock is a crucial factor determined by its oil drainage power. In this particular case, the final effective thickness of the oil drainage zone is measured to be 6.18 m.

In 1990, Zhang Dunxiang and their team conducted a comprehensive analysis in Shengli Oilfield, involving densely sampled intervals of 1–3 m in 19 blocks of well Liang 28. Among these samples, one block consisted of oil sand while the remaining 18 blocks comprised source rock. To determine the effective thickness of the oil drainage zone in the source rock from 2838.9 to 2880 m, various analysis parameters were examined, encompassing porosity, density, TOC content, kerogen microscopy, kerogen elements, vitrinite reflectance, pollen color, extractable bitumen, bulk compositions, alkane gas chromatography, reversed-phase liquid chromatography, and chromatography-mass spectrometry.

The final experimental results indicated that the oil drainage zone's thickness in this section of the source rock was measured to be 6.6 m, which closely aligned with the thickness calculated using the primary migration calculation model mentioned above. This outcome validates the scientific integrity of the model used.

Table 3.5 Parameters for calculating thickness of the effective expulsion zone of well Liang 28

water density ρw	Gravity acceleration g	½ mudstone thickness H	Burial depth of mudstone center	Thickness of sandstone interlayer	Sandstone porosity ϕ_s	Mudstone porosity ϕ_{sh}	Oil-water interfacial tension σ	Pore throat radius r1	Formation pressure coefficient k
1000 kg/m^3	9. 8N/m	20. 55 m	2859 45 m	16. 4 m	19%	10%	19. 7 mN/m	0. 01 μm	1. 2

3.2 Main Controlling Factors of Primary Migration

The primary migration of oil within source rocks, which consist of alternating layers of sandstones and mudstones, is influenced and controlled by various geological factors. These factors include the proportion and thickness of sandstones and mudstones, the arrangement of interbedded formations, physical conditions, pore structure, and formation pressure coefficients. To assess the impact of each factor on the process of primary oil migration, a single-factor analysis method is employed to compare their respective controlling effects. This analysis helps identify the favorable conditions for oil expulsion from the source rock. To facilitate comparative analysis, the efficiency of oil expulsion from the source rock is defined as the ratio between the thickness of the effective oil drainage zone and the thickness of the mudstone. Similarly, in the case of source rocks composed of multiple layers of sandstones and mudstones, the efficiency of oil expulsion is defined as the ratio between the sum of the thicknesses of the effective oil drainage zones in each mudstone layer and the total thickness of the mudstone.

3.2.1 Geologic Factors Related to Effective Oil Expulsion Dynamics

3.2.1.1 Pore Throat Radius

To examine the impact of different pore throat radii in mudstone on the efficiency of oil expulsion from source rock, we conducted a study using a specific sedimentary model. The model consisted of two layers of mudstone and one layer of sandstone, with a total thickness of 72 m. In this study, we varied the pore throat radius of the mudstone from 0.02 to 0.2 μm, while keeping the sandstone thickness at 24 m and the mudstone thickness at 48 m. We then calculated the oil expulsion efficiency of the source rock under these conditions. The results clearly demonstrated that, all else being equal, a larger pore throat radius in the mudstone led to higher oil expulsion efficiency. This relationship is depicted in Fig. 3.7, where the trend shows that as the pore throat radius of the mudstone increases, the oil expulsion efficiency also increases.

3.2.1.2 Formation Pressure Coefficient

To assess the impact of different formation pressure coefficients on the efficiency of oil expulsion from source rocks, we conducted a study using a specific sedimentary model. The model consisted of two layers of mudstone and one layer of sandstone, with a total thickness of 72 m. In this study, we maintained a sandstone thickness of 24 m and a mudstone thickness of 48 m, while varying the formation pressure

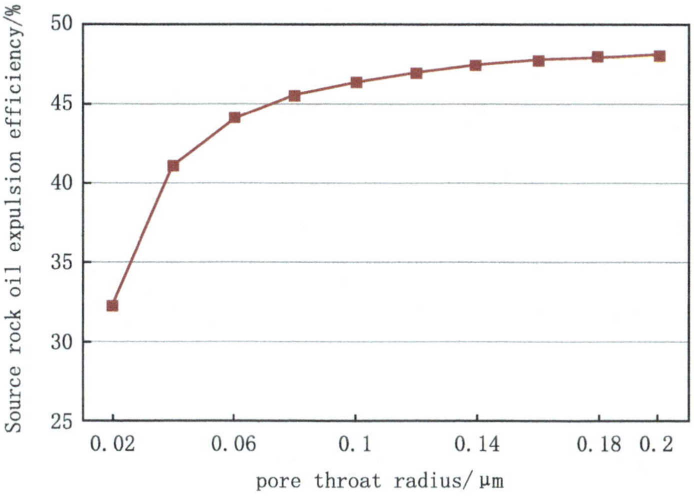

Fig. 3.7 Oil expulsion efficiency of source rocks with different pore-throat radii

coefficient from 1.2 to 1.9. We then calculated the oil expulsion efficiency of the source rock under these conditions. The results clearly indicate that, when all other factors are held constant, a higher formation pressure coefficient corresponds to a greater oil expulsion efficiency. This relationship is illustrated in Fig. 3.8, where the data reveals that as the formation pressure coefficient increases, the efficiency of oil expulsion from the source rock also increases.

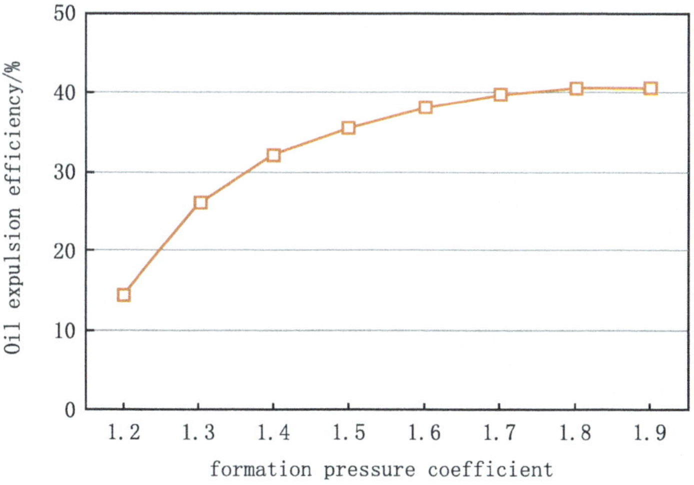

Fig. 3.8 Oil expulsion efficiency at different formation pressure coefficients

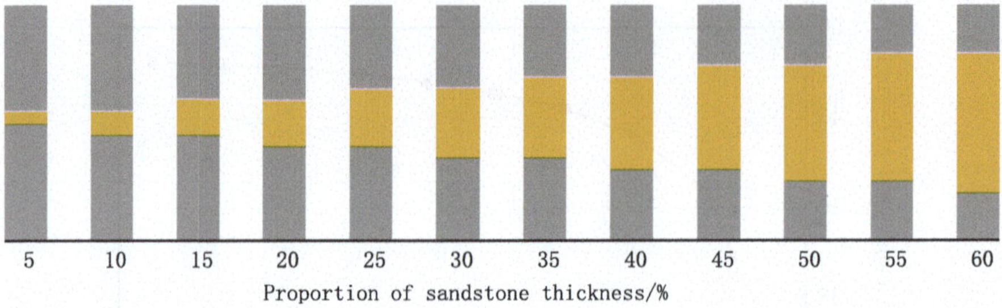

Fig. 3.9 Schematic diagram of source rocks with different sandstone-mudstone thickness ratios

3.2.2 Geological Factors Related to Space

3.2.2.1 Sandstone-Mudstone Thickness Ratio

In order to examine the influence of different sandstone-mudstone thickness ratios on the efficiency of oil expulsion from source rocks, we conducted a study using a specific sedimentary model. The model consisted of two layers of mudstone enclosing a central layer of sandstone, resulting in a total thickness of 72 m. We varied the sandstone thickness percentage, which represents the proportion of sandstone thickness relative to the total thickness of the source rock, ranging from 5 to 60% with equal intervals. This analysis is represented in Fig. 3.9.

Subsequently, we calculated the oil expulsion efficiency of the source rock for each sandstone thickness percentage. The findings indicate that, under comparable conditions, the optimal oil expulsion efficiency was observed when the sandstone thickness percentage was around 30%. This is illustrated in Fig. 3.10. Thus, within geological conditions, it can be concluded that an source rock composed of a sandstone-mudstone ratio of approximately 3:7 tends to be most favorable for the primary migration of oil.

3.2.2.2 Sandstone Porosity

To assess the impact of different sandstone porosities on the efficiency of oil expulsion from source rocks, we conducted a study using a particular sedimentary model. The model consisted of two layers of mudstone enclosing a central layer of sandstone, resulting in a total thickness of 72 m. In this study, we kept the sandstone thickness percentage at 20% and varied the sandstone porosity within the range of 10 to 30%. We then calculated the oil expulsion efficiency of the source rock under these conditions. The findings indicate that, when all other factors are held constant, a higher sandstone porosity corresponds to a higher oil expulsion efficiency. This relationship is depicted in Fig. 3.11, where the data reveals that as the sandstone porosity increases, the efficiency of oil expulsion from the source rock also increases.

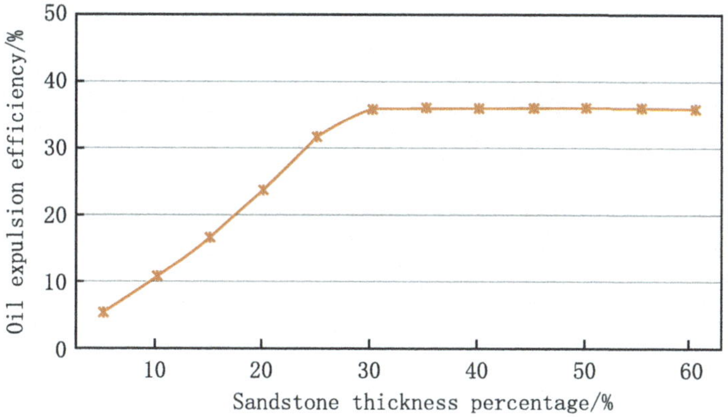

Fig. 3.10 Oil expulsion efficiency of source rocks with different sandstone-mudstone thickness ratios

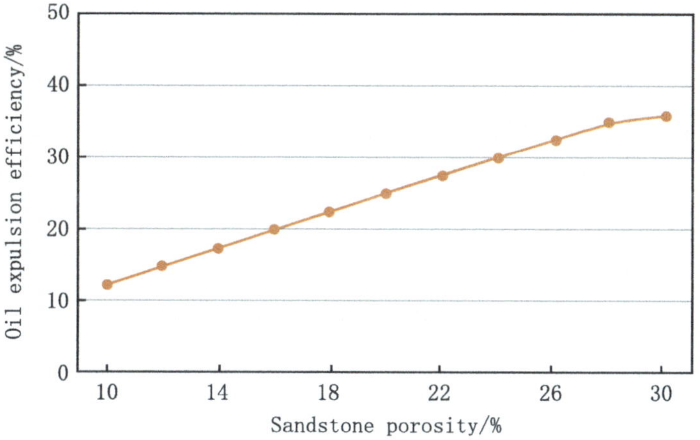

Fig. 3.11 Oil expulsion efficiency of source rocks with different sandstone porosities

3.2.2.3 Single-Layer Thickness of Sandstone-Mudstone

To investigate the impact of different single-layer thicknesses of sandstone and mudstone on the oil expulsion efficiency of source rocks, we conducted a study using a specific example. The source rock in consideration had a total thickness of 72 m, with 24 m allocated to the sandstone layer and 48 m to the mudstone layer. The sandstone-mudstone ratio was fixed at 1:2. In our analysis, we gradually decreased the thickness of the sandstone interlayer from 24 to 1 m while simultaneously increasing the number of layers from 1 to 24 (Fig. 3.12).

Subsequently, we calculated the oil expulsion efficiency of the source rock under these conditions. The findings revealed that, under comparable conditions, thinner thicknesses of the single-layer sandstone and mudstone, as well as an increased

Fig. 3.12 Schematic diagram of source rocks with different sandstone-shale single layer thicknesses

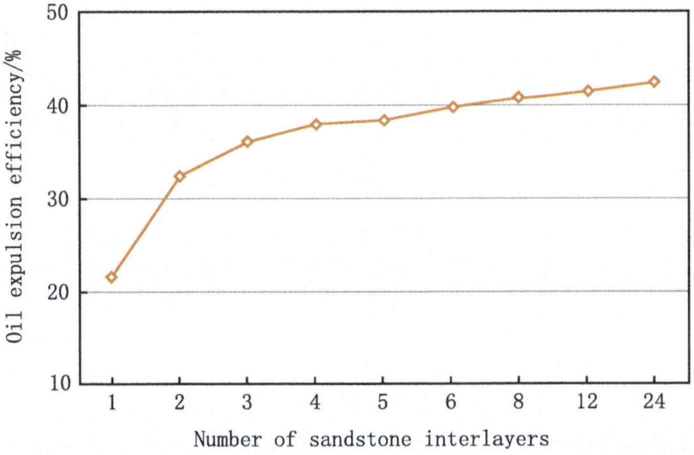

Fig. 3.13 Comparison of oil expulsion efficiency of source rocks with varying sandstone-shale single layer thickness

number of layers, corresponded to higher oil expulsion efficiency. However, it should be noted that once the thickness of the single-layer sandstone and mudstone dipped below 6 m, the rate of increase in oil expulsion efficiency slowed down (Fig. 3.13).

3.2.2.4 Combination Mode

To compare how different combinations of sandstone and mudstone layers impact the oil expulsion efficiency of source rocks, we conducted a study using a specific

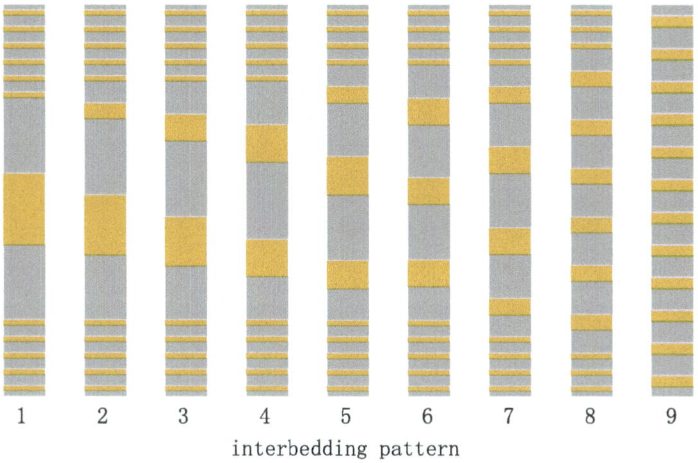

Fig. 3.14 Schematic diagram of source rocks with different interbedding patterns

example. The source rocks analyzed had a total thickness of 72 m, with 24 m designated for the sandstone layer and 48 m for the mudstone layer. The sandstone to mudstone ratio remained constant at 1:2. In our investigation, we used 12 sandstone layers and varied their individual thicknesses unevenly between 1 and 13 m (Fig. 3.14). We then calculated the oil expulsion efficiency and 13 m (Fig. 3.14). We then calculated the oil expulsion efficiency of the source rocks under these conditions. The findings revealed that, when all other factors were kept constant, the oil expulsion efficiency was higher when the individual layer thicknesses of the sandstone and mudstone were closer and more evenly distributed. This relationship is illustrated in Fig. 3.15.

3.3 Patterns of Oil Accumulation

3.3.1 Types of Oil Accumulation

In China's lacustrine basins, the oil generation and accumulation can be broadly classified into two types based on the sedimentary environments and sediments involved. The first type is characterized by oil generation and accumulation occurring within lacustrine and external fluvial-delta (or fan-delta) sedimentary settings. In these settings, the reservoirs consist of shale reservoirs made up of mudstone or shale interspersed with sandstone, as well as conventional reservoirs mainly composed of sandstone. The distribution of these reservoirs is primarily controlled by the deltaic depositional system. An example of this type can be found in the Songliao Basin, where a large river-delta petroleum system has developed in multiple phases during

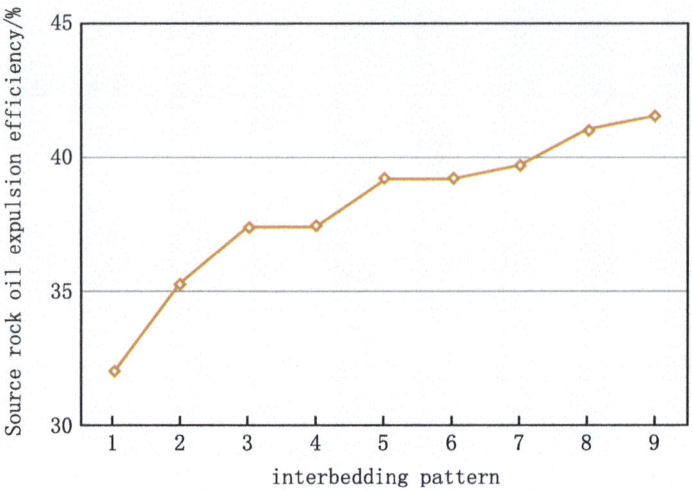

Fig. 3.15 Oil expulsion efficiency of different interlayer combination modes

the Lower Cretaceous. The second type involves oil generation and accumulation taking place within a pure lacustrine sedimentary environment without significant external inputs. In this case, the reservoirs are referred to as autochthonous reservoirs and are composed of black shale or black mudstone interbedded with biogenic limestone or reef limestone. The distribution of these reservoirs is primarily influenced by sedimentary facies and fracture development zones. The Dazhai section of the Jurassic in the Sichuan Basin serves as a typical example of this type of reservoir.

3.3.1.1 Oil Accumulation Mechanisms in Exogenic Sedimentary Systems and lacustrine environments

In the case of the first type of oil accumulation, the basin experiences sustained subsidence and receives continuous inputs of external materials during this phase. These external materials gradually develop into overlapping sedimentary systems such as river-delta, fan-delta, alluvial fan, and lake bottom fan over multiple phases. Through the alternating deposition of delta progradation and lake flooding events, the sandstones and mudstones from these external delta sedimentary systems are vertically interbedded and horizontally superimposed. This complex geological arrangement of interbedded sandstone and mudstone provides favorable conditions for oil generation, migration, and accumulation. The sedimentary facies variations across different parts of the basin lead to regular differences in thickness ratios and interbedding patterns between sandstone and mudstone, resulting in the formation of different types of oil reservoirs.

In the gently sloping zones of the basin, fluvial-delta and fan-delta sedimentary systems commonly develop. Influenced by sediment supply and lake level changes, these deltaic systems may transgress or retreat, leading to a shifting sedimentary pattern. This results in the interdigitation of delta front sandstones and prodelta mudstones, forming vertically stacked interbeds of sandstone and mudstone. The areas near the center of the lake are predominantly composed of prodelta (or lacustrine) mudstones interbedded with sandstones, which are favorable for the development of source rocks. If the oil generated in the mudstones undergoes primary migration within the source rocks, these mudstones interbedded with sandstones can form shale oil reservoirs. In the slope areas near the basin margins, delta front sandstones with better reservoir properties are primarily present, intercalated with thin layers of mudstone. These locations are favorable for the development of reservoir rocks. If these sandstones extend towards the center of the lake and connect with areas of source rock development, or if they are linked through an efficient transport system, the oil generated and expelled from the source rocks can be driven into suitable traps within the delta front sandstones. This leads to the formation of conventional oil reservoirs. The classic example of this phenomenon is the large river-delta sedimentary system that developed during the Lower Cretaceous in the Songliao Inland Lake Basin. This system directly controlled the formation of the main Daqing oil fields.

In the steeply sloping zones of the basin, fan-delta and alluvial fan sedimentary systems commonly form. These systems are characterized by their proximity to the sediment source and the steep gradient. During seasonal floods, coarse detritus is transported directly into the deep lake and rapidly deposited, creating nearshore underwater fans that are constrained by boundary normal faults. These fans consist of a combination of gravely coarse detritus and dark mudstone. Similar to the mechanisms of oil accumulation in the sandstone-mudstone combined geological bodies found in gentle slopes, the sandstone-mudstone interleaved geological bodies in steep slopes have the potential to form shale and conventional oil reservoirs. However, due to the lower structural and compositional maturity of the sediments in steep slopes, as well as less favorable conditions for oil generation and poorer reservoir rock properties, the resulting reservoirs are typically smaller in scale.

In the slope transition areas of the basin or in the distant deep lake regions, delta front (or nearshore underwater fan) sediments can be transported and deposited under the influence of gravity. This process leads to the formation of smaller, thin lake bottom fans. The sandstone deposits from these lake bottom fans occur between mudstones in deep to semi-deep lacustrine environments, creating lithological combinations of mudstones and lens-shaped sandstones. These combinations themselves can serve as self-generating and self-contained lithologic oil accumulations. An example of this type can be seen in the source rock reservoirs developed in the middle part of the third member of the Shahejie Formation, Eocene, in the Liangjialou oil field located in the Dongying depression of the Bohai Bay Basin.

3.3.1.2 Oil Accumulation Mechanisms in Endogenic Lacustrine Sedimentary Environments

The second type of oil generation and accumulation occurs in pure lacustrine sedimentary environments, where there is a lack of external material sources. In this scenario, the sediments consist mainly of chemogenic and biogenic carbonate rocks (such as ostracod shell limestone, reef limestone, oolitic limestone, dolomite, etc.) and mudstones. Carbonate rocks are often found as thin interbeds within mudstones, while certain carbonate rocks can be distributed within black argillaceous rocks or black shales in the form of small mounds, lenses, and pods, without distinct sedimentary architecture. The oil generated in mudstones can migrate a short distance and accumulate in reservoirs composed of carbonate rocks. These reservoirs display karstic cavities and fractures. The combination of these cavity and fracture-developed carbonate rocks and mudstones forms an effective source rock system, as well as a self-generating and self-containing shale oil reservoir. An example of this type can be seen in the Lower Jurassic Da'anzhai section of the Sichuan Basin, where the source rocks exhibit these characteristics.

3.3.2 Mechanism of Petroleum Accumulation

According to the theory of basin formation, oil generation, and oil accumulation, the processes of oil generation, migration, and accumulation occur within the geological evolution of a basin, from its formation to development and subsequent contraction. Therefore, when studying petroleum accumulation mechanisms, it is necessary to consider the entire geological process.

During the continuous subsidence phase of the basin, sedimentary materials steadily accumulate. In a continental oil-bearing basin, clay and sandy deposits are the main components of these sediments. This subsidence and deposition process involves the continuous filling and pressurization of these two types of materials. As the burial depth of the strata increases, the temperature also rises. Under the combined effects of temperature, pressure, and catalytic conditions, organic matter in the oil shale undergoes maturation and transforms into oil. Initially, the oil exists as dispersed droplets within the pores of the mudstone. As oil generation progresses, the abnormal pressure within the mudstone increases, causing these dispersed oil droplets to coalesce into streams. Eventually, driven by pressure gradients, they breakthrough the narrow barriers and infiltrate adjacent sandstone layers, initiating primary migration within the source rock and forming a self-generating and self-containing shale oil reservoir.

During this continuous subsidence phase, the basin acts as a closed system without pressure release conditions between the source rock and external reservoirs. As a result, the oil accumulated in the shale reservoirs during this stage has not undergone long-distance secondary migration. Only a small proportion of the oil migrates a short distance and accumulates in adjacent turbidite sandstone lenses or biogenic limestone (reef limestone)reservoirs, forming lithological oil reservoirs. The majority of the oil remains within the sandstone interlayers of the source rocks, awaiting the establishment of favorable migration conditions. Once these conditions are met, the oil will migrate and accumulate in effective traps outside the source rock.

During the uplift phase of a basin, the strata above the source rock undergo erosion, reducing their thickness and effective stress. This transition in the physical conditions of the basin from loading and pressurization to unloading and depressurization has important implications. Sandstone sediments possess elastic properties, meaning they can rebound towards their original state once the external load is removed, as long as the loading does not exceed their elastic deformation limit. As a result, negative pressure is generated within the pores of the sandstone. The magnitude of this negative pressure is directly related to the amount of rebound, which, in turn, is directly proportional to the thickness of erosion. Typically, the erosion thickness is greatest at the margins of the basin due to structural uplift, gradually decreasing towards the center of the basin. Consequently, the rebound of the sandstone is larger on the slopes of the basin margins, while it is smaller in the areas of the source rock. This discrepancy leads to a more significant pressure drop in the sandstone layers of the basin margin slopes compared to the source rock areas, creating a differential pressure between these two regions. As a result of this pressure differential, oil that was previously stored in the sandstone interlayers of the source rock undergoes secondary migration towards areas of reduced pressure through sandstone systems associated with gentle slope rivers and deltas. Ultimately, the oil accumulates in well-preserved traps, forming conventional reservoirs.

During the comprehensive shrinkage phase of the basin, which typically involves several small cycles of subsidence and uplift erosion, the previously formed oil reservoirs undergo further adjustments and refinements, eventually reaching their final states.

3.3.3 Oil Accumulation Conditions

The formation of petroleum accumulations in subsurface geological conditions is a complex process. It begins with the transformation of organic matter within mudstone into dispersed oil droplets through thermochemical catalysis. However, for these droplets to come together and form commercially viable reservoirs, they must undergo migration from the pores of the mudstone to the pores of the surrounding sandstone layers. This primary migration process occurs within the source rock itself and is integral to the formation of shale reservoirs. In addition, for the oil to accumulate in conventional reservoirs, it needs to undergo secondary migration from the source rock to external sandstone reservoirs. This secondary migration occurs when the front sandstones of an external deltaic sedimentary system effectively connect with the sandstone layers within the source rock. Under the driving forces of secondary migration, the oil that has already accumulated within the sandstone pores can flow into the favorable traps present in the deltaic front sandstones, thereby forming conventional reservoirs.

3.3.3.1 Formation Conditions of Shale Reservoirs

The formation of a shale oil reservoir involves the generation and accumulation of oil within the source rock itself. The conditions necessary for the formation of a shale oil reservoir are directly related to the geological conditions within the source rock. These conditions can be categorized into three aspects: sufficient organic matter, favorable reservoir space, and an effective driving force for oil expulsion between sandstone and mudstone in the source rock.

Adequate Organic Matter Content in Source Rocks for Oil Generation

The formation of a shale oil reservoir relies on two main factors. Firstly, there must be an ample amount of organic matter present in the source rock. The quality and quantity of this organic matter serve as the foundation for oil generation. Secondly, under specific subsurface geological conditions, these organic materials can undergo extensive conversion into oil.

Source Rocks Need to Possess Favorable sandstone (or Carbonate) reservoir Spaces

The generation of oil from organic matter in mudstone occurs within a limited reaction space. If the resulting oil is unable to expel from the mudstone in a timely manner, it hinders the oil generation process. However, this situation changes significantly when sandstone or carbonate intercalations with favorable porosity and permeability conditions are present amidst the mudstones. These intercalations provide ample intergranular pores and fractures, offering significant space for the oil generated from the source rocks and enabling primary oil migration. Additionally, the favorable physical properties of these intercalations facilitate heat conduction, which promotes continuous thermal pyrolysis of the surrounding mudstone or shale. Theoretical calculations based on the primary migration model suggest that a 3:7 sandstone to mudstone thickness ratio in the source rock is optimal for primary oil migration under geological conditions. Moreover, a higher porosity in the sandstone and a more evenly distributed intercalation within the mudstone result in a higher efficiency of oil expulsion from the source rock. In China's continental oil-containing basins, exploration practices have validated that the primary source rocks consist mainly of mudstones or shales intercalated with sandstone or sandstone-comparable carbonate, maintaining a sandstone to mudstone ratio of 3:7 for the most part.

During the development and production stages of low porosity and low permeability tight oil and shale oil reservoirs, sand fracturing techniques are commonly employed to enhance well production. This technique involves injecting a substantial amount of quartz sand or ceramic granules into fractures, improving the stratum's conductive capability. By creating an effective pore pressure differential between the tight and permeable layers, the injected granules promote the extraction of immobile oil within the tight reservoir.

An Efficient Driving Force is Essential to Expel Oil from Source Rock

The process of expelling oil from source rocks relies on a pressure differential that overcomes the resistance caused by capillary forces. When this pressure differential fails to overcome the capillary pressure, the oil generated within the source rock struggles to break through the narrow pathways, resulting in it being trapped within the pores of the mudstone. As the volume of oil increases, these isolated oil droplets gradually merge, forming larger flows and increasing the pressure differential within the source rock. Once the pressure differential surpasses the capillary pressure, it creates an effective force that drives the expulsion of oil. This allows the oil generated in the mudstone to break through the narrow pathways and enter the adjacent sandstone layer until it completely fills the pores of the sandstone.

In summary, if a set of source rocks meets the three aforementioned conditions, it has the potential to form a shale oil reservoir. In simpler terms, a source rock that can produce a significant amount of oil, expel that oil effectively, and accumulate it under geological conditions constitutes a shale oil reservoir.

3.3.3.2 Conventional Reservoir Formation Conditions

Conventional reservoir formation is a process where oil, generated within source rocks, accumulates outside of the source. The conditions for conventional reservoir formation are primarily related to the entrapment of oil outside the source and consist of three essential aspects. Firstly, there must be deltaic sandstones that are effectively connected to the sandstone interlayers within the source rock. This means that there should be a continuous and interconnected network of sandstone reservoirs extending from the source rock. Secondly, an effective pressure difference should exist between the area where the source rock develops and the area outside the source where the reservoir rock is located. This pressure difference helps facilitate the migration of oil from the source rock into the adjacent reservoir rock. Finally, favorable conditions for trapping and preserving the accumulated oil must be present. These conditions can include structural traps, such as folds or faults, or stratigraphic traps, such as pinch-outs or unconformities. These traps prevent the oil from further migrating and help preserve it in the reservoir rock.

Deltaic Sandstones Effectively Communicating with the Sandstone Interlayer in Source Rocks

During the sedimentation process in a basin, the entry of an external river- delta or fan-delta sedimentary system can give rise to the development of delta front sandstones with desirable porosity and permeability on the slope of the basin. The hydrodynamic forces and sedimentary microfacies contribute to the formation of sandstones that extend from the direction of the sediment source to the sandstone interlayers within the source rock. Additionally, other sandstones may become isolated and form separate lenticular bodies. The delta front sandstones that are effectively connected to the sandstone interlayers of the source rock provide ideal locations for oil accumulation outside the source. Moreover, they offer favorable pathways for secondary migration. When the sandstone interlayers within the source rock become saturated with oil, they act as conduits for transporting the oil towards favorable traps within the delta front sandstones. This process leads to the formation of conventional reservoirs due to the forces driving the secondary migration. In contrast, the isolated sandstones can only form lithologic reservoirs through short-distance oil migration. Their mechanisms for reservoir formation and the conditions required are similar to those of shale reservoirs.

Creating a Significant Pressure Difference Between Source Rocks and Reservoirs for Optimal Oil Migration

During the continuous subsidence stage of a basin, it behaves as a predominantly closed system, lacking pressure relief between the area where source rock develops and the reservoir rocks outside of it. Consequently, any oil initially accumulated within the sandstone interlayers of the source rock does not undergo long-distance secondary migration during this stage. When the basin enters an overall uplift stage, the overlying strata above the source rocks erode, resulting in reduced effective stress. The physical state of the basin transitions from a state of loading and pressurization to unloading and depressurization. Due to the elastic nature of sandstone sediments, if the amount of loading does not exceed the limit of elastic deformation, a rebound phenomenon occurs after the external force is released, leading to negative pressure in the sandstone pores. The extent of erosional unloading in the basin determines the magnitude of the sandstone's rebound and, consequently, the degree of pressure reduction in the sandstone pores.

Typically, as a result of basin uplift, the greatest erosion thickness is observed at the basin's edge, gradually decreasing towards the depocenter. As a consequence, there is a more significant pressure drop in the sandstone layers located at the basin's edge compared to the sandstone layers in the source rock development area. This creates an effective pressure difference between these two areas. Under the influence of this pressure difference, the oil present in the sandstone interlayers of the source rocks can undergo secondary migration along the interconnected sandstones. Eventually, it accumulates within favorable traps, giving rise to the formation of conventional reservoirs.

Favorable Trap and Preservation Conditions

In addition to the requirements of suitable delta sandstones and efficient migration dynamics, the formation of oil accumulations outside of the source rock relies on favorable trapping and preservation conditions. These conditions are influenced by two main factors: the lithological attributes, thickness, and continuity of the caprock, and the stability of tectonic movements. Preservation conditions of an oil reservoir depend on the characteristics of the caprock, which is typically composed of mudstone or evaporite formations. These caprocks possess significant thickness, widespread distribution, low porosity and permeability, and stable lateral continuity. These attributes make them ideal for preserving oil reservoirs on a regional scale. However, intense tectonic activity can disrupt the integrity of established oil reservoirs. Conversely, moderate tectonic movements can facilitate the formation of reservoirs. Generally, a stable tectonic environment enhances the preservation of oil reservoirs.

To summarize, if a set of source rocks can generate, expel, and effectively accumulate a substantial amount of oil under specific geological conditions, it qualifies as a shale oil reservoir. Moreover, if these strata contain favorable delta sandstones that connect with the sandstone interlayers in the source rock, and if there are efficient secondary migration dynamics and favorable trap and preservation conditions, multiple conventional reservoirs can form outside of the source. The combined quantity of oil resources stored in both shale and conventional reservoirs represents the actual oil yield of the source rock.

3.3.4 Comparison Between Nonmarine and Marine Oil Reservoirs

The assertion made by Western scholars that oil can only be generated from marine strata, as opposed to lacustrine strata, has been a subject of ongoing debate. Chinese scholars, on the other hand, argue that oil can originate not only from marine strata but also from freshwater lacustrine sediments. Through decades of oil exploration, Chinese petroleum geologists have made significant discoveries of oil reserves in various Mesozoic-Cenozoic lacustrine sedimentary basins across China. Major oil fields such as Daqing, Shengli, and Changqing, each with reserves exceeding a hundred million tons, provide substantial evidence supporting the Chinese scholars'viewpoint that lacustrine strata can indeed generate oil.

However, certain crucial scientific questions related to the debate between oil generation in marine and lacustrine strata still remain unanswered. For instance, if both marine and lacustrine facies have the potential to produce oil, what are the inherent connections and fundamental differences between them? Can the exploration and development techniques employed for marine reservoirs be directly applied to the exploration and development of lacustrine reservoirs in China? In this section, we will conduct a preliminary analysis and discussion to address these scientific questions by comparing the characteristics of marine and lacustrine oil reservoirs.

3.3.4.1 Origin of the Controversy Between Marine-Generated and Nonmarine-Generated Oil

According to Western Scholars, Oil Generation Primarily Occurs from Marine Facies Rather Than Nonmarine Facies

In 1859, Edwin Drake, an American, led the drilling of the world's first oil well in the region surrounding Oil Creek in Titusville Village, Pennsylvania, USA. At a depth of 21.2 m, oil was discovered and, upon pumping, yielded a daily production of 4.1 tons. This oil was sourced from the Paleozoic marine black shale. The successful operation of this well sparked a wave of oil drilling in the surrounding areas of Oil Creek. By 1862, the oil production in the region had reached 410,000 tons. Concurrently, numerous oil wells emerged in western Pennsylvania, southwestern New York, and Ohio, all extracting oil from the Paleozoic marine black shale. This led to a rapid expansion of oil production in the Appalachian Basin, where these regions are located. By 1900, oil production had surged to 4.97 million tons, establishing the basin as a significant Paleozoic oil production area in the United States.

The widespread production of Paleozoic marine oil in the Appalachian Basin generated research interest among geologists. Through comparative research on marine oil and coal, it was gradually discovered that sedimentary organic matter primarily originated from bacteria, phytoplankton, zooplankton, and higher plants. It was noted that higher plants did not appear until after the Middle Devonian period; therefore, the only source of oil in the Paleozoic era of this region was the organic matter provided by marine phytoplankton. Marine phytoplankton is believed to be the main source of all sedimentary organic matter and is closely related to the natural byproducts of marine aquatic plants.

In contrast, coal is derived from terrestrial plants, and the total amount of organic matter transported from land to the sea by rivers accounts for less than 1% of the total organic matter in the ocean. Not only is the quality of organic matter significantly superior in the ocean compared to land, but the quantity of organic matter in the ocean far exceeds that of land. Marine phytoplankton and zooplankton are abundant in proteins and lipids, which are crucial components in the formation of marine oil. In contrast, the nonmarine environment does not provide the necessary conditions for the evolution of marine aquatic plants; it only supports the growth of terrestrial higher plants. These terrestrial plants have extremely low protein and lipid content but are rich in cellulose and lignin, neither of which are primary organic materials for oil generation.

Since 1913, Western oil companies, including those from the United States, have employed geologists to conduct oil exploration and geological surveys in China. However, these efforts did not result in significant discoveries, leading the companies to conclude that China's nonmarine sources lacked substantial oil reserves. The empirical evidence gathered from these explorations eventually led many Western and American geologists to adopt the belief that oil can only be generated from marine deposits.

In essence, Western scholars, particularly those from the United States, support the notion that "marine facies generate oil, whereas nonmarine facies do not." They base their argument on two main factors. Firstly, the early practices of oil exploration in the United States revealed that oil derived from the American Paleozoic era originated from marine sedimentary layers. Secondly, Western and American geologists assert that marine phytoplankton play a crucial role in the production of organic matter, whereas nonmarine environments, which primarily host higher plants, lack the necessary conditions for the presence of marine phytoplankton.

Chinese Scholars Maintain the Perspective that Petroleum Generation in China Can also Take Place in Nonmarine Environments

Starting from the 1930s, a group of Chinese geologists initiated comprehensive petroleum geological surveys in various regions such as Northern Shaanxi, Hexi Corridor, Sichuan, and the Tianshan region. During this period, significant discoveries were made. In 1938 and 1939, the Dushanzi oilfield was found in Xinjiang, located in the nonmarine Paleogene strata, while the Laojunmiao nonmarine oilfield was discovered in Yumen, Gansu, in the Cretaceous strata. One notable figure in this field is Professor Pan Zhongxiang, an expert in Petroleum Geology. After graduating from Peking University in 1931, Mr. Pan conducted multiple petroleum geological surveys, including four visits to Northern Shaanxi and field investigations in Sichuan. He stated that the petroleum resources in Northern Shaanxi stem from nonmarine Triassic and Jurassic strata and highlighted Sichuan's potential for petroleum accumulation due to its favorable geological structures. He even discovered, during his doctoral research at the University of Kansas, that the Paudwas oilfield in Northwestern Colorado, USA, produced oil from nonmarine Paleogene strata. In 1941, during the American Association of Petroleum Geologists (AAPG) conference, Pan Zhongxiang presented his influential paper titled " Nonmarine origin of petroleum in north Shensi, and the Cretaceous of Szechuan, China," thus introducing the concept of "Nonmarine Petroleum Genesis in China" for the first time.

The proposal of nonmarine petroleum genesis had a significant impact on Chinese petroleum geologists, as it instilled confidence in exploring oil resources in China's inland lacustrine basins. This proposition challenged the prevailing belief of "Chinese nonmarine petroleum poverty". Over the course of several decades of exploration, numerous large nonmarine oilfields were discovered one after another, leading to China's oil production exceeding 100 million tons in 1976. These successful outcomes not only demonstrated the abundance of petroleum resources in China's nonmarine lacustrine basins but also indicated the presence of favorable geological conditions for the formation of large oilfields. This has provided valuable empirical data and practical experience for petroleum geologists to conduct further research on the geological characteristics and genesis of nonmarine petroleum in China. Currently, the theory of nonmarine petroleum genesis is gaining acceptance among an increasing number of foreign petroleum geologists.

To summarize, the debate between marine and nonmarine petroleum genesis primarily stems from incomplete conclusions drawn solely based on superficial observations during the early stages of petroleum geological research and exploration. However, global petroleum exploration practices and the results of petroleum geochemical research have shown that both marine and nonmarine strata possess the necessary geological conditions for petroleum and reservoir formation. While there are similarities between the two, there are also essential differences. To effectively apply the successful experiences from marine oilfield exploration and development to guide the exploration and development of China's nonmarine oilfields, it is crucial to understand the intrinsic connections and fundamental distinctions between them.

3.3.4.2 Characteristics of Petroleum Formation and Accumulation in Marine Source Rocks

The United States possesses the world's most abundant reserves of Paleozoic marine sourced oil and gas from shale. By analyzing practical data from marine shale sourced oil and gas exploitation in the U.S. and considering global sedimentary environments of marine source rocks, we can outline the key characteristics of petroleum generation and accumulation in marine source rocks as follows:

The organic matter in marine source rocks primarily originates from marine planktonic algae, particularly thriving in semi-saline and saline environments. These environments with semi-closed reducing conditions facilitate the preservation of organic matter. Marine source rocks worldwide are formed in sedimentary environments consisting of saline water (10–35 salinity) and semi-saline water (1–10 salinity). The organic matter mainly comprises marine phytoplankton such as cyanobacteria, dinoflagellates, coccolithophores, and diatoms. In early Paleozoic marine source rocks of the United States, the organic matter primarily comes from bacteria and cyanobacteria; in the Mesozoic period, it is predominantly derived from coccolithophores and dinoflagellates; and in the Cenozoic era, it mostly originates from flagellates and diatoms. Planktonic algae in semi-saline to saline environments exhibit high potential for hydrocarbon generation. However, due to the semi-closed and reducing marine depositional environment, the preservation of organic matter is not as effective as in completely closed nonmarine deposits.

The reservoir space in marine source rocks typically possesses favorable physical properties. The continental shelves and slopes of oceans provide ideal conditions for the formation and accumulation of hydrocarbons in marine source rocks. River delta sandstones, which have undergone long-distance migration, exhibit high compositional and structural maturity. These sandstones are predominantly composed of quartz with few unstable components. The grains are well-sorted and rounded, and the detrital structure is primarily pore-filling cement, with grains mainly in point contact. Carbonate cement predominantly fills the pores, resulting in relatively good porosity and permeability and providing favorable reservoir conditions for accumulated hydrocarbons.

The geological setting is characterized by stability. The North American craton represents one of the most stable geological structures globally, comprising complete sequences of sedimentary rock strata from the Paleozoic, Mesozoic, and Cenozoic eras, primarily consisting of marine deposits. During the Mesozoic and Cenozoic periods of global plate tectonic shifts and collisions, the North American craton experienced mainly vertical movements without significant folding. Its exceptional stability surpasses that of other global cratons, significantly contributing to the preservation of oil reservoirs. The earliest discoveries and production of conventional oil and gas, as well as shale oil and gas, in the United States originated from the North American craton. The ability to preserve these Paleozoic marine hydrocarbons in their entirety can largely be attributed to the stable geological setting of the North American craton.

3.3.4.3 Characteristics of Petroleum Generation and Accumulation in Nonmarine Source Rocks

(1) The main source of organic matter in lacustrine basins in China's Mesozoic and Cenozoic epochs is planktonic algae thriving in semi-saline to saline environments. These basins, known as nonmarine saline lake basins, provide favorable conditions for the preservation of organic matter. The primary types of algae found in these basins are dinoflagellates and botryococcus. In Jurassic saline lake basins, the predominant organic matter comes from botryococcus, while post-Jurassic formations are primarily composed of dinoflagellates. Due to the enclosed nature of these basins, characterized by deep waters and reducing conditions, the organic matter is well-preserved.
(2) The reservoir spaces in lacustrine basins exhibit significant heterogeneity. The transition zones from basin-edge slopes to deep lake depressions are ideal areas for hydrocarbon generation and accumulation in source rocks. These zones often feature river-delta or fan-delta depositional systems along the basin slopes. The sandstones found in these areas have not undergone long-distance migration and therefore have relatively lower maturity compared to marine reservoirs. They are composed of feldspar, quartz, and rock fragments, with poor sorting and rounding. The structures can be either matrix-supported or grain-supported, with cementation primarily occurring with mudstone. The reservoirs exhibit strong vertical heterogeneity, making lateral continuity poor and phase changes rapid. This poses challenges for exploration and development in lacustrine reservoirs.

(3) The tectonic setting of lacustrine basins in China is unstable. Although these basins formed within stable tectonic units, they have experienced intense tectonic and magmatic activity throughout their sedimentary evolution. Eastern China has undergone significant uplift since the early Paleozoic, resulting in the loss of marine strata. In the Mesozoic era, multiple collisions and subductions between the Pacific and eastern Asian plates caused tectonic alterations and magmatic activity in most parts of China, except for the Ordos Basin. In the Cenozoic era, the collision between the Indian and Tibetan plates (Himalayan orogeny) has further contributed to unstable tectonic conditions. These unstable conditions have a negative impact on the preservation of oil reservoirs in lacustrine basins.

3.3.4.4 Comparative Analysis of Marine and Nonmarine Oil Reservoirs

When comparing the characteristics of petroleum generation and accumulation in marine and nonmarine source rocks, it is observed that both types originate and evolve in semi-saline to saline environments, primarily from planktonic algae. However, nonmarine source rocks, found in fully enclosed reducing environments, offer better conditions for preserving organic matter. In terms of location, marine oil reservoirs typically develop in continental shelf and slope zones adjacent to marine environments, while nonmarine reservoirs form in transitional regions between basin slopes and deep lacustrine depressions. River-delta sandstones in these slope areas provide favorable reservoir space for petroleum generation and accumulation.

While marine reservoirs generally have better physical properties, nonmarine reservoirs exhibit stronger relative heterogeneity in their reservoir spaces, including sandstone interlayers within source rocks and traps on slopes. Nonmarine reservoirs also tend to have poorer physical properties, and some shale reservoirs may have low porosity, permeability, and oil saturation, adding complexity to petroleum exploration and development. In terms of geological stability, marine sedimentary basins have a relatively stable tectonic background, benefiting the preservation of oil reservoirs. In contrast, most nonmarine sedimentary basins have experienced intense tectonic movements and magmatic activities, leading to the destruction of matured reservoirs and significant loss of petroleum components.

In summary, nonmarine oil reservoirs face greater challenges in terms of overall conditions for petroleum generation and accumulation compared to marine reservoirs. Exploration and development of nonmarine reservoirs require a different approach than marine reservoirs. It is essential to analyze and understand the mechanisms and favorable conditions for oil generation in nonmarine source rocks specific to nonmarine oil-bearing basins. This will facilitate the development of exploration and development technologies, methodologies, and experimental apparatus tailored to the unique characteristics of nonmarine reservoirs, advancing the understanding of nonmarine petroleum generation theories.

3.4 Summary

(1) Source rock consists of alternating layers of mudstone or shale and permeable strata, which can include sandstone with good porosity and permeability or carbonate rocks with properties similar to sandy rocks (such as ostracod shell limestone, algal limestone, dolomite, etc.). These sandstone or carbonate layers serve as reservoirs for oil and facilitate ongoing oil generation. The pressure gradient between the sandstone and mudstone helps the oil droplets within the mudstone to aggregate and flow, enabling the primary migration of oil under pressure differences. This interbedded rock formation plays the roles of "source, reservoir, and seal".

(2) In source rock where sandstone and mudstone are interbedded, oil initially exists as dispersed droplets within the mudstone pores. To form economically viable oil reservoirs, these droplets need to migrate from the mudstone pores to the adjacent sandstone pores, initiating in-source migration. The extent of primary migration depends on the natural oil-expulsion force of the mudstone and the storage capacity of the sandstone interbeds.

(3) Primary migration within the source rock occurs when the remaining pressure gradient exceeds the capillary forces. If the remaining pressure differential is not enough to overcome the capillary pressure, the oil generated in the source rock struggles to move through the narrow throats and remains in the mudstone pores. However, as oil generation increases, these retained droplets gradually coalesce, and the remaining pressure gradient within the source rock also rises. Once the remaining pressure differential surpasses the capillary pressure, the oil can overcome the restrictive throats and infiltrate the adjacent sandstone layer until it fills all the sandstone pores. Hence, the capillary pressure difference within the source rock determines the thickness of the retained oil zone in the mudstone. The effective thickness of the oil expulsion zone is calculated by deducting the thickness of the retained oil zones at the top and bottom from the overall thickness of the mudstone.

(4) Various geological factors influence and control the process of primary migration within the source rock, including the proportion, thickness, interbedded arrangement, physical properties, pore structures, and formation pressure coefficient of the sandstone and mudstone. Through single-factor analysis methods, it is evident that a 3:7 ratio of sandstone to mudstone in the source rock promotes primary migration. Larger sandstone porosity, thinner individual sandstone-mudstone layers, and more evenly distributed layers enhance oil expulsion efficiency. Similarly, larger pore throat radius in mudstone and higher formation pressure coefficient result in greater effective oil expulsion force and efficiency.

(5) In China's lacustrine basins, the types of oil generation and accumulation can be broadly classified into two categories. The first type occurs in lacustrine sedimentary environments that are mixed with deposits from external river-deltas or fan-deltas. This results in reservoirs that include shale reservoirs (composed

of interbedded mudstone or shale with sandstone) and conventional sandstone-dominated reservoirs. The second type involves pure lacustrine sedimentary environments without external sources, leading to self-generating and self-contained systems consisting of interbedded black shale or black mudstone with biogenic limestone or reef limestone.

(6) When a group of source rocks is capable of generating and expelling significant amounts of oil and effectively accumulating petroleum under geological conditions, it can itself form a shale oil reservoir. Additionally, if the stratum includes beneficial deltaic sandstones connected to sandstone layers within the source rocks, as well as effective secondary migration forces and favorable preservation conditions, it can give rise to multiple conventional oil reservoirs located outside the source rock. The total volume of petroleum resources accumulated in both the shale and conventional reservoirs represents the actual oil yield of the source rocks.

(7) Generally, the conditions for oil generation and accumulation in nonmarine reservoirs are not as favorable as those in marine reservoirs, making exploration and development more challenging and costly. Therefore, it is not feasible to directly apply the knowledge gained from marine reservoir exploration and development to nonmarine reservoir scenarios. It is crucial to summarize the mechanisms and favorable conditions for petroleum generation and accumulation in nonmarine source rocks based on the specific characteristics of nonmarine oil- bearing basins. This will enable the development of exploration and development techniques, methodologies, and experimental equipment that are specifically suited to nonmarine reservoirs, thereby advancing the theories related to nonmarine oil generation.

References

Dunxiang Z, Fanghou Z, Guanghua L (1990) Primary migration of lacustrine hydrocarbons from mudstone to turbidite in Liangjialou. Oil Gas Geol 11(3):334–344

Zhixin W (2000) Is capillary force the driving force for the primary migration of oil (gas)? Pet Geol 3:195–200

Chapter 4
Tectonic Framework of China

Abstract The Chinese continental landmass is a mosaic of small and micro-lacustrine blocks interspersed with orogenic belts. Major continental blocks include the North China Block and the Yangtze Block, which, after undergoing three stages of tectonic evolution, are bordered by orogenic and fault belts. Within these continental blocks, oil- and gas-bearing basins demonstrate a pattern of multi-phase superimposition, characterized by "zoning from east to west and partitioning from north to south." The crustal thickness of the Chinese mainland decreases progressively from west to east, influencing the geothermal gradient. Geologically, the Chinese landmass is framed by the Siberian, Indian, and Pacific plates. While the northern and central regions remain relatively stable, the Qinghai-Tibet region experiences significant compression from the Indian Plate, resulting in the formation of unique negative-pressure basins. In the east, interactions with the Pacific Plate have led to the development of trench-arc-basin systems and rift basins. Petroleum resources on the Chinese mainland (excluding offshore regions) are predominantly lacustrine. To understand the characteristics of lacustrine source rocks from various geological periods, key oil-bearing basins are analyzed, including the Permian-aged Junggar Basin, the Triassic-aged Ordos Basin, the Jurassic-aged Sichuan Basin, the Cretaceous-aged Songliao Basin, and the Paleogene-aged Bohai Bay Basin. This comprehensive analysis sheds light on the source rock characteristics of oil-bearing basins across different regions, geological periods, and developmental features in China.

China's mainland is composed of multiple small and micro continental blocks intertwined with mountain belts. These continental blocks, although relatively small in size, are numerous. Among the larger continental blocks in China are the North China Block, Yangtze Block, Tarim Block, Junggar Block, Qaidam Block, Songliao Block, and Cathaysia Block (Fig. 4.1).

After undergoing three stages of tectonic evolution during the Proterozoic, Paleozoic, and Mesozoic–Cenozoic eras, these continental blocks in China now feature a tectonic configuration surrounded by numerous active mountain-building zones and fault zones. Within the interior of these blocks, oil and gas basins have developed across the Paleozoic, Mesozoic, and Cenozoic periods, exhibiting a complex

D. Guan et al., *Theory and Simulation Experimental Technology of Oil Generation From Nonmarine Source Rocks in China*, https://doi.org/10.1007/978-981-96-3145-2_4

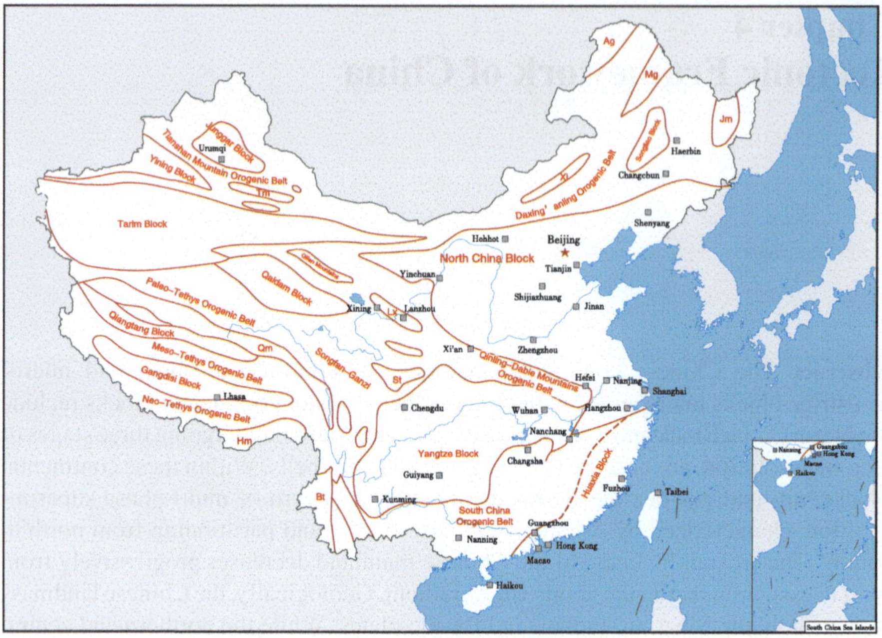

Fig. 4.1 Major tectonic units in China. Hm—Himalayan block; Tm—Turpan-Hami block; Qm—Qiangtang block; Lx—Longxi block; St—Songpan block; Bt—Baoshan-Tengchong block; Ag—Erguna block; Mg—Northern Xing'an block; Jm—Jiamusi block; Xi—Xilinhaote block

and layered structure. These basins, rich in hydrocarbons, are divided by extensive and deep fault zones, resulting in a distinct "east-west zonation and north-south segmentation" pattern (Fig. 4.2).

Across mainland China, the Earth's crust exhibits a distinct zonation from west to east, resembling a stair-step pattern. The western region boasts the thickest crust, measuring up to 70 km in thickness. The central region maintains a relatively consistent crust thickness of over 40 km. In contrast, the eastern region has the thinnest crust, averaging between 35 and 38 km. As we move eastward from the coastal continental shelf towards the Pacific Ocean, the crust gradually transitions into oceanic crust (Fig. 4.3). These variations in crustal thickness contribute to a lower geothermal gradient in the sedimentary basins of the western and central regions, while the eastern region experiences a higher gradient.

The mainland block of China is surrounded by three major tectonic plates—the Siberian, Indian, and Pacific plates (Fig. 4.4). The northern and central areas of the mainland block are relatively stable. However, the Qinghai-Tibet region is continuously affected by compression and tectonic collisions from the Indian plate. This ongoing activity leads to uplift and erosion, making it the only subaerial negative-pressure basin globally. Since the Mesozoic and Cenozoic eras, the eastern part of China has been influenced by subduction resulting from the collisions with the Pacific plate, which primarily consists of oceanic crust. This interaction has given

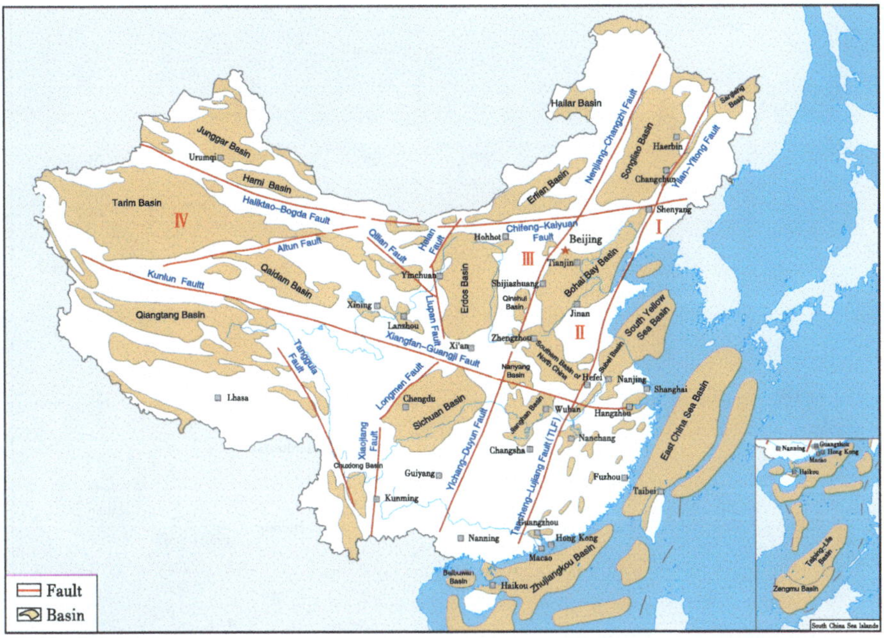

Fig. 4.2 Division of the oil and gas-bearing basin groups in China (after Zhanqian 2003)

rise to a trench-arc-basin system along the western Pacific rim and has contributed to the formation of numerous rift basins and marginal marine basins oriented in a north-easterly and north-northeasterly direction.

Presently, the oil reserves discovered within mainland China, excluding offshore areas, are predominantly nonmarine oil. However, there is an exception in the Tarim Basin, which primarily contains marine oil. To gain a comprehensive understanding of the source rock characteristics in China's nonmarine deposits throughout different geological periods, this study focuses on specific formations as key subjects of investigation. These formations include the Permian source rocks in the Junggar oil-bearing basin located in the western region, the Triassic source rocks in the Ordos oil-bearing basin situated in the central region, the Jurassic source rocks in the Sichuan oil-bearing basin also found in the central region, the Cretaceous source rocks in the Songliao oil-bearing basin located in the eastern region, and the Paleogene source rocks in the Bohai Bay oil-bearing basin also situated in the east. By studying these formations, valuable insights can be obtained regarding the diverse characteristics of source rocks across different regions and geological epochs, as well as the unique developmental features of oil-bearing basins in China.

Fig. 4.3 Distribution of crustal thickness in the Asian continent (after Qianshen et al. 1982) (numerical values in the figure represent the thickness of the crust in km)

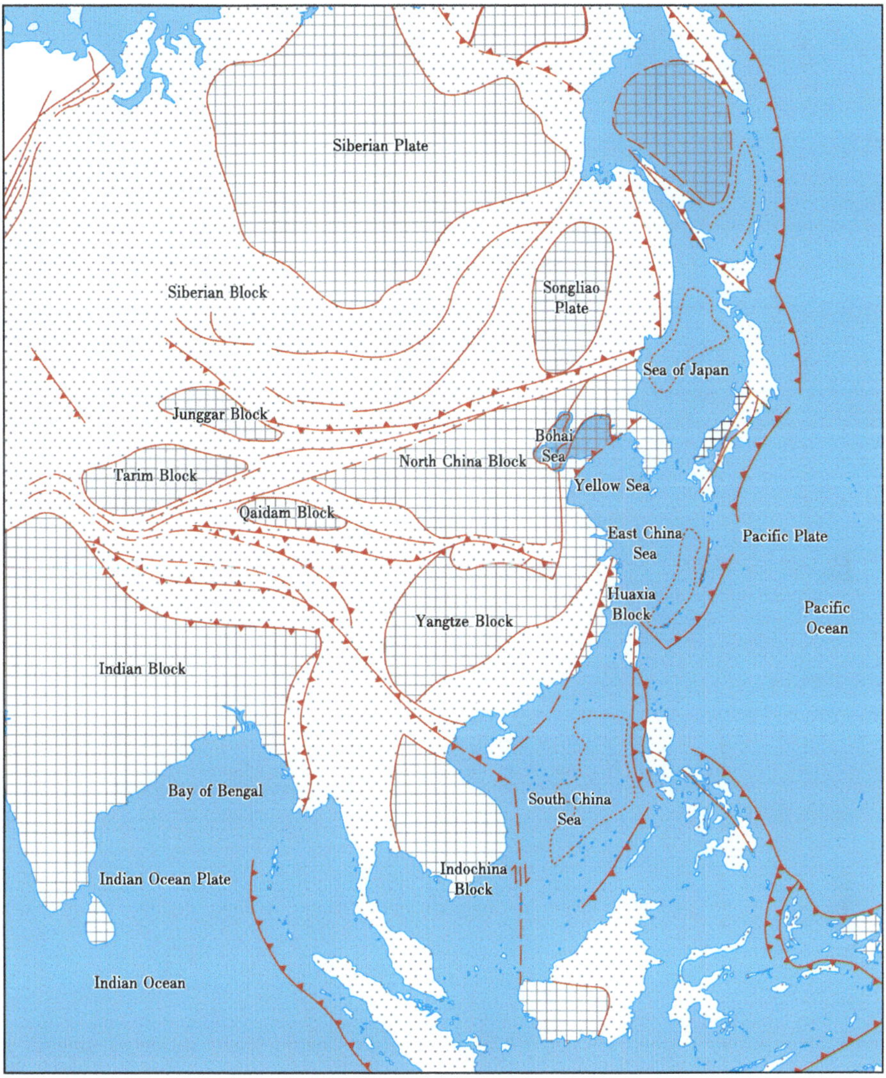

Fig. 4.4 Major tectonic units in China and surrounding regions (after Tao 1997)

References

Qianshen W, Chuanzhen W, Hongchen L et al (1982) Discussion on crustal thickness distribution and crustal structural characteristics in Asia continent. Seismogeology 4(03):1–9

Tao W (1997) Oil and gas reservoir geology of rift basin in eastern China. Petroleum Industry Press

Zhanqian G (2003) Alteration of China petroliferous basin. Xinjiang Pet Geol 24(1):8–12

Chapter 5
Permian Source Rock Characteristics of the Junggar Basin

Abstract The Junggar Basin contains six distinct sets of petroleum source rocks spanning the Carboniferous to the Quaternary periods. These include formations from the Carboniferous, Permian, Triassic, Jurassic, Cretaceous, and Paleogene, distributed across various depressions within the basin. Among them, the Permian period is notable for its widespread lacustrine clastic deposits, present in all depressions of the basin, and serves as the primary source rock, accounting for more than half of the basin's oil and gas resources. In the western region of the basin, the Fengcheng Formation and Lower Wuerhe Formation are key contributors, while in the eastern region, the Pingdiquan Formation dominates. In the southern area, the Lucaogou Formation and Hongyanchi Formation are prominent. These source rocks are composed primarily of mudstone, shale, tuffaceous dolomite, and occasionally coal seams, with depositional environments ranging from semi-deep lakes to fluvial-deltaic systems. The organic matter is predominantly of Type I_1 to Type II_2, characterized by high organic content. Volcanic activity in the region provided heat and catalytically beneficial trace elements, creating favorable conditions for the generation and accumulation of the basin's oil and gas resources.

The Junggar Basin is a significant oil and gas-bearing sedimentary basin and located in the northern part of Xinjiang Uygur Autonomous Region, northwest China. It occupies an area of around 13×10^4 km^2 and is surrounded by the Altai Mountains, Tianshan Mountains, and the Western Junggar Mountains. The basin contains sedimentary strata ranging from the Middle-Late Carboniferous to the Quaternary periods, with the sedimentary sequence reaching a maximum thickness of 15,000 m.

Based on the structural characteristics of the Permian system and the subsequent tectonic modifications within the Junggar Basin, it can be divided into six primary tectonic units: Western Uplift, Eastern Uplift, Luliang Uplift, North Tianshan Fold-Thrust Belt, Central Depression, and Wulungu Depression. In addition, it has 44 secondary tectonic units (Fig. 5.1).

D. Guan et al., *Theory and Simulation Experimental Technology of Oil Generation From Nonmarine Source Rocks in China*, https://doi.org/10.1007/978-981-96-3145-2_5

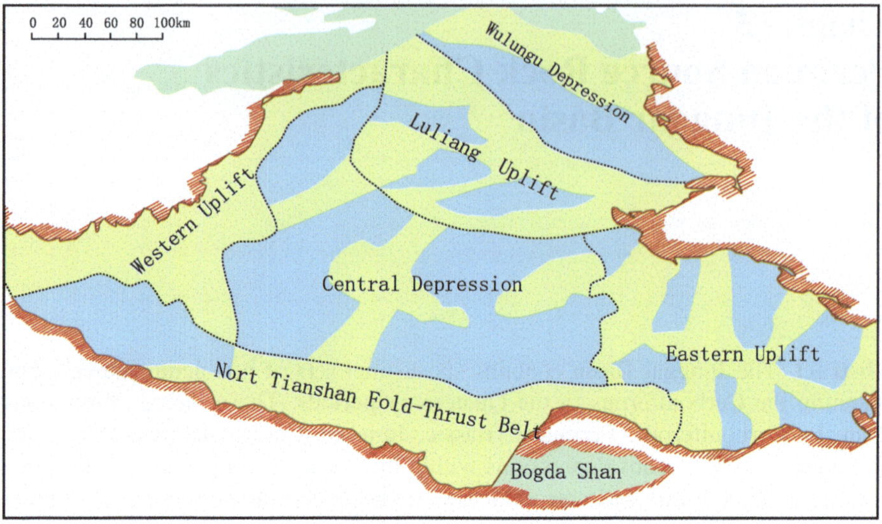

Fig. 5.1 Structural units of the Junggar Basin. Adapted after Jianping et al. (2016)

5.1 Permian Stratigraphic Development Characteristics of the Junggar Basin

The Junggar Basin went through a series of evolutionary events, including basement formation preceding the Carboniferous, transitional development during Carboniferous-Permian, the establishment of Mesozoic-Paleocene inland lake basins, and intense compression during Neogene-Quaternary. In the Triassic-Early Jurassic period, the Junggar Basin experienced its full bloom phase of sedimentary basin development, characterized by the widespread deposition of sedimentary strata that nearly covered the entire basin. This marked a crucial milestone in the transition of the Junggar Basin into a continental basin (Fig. 5.2).

The Permian strata in the Junggar Basin represent the primary sedimentary cover sequence during the early stage of basin formation. The development characteristics of this set of strata vary significantly across different regions due to significant differences in tectonic and sedimentary settings.

The Lower Permian strata in the northwestern and eastern parts of the southern margin exhibit varying degrees of residual marine sedimentation. The northwestern margin is also distinguished by abundant volcanic rock development. The northeastern part of the Junggar Basin is dominated by fluvial settings, potentially accompanied by intense volcanic activity. During the Middle Permian, the basin was still influenced by ocean waters, leading to the deposition of a high-salinity deep-water sedimentary sequence that represents the most significant source rock within the basin. During the Late Permian, the basement of the basin became more homogeneous, with the lithological characteristics of the strata becoming increasingly

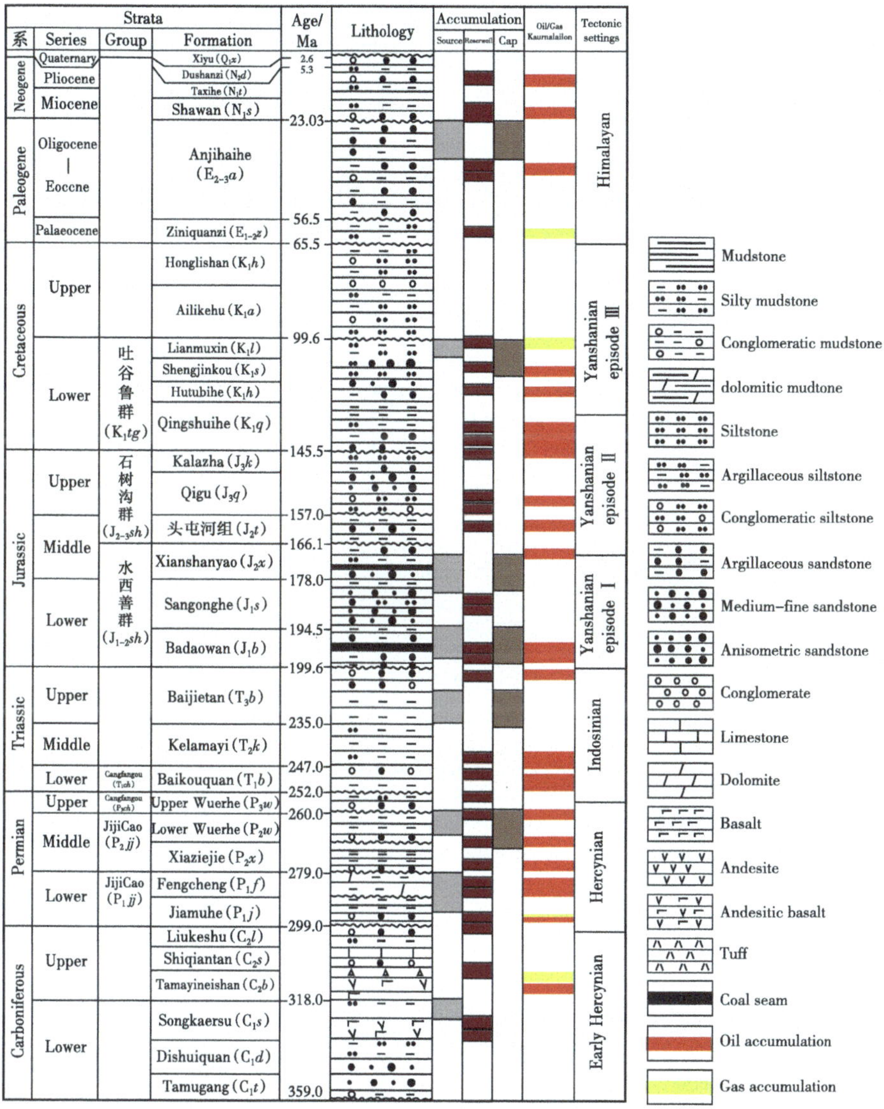

Fig. 5.2 General stratigraphic column of the Junggar Basin. Wenjun2019 Adapted from et al. ()

similar across the basin. However, there are still substantial differences between various places in the development and thickness of this stratigraphic unit (Table 5.1).

Table 5.1 Stratigraphic sequences in different regions of the Junggar Basin (after Wang et al. 2013)

Era	System	Series	Northwestern Junggar Group	Central Junggar Group	Southern Junggar Group	Eastern Junggar Group
Cenozoic	Quaternary	Pleistocene	Xiyu (Q_1x)	Xiyu Q_1x	Xiyu Q_1x	Xiyu (Q_1x)
	Paleogene	Pliocene	Dushanzi (N_2d)	Dushanzi (N_2d)	Dushanzi (N_2d)	Dushanzi (N_2d)
		Miocene	Taxihe (N_1t)	Taxihe (N_1t)	Taxihe (N_1t)	Taxihe (N_1t)
			Shawan (N_1s)	Shawan (N_1s)	Shawan (N_1s)	Shawan (N_1s)
		Eocene-Oligocene	Wulunhe ($E_{2-3}a$)	Anjihehai ($E_{2-3}a$)	Anjihehai ($E_{2-3}a$)	Anjihehai ($E_{2-3}a$)
		Palaeocene-Eocene		Ziniquanzi ($E_{1-2}z$)	Ziniquanzi ($E_{1-2}z$)	Ziniquanzi ($E_{1-2}z$)
Mesozoic	Cretaceous	Upper	Honglishan (K_2h)	Donggou (K_2d)	Donggou (K_2d)	Hongshaquan (K_2h)
			Ailikehu (K_2q)			
		Lower	Lianmuqin (K_1l)	Lianmuqin (K_1l)	Lianmuqin (K_1l)	Lianmuqin (K_1l)
			Shengjinkou (K_1s)	Shengjinkou (K_1s)	Shengjinkou (K_1s)	Shengjinkou (K_1s)
			Hutubihe (K_1h)	Hutubihe (K_1h)	Hutubihe (K_1h)	Hutubihe (K_1h)
			Qingshuihe (K_1q)	Qingshuihe (K_1q)	Qingshuihe (K_1q)	Qingshuihe (K_1q)
	Jurassic	Upper			Kelazha (J_3k)	Kelazha (J_3k)
			Qigu (J_3q)		Qigu (J_3q)	Qigu (J_3q)
		Middle	Toutunhe (J_2t)	Toutunhe (J_2t)	Toutunhe (J_2t)	Toutunhe (J_2t)
			Xiashanyao (J_2x)	Xiashanyao (J_2x)	Xiashanyao (J_2x)	Xiashanyao (J_2x)
		Lower	Sangonghe (J_1s)	Sangonghe (J_1s)	Sangonghe (J_1s)	Sangonghe (J_1s)
			Badaowan (J_1b)	Badaowan (J_1b)	Badaowan (J_1b)	Badaowan (J_1b)
	Triassic	Upper	Baijietan (T_3b)	Baijietan (T_3b)		
					Huangshanjie (T_3h)	Huangshanjie (T_3h)
		Middle	Kelamayi (T_2k)	Kelamayi (T_2k)	Kelamayi (T_2k)	Kelamayi (T_2k)
		Lower	Baiquankou (T_1b)	Baiquankou (T_1b)	Shaofanggou (T_1s)	Shaofanggou (T_1s)
					Jiucaiyuan (T_1j)	Jiucaiyuan (T_1j)
Palaeozoic	Permian	Upper	Wuerhe (P_3w)	Wuerhe (P_3w)	Wutonggou (P_3wt)	Wutonggou (P_3wt)
					Quanzijie (P_3q)	Quanzijie (P_3q)
		Middle	Wuerhe (P_2w)	Wuerhe (P_2w)	Hongyanchi (P_2h)	Qingdiquan (P_2q)
					Lucaogou (P_2l)	
			Xiazijie (P_2x)	Xiazijie (P_2x)	Jingjingzigou (P_2j)	Jiangjunmiao (P_2j)
					Wulabo (P_2w)	
		Lower	Fengcheng (P_1f)	Fengcheng (P_1f)	Tashikula (P_1t)	Jingou (P_1j)
			Jiamuhe (P_1j)	Jiamuhe (P_1j)	Shirenzi (P_1s)	
	Carboniferous	Upper				Shiqietan (C_2s)
					Qijiagou (C_2q)	
			Molaoba (C_2m)		Bogeda (C_1b)	Dishuiquan (C_1d)
		Lower				

5.1.1 Lower Permian

In the northwestern to central regions, the Lower Permian comprises the Jiamuhe Formation (P_1j) and the Fengcheng Formation (P_1f). The Jiamuhe Formation is primarily distributed within the Mahu sag, the Pen 1 Xi sag, and the Shawan sag of the western part of the basin. It exhibits significant thickness variations, with a maximum thickness exceeding 3000 m. The lithology of the Jiamuhe Formation primarily consists of conglomerates and sandstones interbedded with gray-green debris sandstones. The base of the formation is composed of tuff, while the upper part consists of andesite and thick layers of tuffaceous volcanic rocks. The lower section of the Fengcheng Formation consists of deep gray dolomitic tuff, tuffaceous dolomite, tuff, silty mudstone. The upper section consists of light gray dolomitic tuff and gray-black argillaceous dolomite. These are interbedded with thin layers of mudstone and sandy mudstone, suggesting that sedimentation occurred in stagnant bays or lagoons. The Fengcheng Formation is considered a major source rock interval within the Permian sequence of the Junggar Basin.

In the southern margin of the basin, the Lower Permian includes the Shirenzigou Formation (P_1s) and the Tashikula Formation (P_1t), which belong to a marine depositional environment. The lower part of the Shirenzigou Formation is characterized by thick beds of gray-green calcareous conglomerates, sandstones, siltstone, and sandy mudstones. The upper part consists of thin layers of gray-black sandy mudstone interbedded with dark gray fine sandstone, fragmentary gray limestone, and calcareous sandstone. The formation, which has a thickness that ranges from 240 to 1200 m, contains visible plant fossils. The Tashikula Formation comprises rhythmic and frequent alternations of gray sandstones, siltstones, and black mudstones, along with intercalations of sandy limestone, oolitic limestone, and thin layers of volcanic ash deposits. With a thickness that ranges from 1200 to 2200 m. The Tashikula Formation contains spores and plant remains.

The Jingou Formation (P_1j) represents the Lower Permian in the eastern region of the basin. It is mainly exposed in the easternmost part and has a thickness of around 670 m, consisting of conglomerates, sandstones interbedded with gray sandy mudstone within the lower part of the formation. The middle to upper part is characterized by purple-red sandy mudstones interbedded with gray-green sandstones and conglomerates, containing foraminifera and ostracod fossils. This section ranges from 450 to 1100 m thick.

5.1.2 Middle Permian

In the northwestern to central parts of the basin, the Middle Permian is composed of the Xiazijie Formation (P_2x) and the lower Wuerhe Formation (P_2w). The Xiazijie Formation mainly consists of alluvial fan deposits, withgray-brown and brown sandstones and conglomerates containing pebbly sandstone and sandy mudstone characterizing its lower part. The upper part consists of brown conglomerates. The lower Wuerhe Formation (P_2w) is composed of various thicknesses of dark gray, gray, and gray-green conglomerates, mudstones, sandy mudstones, siltstones as well as multiple layers of thin coal seams. These deposits represent fluvial fan and lacustrine swamp deposits.

At the southern margin of the basin, the Middle Permian strata are composed of the following formations from bottom to top: the Wulapo Formation (P_2w), the Jingjingzigou Formation (P_2j), the Lucaogou Formation (P_2l), and the Hongyanchi Formation (P_2h). The Wulapo Formation is primarily composed of a sandstone sequence that includes conglomerates, sandstones, siltstones, and mudstones. The predominant colors are gray green with intercalations of purple-red. The thickness of the formation ranges from 440 to 1300 m. The Jingjingzigou Formation is composed of interbeds of sandy mudstone, siltstone, tuffaceous sandstone, tuffaceous mudstone, and grayish blocky tuff. It has a thickness ranging from 500 to 900 m. The Lucaogou Formation comprises thin to medium-thick layers of oil shale in gray-black, black, and deep gray colors. Additionally, it contains thin interbeds of sandy limestone and siltstone. It contains fossils such as plant spores and insect burrows. The formation has a thickness of 800–1000 m and represents the lacustrine sedimentation that occurred in a stagnant and isolated basin.

In the eastern part of the Junggar Basin, the Middle Permian is stratigraphically subdivided into two units: the Jiangjunmiao Formation (P_2j) and the Pingdiquan Formation (P_2p). The Jiangjunmiao Formation is primarily found in the eastern region, including Zhangpenggou-Shiqiantan area. It also occurs in intermountain depressions located north of Wucaiwan depression and Laojunmiao. This formation mainly consists of fluvial facies characterized by red or mottled mudstone interbedded with sandstone and conglomerate. The Pingdiquan Formation widely extends across the foreland of Kalameili Mountains. It mainly comprises dark-colored lacustrine mudstone deposits with notable lateral variations. In depressional areas like Jimusaer Depression, it exhibits lacustrine facies featuring deep gray mudstone interbedded with laminated sandstone. Importantly, this formation represents a major source rock unit in the eastern part of Junggar Basin.

5.1.3 Upper Permian

In the northwestern to central regions of the Junggar Basin, the Upper Permian is represented by the upper Wuerhe Formation (P_3w). It predominantly consists of brownish conglomerates interbedded with sandy mudstones, reflecting deposition in fluvial-aeolian and alluvial fan settings. In the southern margin area, the Upper Permian can be divided into two formations: the Quanzijie Formation (P_3q) and the Wutonggou Formation (P_3wt). The Quanzijie Formation is mainly composed of conglomerates with minor occurrences of gravelly mudstones, carbonaceous mudstones, and thin coal seams. Sandstones are relatively limited in this formation which has a thickness ranging from 200 to 300 m. These lithological characteristics suggest that deposition was related to debris flows and fluvial processes within alluvial fan environments. The Wutonggou Formation displays distinct lithological variations. The lower units are composed of greenish conglomerates, sandstones, mudstones, and coal seams that exhibit regular cyclic alternations. These units were deposited in meandering river environments and have a thickness ranging from 120 to 220 m. On the other hand, the upper units of the Wutonggou Formation consist of banded mudstones with dark red, purple, and green colors interbedded with thin layers of fine sandstone, mudstone lenses, and carbonaceous mudstones. These units, with a thickness of around 68–128 m, represent shallow-water marsh and floodplain deposits. In the eastern part of the basin, the Upper Permian formations, collectively referred to as the Xiacangfanggou Group (P_3ch), include the Quanzijie Formation (P_3q) and the Wutonggou Formation (P_3wt). These formations are characterized by a combination of fluvial conglomerates, sandstones, and carbonaceous mudstones.

5.2 Characteristics of the Permian Source Rocks in the Junggar Basin

The Junggar Basin has six sets of source rocks that developed during the Carboniferous, Permian, Triassic, Jurassic, Cretaceous and Paleogene systems. These source rocks are distributed throughout the basin with some being found in specific regions. They have provided a strong basis for abundant hydrocarbon resources in the Junggar Basin. The Permian source rocks are considered to be the most significant among them as they contribute to over half of the oil and gas resources in this basin.

5.2.1 Depositional Environments and Paleontological Features

The Permian strata represents the primary source rock succession in the Junggar Basin. The Lower Permian Fengcheng Formation and the Middle Permian lower Wuerhe Formation dominate the source rocks of the western part of the basin. In the eastern part of the basin, primary source rock development occurs in the Middle Permian Pingdiquan Formation. Meanwhile, in the southern part of the basin, source rock development can be found primarily in the Middle Permian Lucaogou Formation and Hongyanchi Formation.

5.2.1.1 Environmental Characteristics of Source Rocks

Fengcheng Formation (P_1f)

The Fengcheng Formation is primarily composed of dark mudstone, dolomitic mudstone, and argillaceous dolostone, interbedded with sandstone, siltstone, and thin layers of limestone. Calcite, dolomite, and pyrite can be observed in some intervals. The average TOC content is around 1.33%. Additionally, horizontal bedding is well-developed in the formation. In the central and western parts of the basin, the thickness of the Fengcheng Formation ranges from 200 to 700 m, with an average thickness of approximately 300 m. In the Pen 1 Xi sag and Mahu sag the thickness of the dark mudstone can reach 500–600 m, and in the Changji sag, it can reach up to 700 m. The dark mudstone shows a gradual thinning trend from northwest to southeast. The lithology and sedimentary characteristics suggest that there was intense crustal subsidence and significant deepening of water bodies on the northwestern margin during deposition of the Fengcheng Formation. Abundant deposition of dolomite and dolomitic limestone indicates the Fengcheng Formation developed in a high salinity water bodies and semi-deep lake or deep lake sedimentary environments. Thick deposits of dark mudstone with well-developed horizontal bedding reflect predominantly stable depositional conditions. The water conditon is characterized by alkaline ($pH > 8$) and strong reducing conditions (well developed sulfide minerals).

Lower Wuerhe Formation (P_2w)

The lower Wuerhe Formation (P_2w) predominantly consists of dark gray and gray-black mudstone and shale interbedded with gray and gray-green sandstone and conglomerate. Coal seams are also present within the formation. The thickness of the dark mudstone ranges from 50 to 250 m, reaching up to 300 m in the Mahu sag and a maximum of 450 m in the Pen 1 Xi sag. The mudstone exhibits horizontal bedding, and evidence of erosional structures can be observed in the conglomerates. Thin coal seams are commonly found, and the mudstone contains carbonized plant remains and pyrite. In certain areas, dolomite and dolomitic limestone are present, while the average TOC content is 0.95%. Based on lithological characteristics, it can be inferred that during the deposition of the Wuerhe Formation, water bodies expanded larger compared to those during deposition of the Xiazijie Formation. Stable lake basins were primarily located to the east of Wuerhe, including areas such as Mahu and the Pen 1 Xi sags, while other regions were dominated by river, delta, and swamp environments. The water medium is weakly alkaline ($pH = 7–8$) and exhibits weak sulfide conditions (weak reducing environment).

Pingdiquan Formation (P_2p)

The Pingdiquan Formation (P_2p) is characterized by gray-green and dark gray mudstone, siltaceous mudstone, gray-black oil shale interbedded with mudstone, oolitic limestone, stacked cone limestone, and gray or gray-green sandstone. Yellow-green conglomerate and coarse-grained sandstone with gravel are commonly found at the base of the formation while thin coal seams or layers can be observed in the upper parts. The formation contains minerals such as calcite, dolomite, siderite, and rhodochrosite. Some layers also contain minor amounts of primary pyrite crystals. The dark mudstone and oil shale exhibit horizontal bedding along with fine-scale horizontal lamination and seasonal laminae. Fine-scale ripple bedding is visible in siltstone and siltaceous mudstones whereas medium to large-scale crossbedding is present in sandstones and conglomerates which also show evidence of scouring structures. The average TOC content in Subsurface samples reaches 3.36%, while surface samples average at 1.22%.

Based on the sedimentary characteristics, it can be inferred that the Pingdiquan Formation represents a transgressive-regressive depositional sequence. During the early stage, the water body was shallow, and rivers were predominant in most areas. In the middle stage (from the upper part of Member S3 to the lower-middle part of Member S2), a widespread lacustrine environment developed, indicating a semi-deep lake setting. In the late stage, water regression occurred, leading to the formation of localized marshes. The dominant sediment consists of dark mudstone, with the presence of minerals such as calcite, dolomite, siderite, and rhodochrosite. Some layers contain minor amounts of primary pyrite crystals. Overall, the water medium is characterized by alkaline conditions ($pH > 8$) and sulfide-rich environments (strongly reducing conditions).

Lucaogou Formation (P_2l)

The Lucaogou Formation is mainly composed of dark gray sandstone, sandy shale, black oil shale interbedded with limestone and dolomitic limestone. Minerals such as calcite, dolomite, siderite, and minor amounts of pyrite are present. The average TOC content in the Subsurface black mudstone is 4.5%, while the surface mudstone (shale) and oil shale have a significantly higher average TOC content of 7.04%.

Hongyanchi Formation (P_2h)

The Hongyanchi Formation (P_2h) consists mainly of dark gray and gray-green mudstone, sandy shale interbedded with yellow-green siltstone and mudstone, with localized carbonaceous mudstone and minor occurrences of calcite. The average TOC content in the shale is 1.77%. The source rocks exhibit horizontal bedding. Sedimentary characteristics indicate a shallower water body and increased hydrodynamic processes compared to the Lucaogou Formation. The lower part of the unit is characterized by the development of thick deltaic deposits. The fan-delta system is mainly composed of subaqueous distributary channels at the front of the fan-delta, with localized swamp environments. Overall, it exhibits depositional characteristics of a shallow lake and shoreline environment. The water medium is weakly alkaline (pH = 7–8) and non-sulfidic (weakly reducing conditions).

5.2.1.2 Paleontological Characteristics of Source Rocks

Fengcheng Formation (P_1f)

The Fengcheng Formation is characterized by a relatively abundant presence of spore pollen fossils, with no animal fossils discovered to date. The spore pollen assemblage is composed exclusively of gymnosperm pollen, with the most developed being the ribbed bisaccate pollen. Monosaccate pollen, multigrooved ribbed pollen, and non-ribbed bisaccate pollen also constitute a certain proportion of the assemblage. Occasional or small amounts of monosulcate bisaccate pollen and trisaccate ribbed pollen are also observed.

Based on the analysis of lithological characteristics, it can be observed that the deposition of the Fengcheng Formation covered a larger extent within the lake compared to the deposition of the Jiamuhe Formation. Moreover, there is a notable occurrence of tuff within the Fengcheng Formation which suggests volcanic activity during that period. Large amount of volcanic ash was erupted and deposited during the deposition of the Fengcheng Formation. This episodic influx of volcanic ash into the water bodies created intermittent spatial isolation, leading to the development of oxygen-depleted and strongly reducing environments.

Lower Wuerhe Formation (P_2w)

Abundant spore fossils, which forming the Raistrickia-Cordaitina-Protohaploxypinus-Hamiapollenites spore assemblage, have been discovered within the Lower Wuerhe Formation. This particular assemblage is characterized by the significant presence of sugarcane spores, well-developed Cordaitina pollen in gymnosperms, and a considerable amount of multi-groove and single-groove pollen.

According to the spore assemblage characteristics and paleogeographic analysis, there is a central uplift zone that acts as a barrier between the Lower Wuerhe Formation and the northeastern margin of the basin. This means that it is highly unlikely for seawater to have any influence in this area.

According to the current drilling data, the formation is primarily developed in the depositional environments of alluvial fans, rivers, and marshes, with no evidence of stable lake deposits. Therefore, the sedimentary area of the Lower Wuerhe Formation had a relatively small-scale lake, mainly consisting of freshwater. The lake water was primarily sourced from rivers.

Pingdiquan Formation (P_2p)

The Pingdiquan Formation exhibits a vertical alternation of salt-tolerant and fresh-water fossil assemblages, with a relatively rich presence of fish fossils and varying degrees of development of invertebrate fossils in different regions. Based on the vertical distribution patterns of fossils, two distinct fossil assemblages can be identified.

(1) Anthraconauta-Pseudomodiodus-Senderzoniella-Mrassiella assemblage, occurring in the lower and middle parts of the Pingdiquan Formation in the Zhangpenggou area. The fossil assemblage is abundant, often concentrated in beds, with high richness and diversity, comprising a total of 6 genera and 21 species, indicating a brackish water environment.
(2) The Palaeomutela-Palaeanodonta-Abiella assemblage can be found in the lower part of the first member of the Pingdiquan Formation located in Zhangpenggou area. The fossil preservation is excellent, with large individuals and high abundance, exhibiting low diversity, indicating a freshwater environment.

Lucaogou Formation (P_2l)

The Lucaogou Formation is the most fossil-rich stratigraphic unit in the southern margin of the basin, containing a variety of fossils such as crustaceans, bivalves, fish, amphibians, spores, and plants.

The Lucaogou Formation contains a rich and diverse fossil assemblage of ostracods, with a predominance of local genera and species. This assemblage is characterized by the presence of Darwinula-Kelameilina-Tomiella-Hongyanchilella.

Additionally, the formation contains numerous bivalve fossils that occasionally accumulate to form shell beds. Researchers have identified 15 species and 5 genera (including indeterminate species) in the area, with the Antraconauta and Pseudomodiolus-Mrassiella assemblages being particularly prominent. Notably, both Antraconauta and Pseudomodiolus are well-represented in this formation.

The vertebrate fossils in the Lucaogou Formation consist only of ancient codfish and amphibians. The fish fossil (Turfanisvartus) is regarded as the typical Middle Permian fossil found in the northern Tianshan Mountains of the Junggar Basin and serves as an important basis for regional stratigraphic correlation. The spore fossil assemblage is highly developed and comprises the Cordaitina-Vittatina-Striatoabieites-Hamiapollenites assemblage.

Plant fossils are commonly found in the lower section of the Lucaogou Formation.

Based on the developmental characteristics of various groups of fossil organisms, it is believed that during the deposition of the Lucaogou Formation, the southern margin of the basin had extensive lakes with deep water and high salinity. The fossil record of ancient organisms is exceptionally rich. Among bivalve fossils, the salt-tolerant types such as Anthraconauta and Pseudomodiolus are well-developed in the formation. Local species of ostracods make up a significant proportion, suggesting deep water conditions at that time. In addition, the abundance of salt tolerant Permiana and Tomiella species is the highest among different regions. The presence of abundant planktonic fish fossils and the development of oil shales further indicate a deepwater environment with predominantly reducing conditions in the southern margin of the basin during deposition of the Lucaogou Formation.

Hongyanchi Formation (P_2h)

The Hongyanchi Formation is rich in fossils, including bivalves, ostracods, spores, plant fossils, and fragmented vertebrate fossils. Bivalve fossils are abundant in this formation. Fossils in the lower to middle part of the Hongyanchi Formation are similar to those found in the Lucaogou Formationn. In the upper part, a combination of six genera and twelve species is found, including Senderzoniella, Abiella, and Palaeomutela. This combination is characterized by a significant decrease in Anthraconauta, the extinction of Pseudomodiolus and Mrassiella, and the emergence of abundant freshwater bivalve fossils such as Palaeomutela, Palaeanodonta, and Senderzoiella along the southern margin of the basin.

Ostracod fossils are plentiful in the Hongyanchi Formation and share similarities with those discovered in the underlying Lucaogou Formation. As a result, they can be classified into the same fossil assemblage or group. The Hongyanchi Formation is distinguished from the Lucaogou Formation by the absence of Kelameila Plena and Qitainaoblonga, while new species like T. ventricostata, P. ermianacompta, P. tubercarinata, Xinjiangella antiquata, and X. insolita have emerged in this formation as replacements. Pollen fossils within the Hongyanchi Formation form the Cylogranisporites-Cycadopites-Cordaitina pollen assemblage. Among the different types of pollen, gymnosperm pollen is the most prevalent. This type of pollen is characterized by ribbed double-sac structures, but there are also significant amounts of single-sac, single-groove, and multi-groove ribbed pollen present. A small amount of non-ribbed double-sac pollen is also present, and spores of sugarcane plants account for 7–10.5% of the assemblage. The presence of Cycadopites and Urmits, which are single-groove pollen, differentiates this pollen assemblage from that of the Lucaogou Formation.

The plant fossils in the Hongyanchi Formation are dominated by Paracalamites sp., which is a common genus in the Permian period. The combination of these biological fossils reflects that during the sedimentation of the Hongyanchi Formation, the overall depositional environment remained similar to that of the Lucaogou Formation. However, in the later stages of deposition in the Hongyanchi Formation, the water changed from saline to freshwater, resulting in a fossil assemblage dominated by freshwater organisms. Additionally, the size of the lake basin gradually reduced, and the water depth decreased.

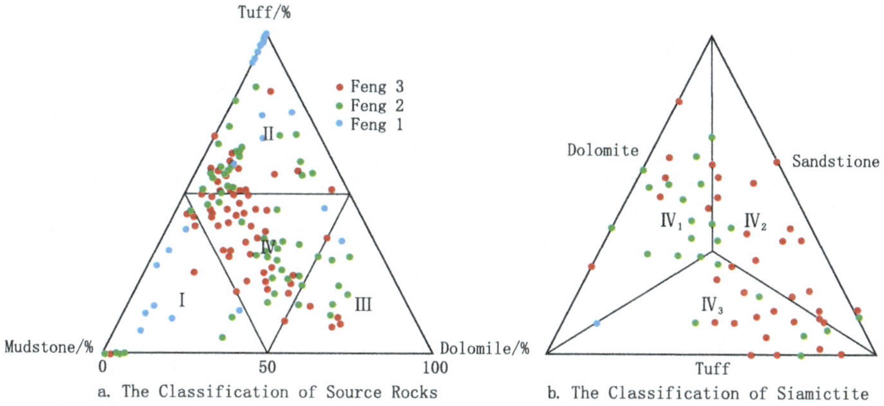

Fig. 5.3 Ternary diagram of source rock composition in the Fengcheng Formation, Mahu Sag (after Xiaojun et al. 2018)

5.2.2 Main Characteristics of the Permian Source Rocks

The Permian strata represent the most significant intervals for source rock development in the Junggar Basin. In the western part of the basin, including the Mahu Sag and the Pen 1 Xi Sag, the source rocks are mainly developed in the lower Permian Fengcheng Formation and the Middle Permian lower Wuerhe Formation. In the eastern part of the basin, the source rocks are primarily developed in the Middle Permian Pingdiquan Formation. In the southern part, they are mainly developed in the Middle Permian Lucaohe Formation and Hongyanchi Formation.

5.2.2.1 Mineral Composition of Source Rocks

Fengcheng Formation (P_1f)

The mineral composition of the Fengcheng Formation source rocks is quite complex, consisting mainly of terrigenous clastics, carbonate components, and volcanic materials that are mixed in varying proportions (Fig. 5.3). These can be further subdivided into four major categories based on the proportion of different components: argillite, argillaceous dolomite, tuff, and diamictite.

(1) Argillite: it has low clay mineral content and appears bright under crossed polarized light, with no interference colors observed. The matrix primarily consists of fine-grained alkaline feldspar and some carbonate minerals. Coarser quartz and feldspar grains are transparent and colorless when viewed under plane-polarized light but display first-order gray-white interference colors when viewed under crossed polarized light. Although there is a certain degree of sorting, the grains

appear angular and do not exhibit contact with each other, displaying distinct aeolian characteristics.

(2) Tuff: This rock is mainly made up of tuffaceous materials, with some continental clastic and carbonate components. When viewed under crossed polarized light, the tuffaceous components show complete extinction. The carbonate components appear as clasts and display iridescent high-order white interference colors that resemble a pearl-like appearance.

(3) Argillaceous dolomite: It is primarily composed of carbonate minerals, particularly dolomite, and some samples contain specific alkali minerals. The dolomite grains formed through primary sedimentation are fine and uniform. However, in some samples, subsequent processes have caused partial or complete recrystallization, leading to the formation of larger grains of carbonate minerals.

(4) Diamictite: It constitutes the largest proportion among all rock types in the Fengcheng Formation. It can be further divided into three subfacies based on the proportions of dolomitic, argillaceous, and tuffaceous components. This rock contains a significant amount of volcanic detritus, indicating frequent volcanic activities during deposition of the formation. Compared to argillaceous rocks, it has relatively lower amounts of quartz and feldspar detritus which exhibit sharp angular shapes. Carbonate minerals are often affected by diagenetic processes and form larger clusters or bands. Organic matter found within this type of rock can be classified into two major categories: asphaltic and particulate. Asphaltic organic matter may be related to microbial activity and exhibits little to no fluorescence under blue light excitation while particulate organic matter shows yellow fluorescence but overall fluorescence is not very strong. The mineral characteristics of the source rocks in the Fengcheng Formation are summarized in Tables 5.2 and 5.3.

Lucaogou Formation (P_2l)

The quantitative analysis of X-ray diffraction indicates the distribution of clay minerals in the Lucaogou Formation ranges from 2 to 62%, with an average of 33.02% (Table 5.4). The content of quartz and feldspar ranges from 25 to 84%, with an average of 53.86%. The carbonate minerals range from 0 to 68%, with an average of 13.1%.

5.2.2.2 Elemental Composition of Source Rocks

Fengcheng Formation (P_1f)

The elemental composition of the Fengcheng Formation is shown in Table 5.5

Trace element compositions from 63 samples taken from the Fengcheng Formation in the Wuerhe region were analyzed and plotted on a standardized spider diagram

Table 5.2 Mineral composition of source rocks in the lower Permian Fengcheng Formation, Junggar Basin (after Shifa et al. 2014)

Mineral composition/%												
Quartz		Feldspar		Calcite		Dolomite		Pyrite		Clay		Other
Range	Average	Range	Average	Range	Average	Range	Average	Range	Average	Range	Average	
3–45	30.4	6–53	26.5	2–48	11.2	2–80	20.6	0–10	3.6	0–9	3.8	3.9

Table 5.3 Clay mineral composition in source rocks of the Fengcheng Formation, Fengcheng Area, Junggar Basin (after Xuelei et al. 2022)

Relative content of clay minerals (%)									
Montmorillonite		Illite–smectite mixed		Illite		Kaolinite		Chlorite–smectite mixed	
Range	Average	Range	Average	Range	Average	Range	Average	Range	Average
0–81	43.5	0–85	30.1	4–57	14.5	3–15	6.5	0–38	5.4

Table 5.4 Mineral composition of source rocks in Lucaogou Formation, southeastern margin of Junggar Basin (after Yang et al. 2018)

Sample position	Clay/%		Quartz + feldspar/%		Carbonate/%	
	Range	Average	Range	Average	Range	Average
Upper Lucaogou	2.00–62.00	36.82 (73)	25.00–82.00	51.52 (73)	0.00–68.00	11.66 (73)
Middle Lucaogou	2.00–41.00	12.14 (7)	34.00–84.00	70.86 (7)	0.00–61.00	17.00 (7)
Lower Lucaogou	3.00–28.00	14.29 (7)	52.00–75.00	61.29 (7)	5.00–41.00	24.43 (7)
Lucaogou	2.00–62.00	33.02 (87)	25.00–84.00	53.86	0.00–68.00	13.11 (87)

(Figs. 5.4, 5.5 and 5.6). The analysis revealed significant variations in trace element enrichment patterns among different rock types.

In Fig. 5.4, the spider diagram of trace elements of tuff samples drilled from the Fengcheng Formation reveals that different types of tuffs display a strong enrichment of elements such as Li, Mo, and W compared to the average values of the upper continental crust. However, high-field strength elements like Nb, Ta, and Th are noticeably depleted. Oxygen-sensitive elements N, V, and Cr exhibit weak enrichment. The tuff samples have similar trace element compositions to intercalated andesite layers within the same stratigraphic level but differ significantly in their significant enrichment of Mo and W. This suggests that during deposition, deep-sourced hydrothermal fluids enriched with W and Mo influenced sedimentation processes in the tuff samples.

Figure 5.5 shows the spider diagram of trace elements of dolomite rocks in the Fengcheng Formation. The dolomite samples exhibit similar enrichment of Li, Mo, and W as observed in tuff samples, however, the dolomite rocks also exhibit strong enrichment of Ni. High-field strength elements such as Hf, Ta, Th, Zr, and Nb are significantly depleted in all types of dolomite rocks except siliceous dolomite. This suggests that there may be a common or similar source of material for both tuff and dolomite rock types. Siliceous dolomite displays relatively higher enrichment of Cu, Y, Th, and U, indicating that the fluids during the deposition of dolomite rocks in the Fengcheng Formation not only exhibit siliceous characteristics but also display a multi-source nature.

Table 5.5 Content of major elements of source rocks in the Fengcheng Formation, Junggar Basin (after Guo et al. 2018)

Rock type		Major elemental oxide content/%										
		SiO_2	Al_2O_3	CaO	FeO	K_2O	MgO	MnO	Na_2O	P_2O_5	TiO_2	LOI
Tuffs	Range	38.00–80.12	3.43–14.21	2.3–11.43	2.33–5.42	1.81–6.58	7.69–10.5	0.05–0.143	0.423–12.1	0.003–0.174	0.003–0.174	5.12–17.52
	Average	54.29	9.67	6.23	4.10	4.32	8.93	0.073	4.62	0.051	0.051	11.38
Dolomites	Range	34.68–47.94	2.61–8.84	7.8–14.73	2.73–4.17	0.24–5.51	2.21–9.01	0.018–0.13	2.4–3.58	0.005–0.059	0.131–0.393	16.57–26.28
	Average	42.89	6.82	10.37	3.05	4.32	4.61	0.065	2.91	0.019	0.29	19.98
Reedmergnerite	Range	45.00–74.42	0.09–2.71	0.24–11.63	0.06–1.81	0.05–1.83	0.16–8.41	0.001–0.054	6.31–12.2	0.008–0.025	0.04–0.128	1.1–18.44
	Average	62.71	1.28	3.81	0.81	0.71	3.09	0.021	10.08	0.016	0.055	8.66
Average		51.48	6.22	6.64	3.07	2.98	5.85	0.058	5.44	0.032	0.141	14.04

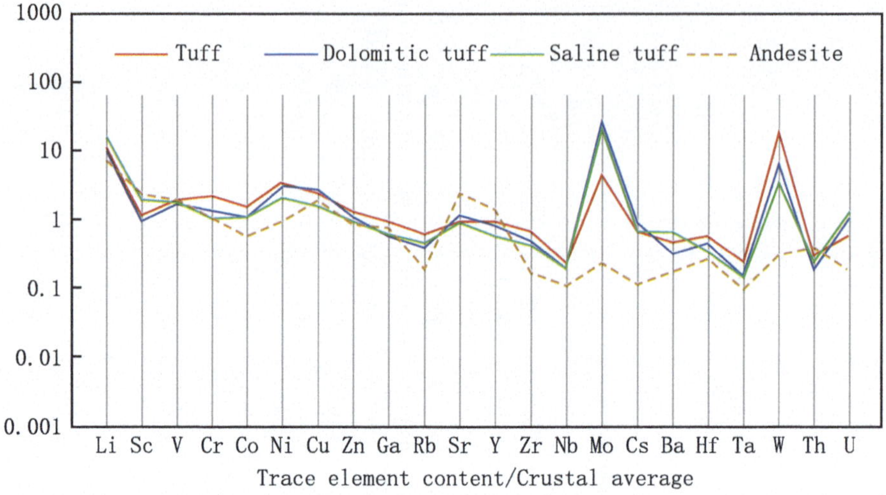

Fig. 5.4 Spider diagram of trace elements in the Fengcheng Formation source rocks (tuffaceous rocks), Wuerhe region (after Chang et al. 2017)

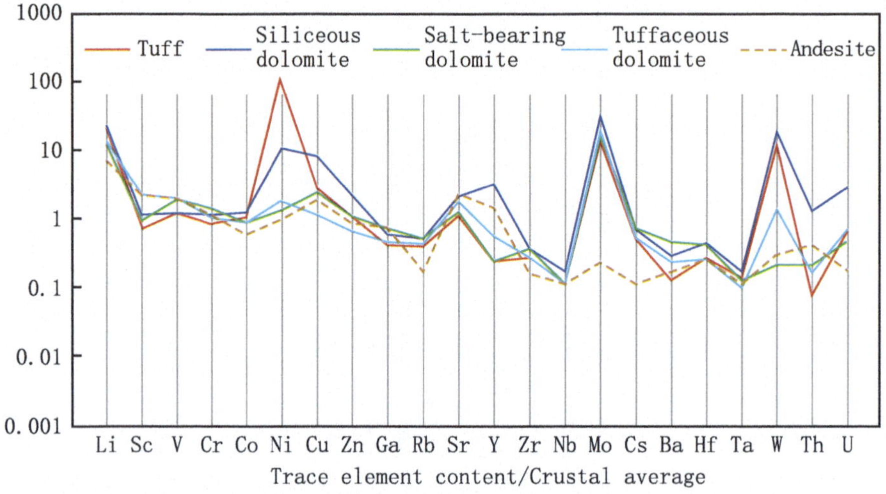

Fig. 5.5 Spider diagram of trace elements in the Fengcheng Formation source rocks (dolomitic rocks), Wuerhe region (after Chang et al. 2017)

Figure 5.6, which depicts the spider diagram of trace elements of salt rocks, shows that the trace element content is relatively low compared to average crustal values. Nevertheless, there are still some notable features worth noting. In addition to the enrichment of Li, Ni, W, and Mo elements commonly observed in tuff and dolomite rocks, the reedmergnerite bands also exhibits enrichment of Cu. Although the overall content of trace elements in the mineral reedmergnerite is noticeably

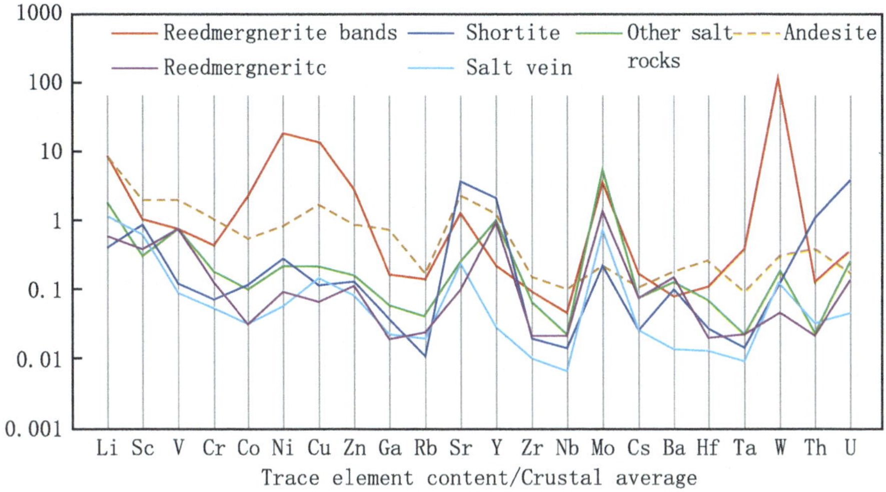

Fig. 5.6 Spider diagram of trace elements in the Fengcheng Formation source rocks (salt rocks), Wuerhe region (after Chang et al. 2017)

depleted compared to the reedmergnerite bands, it still exhibits relative enrichment in Li, Cu, Mo, and W. Given the general enrichment of trace elements in the volcanic ash and limestone, it is considered that the enrichment of trace elements in the reedmergnerite bands is primarily attributed to their higher content of volcanic ash or limestone components.

Lucaogou Formation

Rock samples of the Lucaogou Formation have high concentrations of Si, Fe, and P elements. Of these elements, P is a crucial nutrient that plays a significant role in controlling primary productivity. It creates a favorable nutritional environment for aquatic organisms, especially algae, which promotes the formation and development of oil-prone source rocks within the Lucaogou Formation.

5.2.2.3 Organic Matter Abundance of Source Rocks

The Lower Permian Fengcheng Formation source rocks have an average TOC content of 1.28%. The content of bitumen "A" is 1.492‰. The TOC content is relatively high, but the content of soluble organic matter is low within source rocks of the lower Wuerhe Formation. The content of bitumen "A" ranges from 0.027 to 0.570‰, with an average of 0.177‰. The total hydrocarbon content ranges from 0.012 to 0.250‰, with an average of 0.080‰.

Organic Matter Types of Source Rocks

The type of organic matter can be identified through various indicators such as the microscopic composition of kerogen, elemental composition, Rock-Eval pyrolysis parameters, and carbon isotope composition. Organic matters of the Fengcheng Formation are characterized by a predominant presence of sapropel, a relatively high content of exinite, and a low content of inertinite. Based on the aforementioned observations, it has been identified that the organic matter types present in the source rocks are type I, type II_1, and type II_2. The organic matter in the Pingdiquan Formation is characterized as sapropelic and sapropel-humic, while the organic matter within the Lucaogou Formation is predominantly sapropelic and sapropel-humic and is classified as type I and type II_1.

Thermal Maturity of Organic Matter

The vitrinite macerals of the Fengcheng Formation source rocks in the northwest margin region have a reflectance (Ro) range of 0.85–1.12%, indicating organic matters are currently at their peak oil generation stage. The vitrinite reflectance of the lower Wuerhe Formation (P_2w) source rocks correlates well with burial depth. At a depth of 1200 m, the vitrinite reflectance is 0.7%, increasing to 1.3% at around 3300 m and reaching 2.0% at approximately 5500 m. Source rocks of the Permian strata in the central part of the Junggar Basin have generally reached mature-high mature stage.

In the eastern part of the Junggar Basin, organic matter within the Pingdiquan Formation displays varying maturity levels. It is immature at depths shallower than 900 m, low maturity between 900 and 1483.71 m, and reaches the peak of mature oil generation beyond 1483.71 m. In areas such as Huoshaoshan, Shanan, Beisantai, and Santai, where the burial depth of the source rocks ranges from 1800 to 2500 m, the vitrinite reflectance ranges from 0.6 to 0.75%, indicating that the source rocks have entered the mature stage.

5.2.3 Accumulation Conditions of the Permian Source Rocks

5.2.3.1 Favorable Hydrocarbon Generation Space Promoting Organic Matter Transformation

The Junggar Basin in the Permian period witnessed the development of organic-rich sedimentary systems, and the unique evolutionary history of the Mahu Sag has led to the formation of distinctive source rocks. Research confirms that the Fengcheng Formation serves as the primary source rock in the Mahu Sag. It covers an area of approximately 500 km^2 with a thickness ranging from 50 to 300 m. The Fengcheng Formation is predominantly distributed in the central-western part of the Mahu Sag and near the Wuxia Fault Zone. Within the vertical succession, most of the source rocks are concentrated in the lower section of the Fengcheng Formation, exhibiting significant thickness variations ranging from 30 to 150 m. The typical petrological characteristics of rocks within the Fengcheng Formation are primarily composed of three endmembers: continental clastics, carbonate components, and volcanic materials, which are mixed in varying proportions. This mixture gives rise to various rock types such as conglomerates, sandstones, limestones, mudstones, tuffaceous rocks, and evaporites, as well as transitional lithologies. Frequent interbedding and distinct textures can be observed in vertical direction within the Fengcheng Formation. In the formation, it is common to observe interbeds that appear as thin layers, lenticular and laminated structures. Furthermore, some of these interbeds consist of extremely thin layers of algal laminated carbonate rock. The composition of rocks can vary significantly at a small scale, sometimes as little as centimeters or millimeters. As a result, one rock sample may contain multiple types of lithology.

The upper part of the 1st to 3rd members of the Fengcheng Formation is characterized by frequent alternation between limestone and shale layers, with individual layer thickness generally less than 0.5 m. The porosity is generally less than 10%, with an average of 5.79%, and the permeability is generally less than 0.1 mD. However, there are also a few samples with porosity exceeding 10% and permeability exceeding 0.1 mD, which correspond to core scans showing the development of microfractures.

Source rocks with high porosity offer more room for the generation of organic matter and facilitate its transformation. As the organic matter undergoes continuous conversion, hydrocarbon-generating overpressure gradually increases (Fig. 5.7). Due to the relatively high permeability of the source rock, hydrocarbons in the pores migrate to adjacent layers with larger pores under the influence of hydrocarbon-generating overpressure. This process is known as primary migration. As hydrocarbons are expelled, pore pressure decreases, initiating the hydrocarbon generation process once again. This cyclic process efficiently led to the transformation of organic matter within source rocks and results in widespread occurrence of mature oil within it.

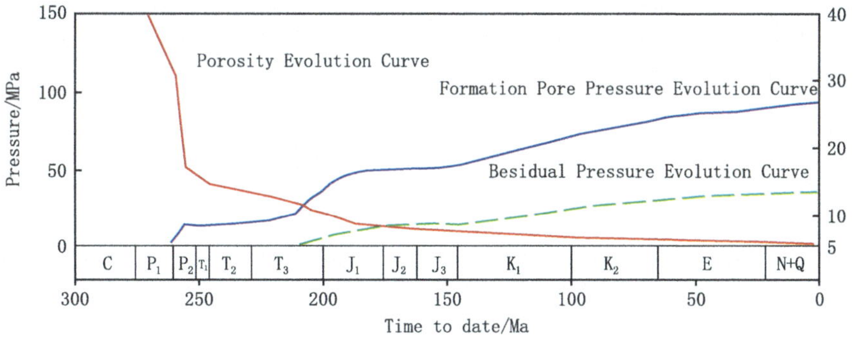

Fig. 5.7 Formation pressure and pore evolution profile in the Mahu Sag (after Dongming et al. 2021)

Based on the spatial distribution of lithofacies, it can be inferred that there is favorable storage space for the initial oil expulsion from the source rocks. Research indicates that the source rocks of the Fengcheng Formation were deposited in a fan-delta system. Influenced by factors such as provenance, basin water salinity, paleoclimate, terrigenous sediment supply, and volcanic activity, a delta front belt characterized by dolomitic sandstones is formed in the area towards the land. The clastic particles in this belt are relatively coarse, mainly fine to medium sandstones, but they undergo significant diagenesis, resulting in dense lithostratigraphic layers. Towards the basin, the source rocks are characterized by thin interbeds of dolomitic silt-fine sandstones, dolomitic mudstones, and mudstones. This has created a highly typical source-reservoir integrated shale oil play within the basin-slope area.

5.2.3.2 Catalytic Effects of Various Trace Elements on Oil Generation

Source rocks within the Fengcheng Formation in the Mahu sag is formed in an alkaline lake environment under the influence of volcanic activity. The dominant minerals of the source rocks are quartz, feldspar, calcite, and dolomite. Additionally, some uncommon minerals such as reedmergnerite and sodalite can been observed. The clay content is minimal, with smectite being the primary clay mineral composition. The alkaline nature of the Fengcheng Formation is a result of volcanic activity and sedimentation in an environment that is rich in sodium but poor in potassium. The presence of sodalite indicates an alkaline environment, while reedmergnerite and suggests the presence of hydrothermal activity. The high feldspar content and low clay content (<10%) suggest that physical weathering is dominant over chemical weathering in the Fengcheng Formation. Under the influence of volcanic and hydrothermal activities, certain trace elements in source rocks of the Fengcheng Formation show significant enrichment compared to the average crustal values. Trace elements such as Li, Mo, and W are strongly enriched in tuff samples. Dolomite samples exhibit enrichment of trace elements including Li, Mo, W, Ni, Cu, Y, Th, and U. Similarly, salt rock samples demonstrate enrichment of trace elements like Li, Ni, W, Mo and Cu. Metallic elements in hydrothermal fluids are absorbed by the sediment and adsorbed onto organic matter, catalyzing the process of hydrocarbon generation and significantly enhancing oil generation efficiency.

References

Chang H (2017) The genesis of exhalative rocks of the lower permian fengcheng formation in Urho area, Northwestern Junggar Basin[D]. Chengdu: Chengdu University of Technology

Chunli G (2016) Material composition and geochemical characteristics of the lower Permian Fengcheng Formation jet rocks in the northwestern margin of Junggar Basin. Doctoral dissertation, Chengdu University of Technology

Dongming Z, Yong T, Wenjun H et al (2021) Conventional-unconventional oil-gas orderly symbiosis and reservoir-forming model of all-oil-gas system in Fengcheng Formation, Mahu sag, Junggar Basin. Pet Explor Dev 48(1):38–51

Guo C (2018) The characteristics of compositions and geochemical of exhalative rocks of the lower permian fengcheng formation in Urho area, Northwestern Junggar Basin[D]. Chengdu:Chengdu University of Technology

Hailiang C (2016) The genesis of exhalative rocks of the Lower Permian Fengcheng Formation in Urho area, Northwestern Junggar Basin. Doctoral dissertation, Chengdu University of Technology

Han Y (2017) Geological characteristics of shale gas in Lucaogou Formation in Southeast Margin of Junggar Basin. Master dissertation, Chengdu University of Technology

Jianping C, Wang X, Deng C et al (2016) Geochemical characteristics of source rocks and crude oil in Junggar Basin. Acta Geol Sin 90(1):37–67

Shifa Z et al (2014) Petrology and lithogeochemical characteristics of dolomites of the Lower Permian Fengcheng Formation in the northeastern part of Junggar Basin. Geol Rev 60(05):1113–1122

Wang X, Zhi D, et al (2013) Source rocks and crude oil geochemistry in the Junggar Basin[M]. Beijing: Petroleum Industry Press

Wenjun H, Xulong W, Yang Z et al (2019) Petroleum geological conditions, resource potential and exploration direction of Junggar Basin. Mar Pet Geol 24(2):75–84

Xiaojun W, Wang T, Cao J (2018) Basic characteristics and efficient hydrocarbon generation of alkali lake source rocks of Fengcheng Formation in Mahu Depression. Xinjiang Pet Geol 39(1):7

Xuelei Z (2022) Clay mineral composition characteristics and genesis of Fengcheng Formation in Mahu Depression. Xinjiang Pet Geol 43(01)

Yang H (2018) Shale gas geological characteristics of lucaogou formation in the southeastern margin of Junggar Basin[D]. Chengdu: Chengdu University of Technology

Chapter 6
Characteristics of Triassic Source Rocks in the Ordos Basin

Abstract The Ordos Basin, situated in the western part of the North China Block, originated as a large deep-water lacustrine basin during the Mesozoic, shaped by plate collision dynamics. Its geological evolution can be divided into three stages: overall subsidence, uplift and erosion, and subsequent shrinkage and adjustment. The basin contains two primary source rock formations: the Triassic Yanchang Formation and the Jurassic Yan'an Formation. This chapter provides an in-depth analysis of the Triassic Yanchang Formation, focusing on its role as the primary source rock. It examines the depositional environment, paleontological characteristics, stratigraphy, sedimentary facies, and petroleum geological properties of the formation. Additionally, the study evaluates the petroleum accumulation conditions of the Yanchang Formation, offering insights into its significance within the Ordos Basin's petroleum system.

The Ordos Basin is located in the western part of the North China Block, spanning across five provinces: Shaanxi, Gansu, Ningxia, Inner Mongolia, and Shanxi, with an area of 25×10^4 km^2. The basin has a rectangular shape and is surrounded by mountain ranges. During its early stage, it underwent stable subsidence, followed by the migration of the depocenter, resulting in a large superposed sedimentary basin. The basement of the basin consists of the Archean and Proterozoic metamorphic rock series. The sedimentary cover comprises various systems, including the Changcheng System, Jixian System, Zhenning System, Cambrian System, Ordovician System, Carboniferous System, Permian System, Triassic System, Jurassic System, Cretaceous System, Paleogene System and Neogene system. The stratigraphy is well-developed, and sedimentary cycles are clearly observed.

The Ordos Basin is divided into six primary structural units, the Yimneg Uplift, the Weibei Uplift, the Western Overthrust Belt, the Tianhuan Depression, the Yishan Slope, and the Jinxi Flexure Belt, based on its present structural morphoFiglogy and evolutionary history (Table 6.1 and . 6.1).

The Yimeng Uplift is situated in the northern region of the Ordos Basin and has been uplifted since the Paleozoic era. As a result, strata within this primary unit have thinned or pinched out towards this area. During the Neogene period, subsidence

D. Guan et al., *Theory and Simulation Experimental Technology of Oil Generation From Nonmarine Source Rocks in China*, https://doi.org/10.1007/978-981-96-3145-2_6

Table 6.1 Division of structural units in the Ordos Basin

No.	Primary structural units	Area/km^2	Layer thickness/m
1	Yimneg uplift	42,000	3500
2	Yishan slope	110,000	6000
3	Weibei uplift	19,000	7500
4	Jinxi flexure belt	26,000	4000
5	Tianhuan depression	28,000	11,000
6	Western overthrust belt	24,000	8000

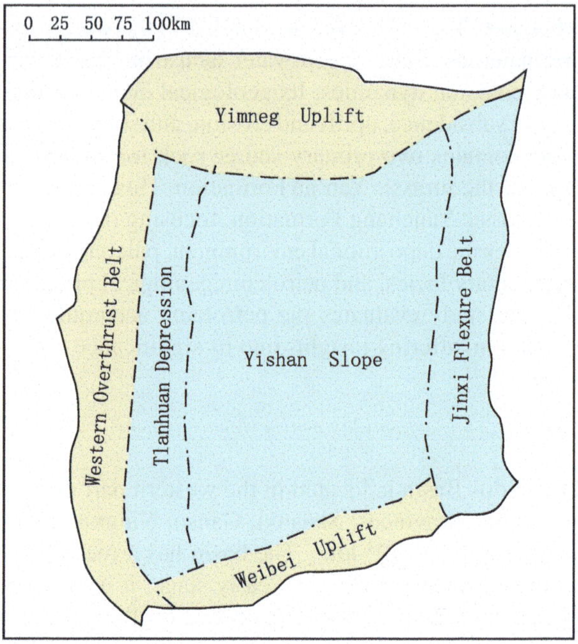

Fig. 6.1 Structural map of the Ordos Basin. Modified after Zhanli et al. (2017)

occurred in the Hetao Basin which caused it to separate from both Yinshan Mountains and Yimeng Uplift, leading to its current structural configuration.

During early Paleozoic era, the Yishan Slope was an elevated region without well-formed sedimentary layers. During the Cambrian and Early Ordovician periods, the Yishan Slope experienced a marine depositional environment. From the late Paleozoic onwards, it began to receive contiental sedimentation and gradually formed a gentle westward-dipping monocline with a dip angle of less than 1°. The Yishan Slope occupies a vast area in the central part of the basin, where it is characterized by well-developed nosing structures.

The Weibei Uplift is situated at the southern edge of the Ordos Basin. During the middle to late Paleozoic, it was a slope that dipped towards the south and experienced relative subsidence on its eastern side with the deposition of the Benxi Formation.

In Cenozoic times, while faulting and subsidence occurred in the Weihe River area, there was an uplift and tilting of Weibei Uplift which resulted in its present-day structural configuration.

The Jinxi Flexure Belt underwent uplift during the Middle to late Paleozoic period. Thin layers of sedimentary strata can only been observed within the Cambrian, early Ordovician, late Carboniferous, and early Permian formation. The zone underwent uplift until the late Jurassic, separating from the North China Platform and becoming the eastern margin of the Ordos Basin.

The Tianhuan Depression had a slope that dipped westward in the Paleozoic era. It began to subside and form a depression belt only during the late Triassic period. The subsidence continued throughout the Jurassic and Cretaceous periods, with the center of subsidence gradually moving eastward. The sedimentary zone showed an asymmetrical syncline structure with steep western limbs and gentle eastern limbs.

During the early Paleozoic, the Helan Rift was located in the northern segment of the Western Overthrust Belt. Meanwhile, the central and southern segments were characterized by marginal subsidence as part of the Ordos platform's edge. In the late Paleozoic, these same segments underwent foreland subsidence. From the middle Permian to the Jurassic, they formed a distinctly segmented and discontinuous deep trough.

6.1 Petroleum Geological Evolution of the Ordos Basin

During the early Paleozoic era, the Ordos Basin was a part of the North China Basin. However, uplift in the mantle led to the formation of Qinling-Qilian Trough which caused gradual subsidence of the Ordos Block. Consequently, it became a broad continental shelf facing the Qin-Qi Sea in the south. The basin experienced overall uplift from the late Ordovician, leading to the erosion and absence of the middle Ordovician to the lower Carboniferous stratigraphic units. During the Late Carboniferous, the Ordos Basin underwent another phase of subsidence, resulting in the deposition of a series of coal-bearing strata developed within marine-continental alternating facies (Linsen 2011). During the Permian period, the Ordos platform experienced a shift from marine to continental environments. As a result, the sea receded from the basin and a lacustrine depression basin began to develop.

During the Mesozoic, the Ordos Basin experienced rapid subsidence under the tectonic collision between the North China Block and the Yangtze Block. This led to the formation of a large deep-water depression basin, providing favorable regional tectonic conditions for the development of source rocks. From the Triassic to the Cretaceous, the basin evolution can be broadly divided into three stages: overall subsidence, uplift and erosion, and subsequent contraction and adjustment (Table 6.2).

Table 6.2 Petroleum geological evolution stages and features of the Ordos Basin

Era	Development stage	Physical field properties	Petroleum geology characteristics	Geotherm features
Quaternary Palaeo-Neogene Late Cretaceous	Basin shrink and adjusted development stage	Subsidence (pore pressure increase) Uplift (pore pressure decrease)	Continuous charge of oil and gas Forming new accumulations	
Late Jurassic–early Cretaceous	Basin uplift and erosion stage	Sediments Unloading and pressure relief	Process of hydrocarbon accumulation	
Yanan Formation depositional period Fuxian Formation deposition period Yanchang Formation deposition period	Basin continuous subsidence stage	Sediments loading and pressure increase	Process of hydrocarbon generation	Average 4–4.5 °C/100 m

6.1.1 *Development Stage of Continuous Subsidence in the Basin—Energy Accumulation Process*

During the Late Triassic to Middle Jurassic period, the Ordos Basin experienced continuous subsidence, leading to development of layers of lacustrine dark mudstones and deltaic sandstones within the basin, which provided abundant material basis for the formation and development of source rocks. During the sedimentation period from Chang 9 member to Chang 4 + 5 member, the basin underwent continuous subsidence. The surrounding provenance regions continuously supplied rich material that filled up the basin, resulting in a thick sequence of semi-deep to deep lacustrine mudstones and oil shales with high organic content. These mudstones and shales are interbedded with delta front sand bodies that are widely distributed in the basin, creating favorable spatial conditions for source rock and organic matter development in Triassic Yanchang Formation. During late Yanchang period, there were multiple episodes of thrusting and strong intracontinental deformation along the southwestern margin of the basin, which caused erosion in the region. During this period, the basin had a vast swamp environment that led to the widespread occurrence of coal seams in the upper Yangchang Formation. During the Early Jurassic Fuxian period, the basin inherited its paleogeomorphology from the late Triassic. The dominant sedimentary environment in the basin during this time was fluvial and lacustrine, with a relatively limited extent. During the deposition of the Yan'an Formation, the basin underwent another phase of subsidence with its center shifting eastward. However, the magnitude of subsidence was smaller compared to the late Triassic. The sedimentation during this time was characterized by small-scale fluvial, lacustrine, and swamp facies which led to the formation of coal-bearing strata interbedded with gray-white sandstones, black shales, and mudstones. These strata are considered as the primary source rocks within the Yan'an Formation in Ordos Basin. During the continuous subsidence of the basin, sediment continues to accumulate, leading to an increase in the lithostatic pressure exerted on the sediment. By the end of the Middle Jurassic, as the lithostatic pressure on the sediment continued to increase, the organic-rich mudstone and oil shale in the southern part of the Ordos Basin, gradually began to undergo maturation. At this stage, the burial depth of the Chang 7 oil-bearing formation is around 1400 m, with a geothermal gradient of approximately 4–4.5 °C per 100 m. The vitrinite reflectance (R_o) value is greater than 0.6%, indicating that the organic matter has reached early maturity. The paleogeothermal temperature is estimated to be around 70 °C. During the later stage of overall basin subsidence in the late Middle Jurassic, the Chang 7 Member, which is the main source rock of the basin, entered an early maturity stage. Additionally, influenced by the Yanshanian movement, the basin experienced intensified faulting and volcanic activities. Consequently, a significant amount of heat energy accumulated within rock layers leading to thermal catalysis that initiated oil generation and conversion from organic matter. As this process continued progressively, abnormally high pressure gradually formed within mudstone of Chang 7 Member. This pressure difference between mudstone and sandstone layers created favorable conditions for primary oil migration.

6.1.2 Stage of Basin Uplift and Erosion—Energy Release Process

During the late Jurassic to early Cretaceous period, the basin underwent an overall uplift and erosion. This stage occurred amidst the Yanshanian Orogeny, which caused strong compression along all margins of the basin—eastern, western, southern, and northern. The western margin continued experiencing eastward thrusting that led to the advancement of late Jurassic subsidence belt towards east forming present-day Tianhuan syncline. At the same time, there was a westward advancement of eastern uplift belt resulting in Shanxi block's uplift and formation of a west-dipping monocline within Ordos Basin. During this uplift and erosion process, the Ordos Basin developed into an independent basin with boundaries like its present-day extent. In this phase, the physical properties of the basin transitioned from the continuous subsidence and loading-pressure phase to the uplift and erosion phase, resulting in unloading and pressure release. This process played a crucial role in the primary migration, secondary migration, and petroleum accumulation within the basin, inducing favorable conditions. The source rocks of the Yanchang Formation entered the peak period of hydrocarbon generation and expulsion.

6.1.3 Comprehensive Contraction Stage of the Basin—Energy Adjustment Process

During the late Cretaceous period, the intense Yanshanian Orogeny led to ultimately concluded the development of the large-scale Mesozoic inland Ordos Basin. This marked the beginning of a stage where basin shrinkage and adjustment occurred. During this stage, the entire basin experienced widespread uplift, with the absence of Late Cretaceous sedimentation. The lake basins gradually became shallower, diminished, and eventually disappeared. The tectonic framework of the basin stabilized, completing most processes of petroleum generation, migration, and accumulation. The lake basins gradually became shallower, contracted, and eventually disappeared. The tectonic framework of the basin stabilized, completing most processes of petroleum generation, migration, and accumulation. Simultaneously, westward subduction of the Pacific Plate led to overall uplift of Ordos Basin which formed present-day plateau landscape.

6.2 Characteristics of Triassic Source Rocks in the Ordos Basin

Two main sets of source rocks developed in the Ordos Basin during the Mesozoic era. The Triassic Yanchang Formation is the primary source rock of the basin, with hydrocarbon sources mainly distributed in the Chang 9 to Chang 4 + 5 members. Another significant set of source rocks is the Jurassic Yan'an Formation, primarily distributed in the Yan 8 to Yan 6 members. The focus of this study is primarily on analyzing the characteristics of the Triassic Yanchang Formation.

6.2.1 Sedimentary Environment and Paleontological Features

The Ordos Basin was once a part of the North China Basin. The basement of the basin consists of the Archean crystalline metamorphic rocks and Paleoproterozoic metamorphic rocks. In the Mesoproterozoic and Neoproterozoic periods, the basin underwent uplifts and experienced marine transgressions towards its southwest direction. During the Mesoproterozoic and Neoproterozoic eras, the Ordos syncline was generally in an uplifted state, while the southwestern region of the syncline experienced extensive marine transgressions. During the early Paleozoic era, the syncline gradually subsided and marine transgressions continued to expand. The Ordos Basin transformed into a wide continental shelf that faced the Qin-Qi Sea in its southern margin. From the south and west, the Qinling, Helan, and Qilian oceanic troughs encroached the basin leading to extensive coverage of an epeiric sea environment in many areas of the basin. As a result of this transformation, shallow marine carbonate rocks were developed in basin. During the late Ordovician, the basin experienced a period of overall uplift, leading to the absence of several stratigraphic units from the middle Ordovician to the lower Carboniferous. The Ordos Platform subsided again in the late Carboniferous, resulting in the deposition of a succession of coal-bearing strata represented by the Taiyuan Formation and the Shanxi Formation, both of which exhibit marine—continental transitional facies. The sea gradually receded from the early Permian and the basin underwent a phase of continental deposition. The topography displayed a pattern of high in the north and low in the south, with steep slopes in the west and gentle slopes in the east. During its early stage, fluvial deposition was predominant while lake deposits became dominant during its later stage within the basin. During the Late Triassic period, the basin experienced a continuous subsidence and stable deposition which led to its peak development. This resulted in the formation of river-lacustrine deltaic facies within the basin. The tectonic movements at the end of the Triassic period caused uplift and subsequent erosion of the basin, forming a landscape characterized by numerous valleys. In the Jurassic period, sedimentation occurred in these river valleys during their early stages, followed by lacustrine deposition later. Early in the Cretaceous, the western margin of the basin experienced eastward thrusting, leading to the development of the Tianhuan Syncline. Simultaneously, the eastern uplifted belt advanced westward, creating a large west-dipping monocline within the Ordos Basin. This tectonic activity resulted in the deposition of fluvial sediments in the area. The basin remained uplifted from early Cretaceous, resulting in further erosion of the Mesozoic strata and the formation of the present-day plateau topography. Within the basin, only certain areas retain thin deposits of the Paleogene to Neogene sediments.

Comprehensive analysis of geology, geochemistry, paleobotany, and palynology suggests that the Ordos Basin underwent a transition from a warm and humid climate to a dry one during the Mesozoic (Table 6.3). During the late Triassic, the basin continuously subsided under a warm and humid climate. The deep lake waters provided ample accommodation space for sedimentation, and the favorable climatic conditions led to abundant and widespread aquatic life, resulting in high productivity in the lake. This environment facilitated the development of various ancient biological fossils, including bivalves, ostracods, branchiopods, and vertebrate fish (Hua et al. 2016), providing a rich material basis for the organic matter accumulation in the Yanchang Formation.

In semi-deep to deep lake environment, the bottom of the lake is characterized by oxygen-deficient conditions. This made it challenging for benthic organisms to survive. The lake ecosystem consisted mainly of fish and plankton, including various types of algae and zooplankton. The predominant rock types were horizontally bedded gray-black mudy shale and oil shale, which often contained scattered occurrences of pyrite within these layers.

In the shallow lake zone, the water bottom is characterized by an oxidizing to weakly reducing condition. The water body is open and rich in organic matter, which provides an optimal environment for the growth and reproduction of various taxonomic groups. As a result, the preserved biota in the sedimentary layers of this zone is

Table 6.3 Sedimentary environment characteristics of the Yanchang formation in Ordos Basin

Strata	Oil layers	Sequence sequence	Sequence stratigraphy		Lake level	Paleo-climate	Provenance	Sediment environment	Tectonic settings	Tectonic events
Yanchang Formation	Chang 1	I	IV	Lacustrine deltaic	Lake transgression	Semi-humid and semi-arid	Southwest-dominant provenance, weakening of northern provenance	Meandering river delta		Under the influence of the Indosinian orogeny, there is a structural differentiation with the east being higher and the being lower in terms of paleotopography
	Chang 2 Chang 3				Lake regression				Overall uplift	
	Chang 4 + 5		III	Lacustrine deltaic	Lake transgression	Humid and warm	Predominantly supplied from north and southwest regions	Delta plain swamp facies		
	Chang 6				Lake regression			Delta front	Steady	
	Chang 7		II	Lacustrine				Semi deep–deep lake	Maximum subsidence	
	Chang 8				Lake maximum transgression			Braided river delta		
	Chang 9		I	River	Lake transgression				Subsidence	
	Chang 10							River	Initial subsidence	

exceptionally abundant, leading to a rich fossil record. Shaanxiconcha is the dominant bivalve species, exhibiting a high abundance and numerous individuals. It is densely distributed in the lower part of the Chang 7 Member, and also occurs in high density within the silty mudstone of the Chang 6 and Chang 4 + 5 members. The predominant ostracod species is Tungchania, mainly concentrated in the black shale and silty mudstone of the Chang 7 Member, forming dense layers and coexisting with fish scales. Additionally, there are considerable quantities of Darwinula, Gomphocythere, Lutkevichinlla, and leaf-limbed crustacean Euestheria (larvae) preserved. The fossil-bearing strata predominantly consist of dark-colored mudstone or silty mudstone with horizontal bedding, containing siderite nodules.

In the shore-deltaic transition zone, the water bottom is characterized by abundant oxygen, turbulent water, and rich organic matter. This environment is particularly suitable for the growth of organisms that are resistant to waves, feed on muddy sediment, and have thick shells and large body sizes. The Unio species, particularly U. xuefeng-chuanensis, U. ningxiaensis, U. huangbogouensis, and U. wayaopuensis are the dominant large organism individuals in the area. The fossil assemblage also includes ostracods such as Darwinula, Lutkevicninella and Euestheria. In the upper part of the Yanchang Formation, there is a coexistence of Darwinula, Lutkevicninella and Tungchania with multiple layers indicating a nutrient-rich fresh to brackish water environment.

6.2.2 *Stratigraphy and Depositional Facies Characteristics of the Yanchang Formation*

The Upper Yanchang Formation in the Ordos Basin represents a complete continental depositional system composed of fluvial, deltaic, and lacustrine sediments. The sediment grains are relatively fine, making it the primary source rock sequence in the basin. This formation is widely distributed in the Ordos Basin and its surrounding areas. Based on the characteristics of sedimentary cycles and lithological compositions, the Yanchang Formation can be divided into five lithological units from bottom to top: feldspar sandstone zone (T_3y_1), oil shale zone (T_3y_2), oil-bearing zone (T_3y_3), conglomerate sandstone zone (T_3y_4), and the Wanyaobao coal series (T_3y_5). Furthermore, it comprises ten oil layers (Fig. 6.2 and Table 6.4).

During the deposition of the Chang 10 Member, the long axis of the basin extended in a northwest-southeast direction. The shoreline of the lake varied, being narrower in the north and wider in the south (Fig. 6.3a). The central region represented a shallow lake subfacies, while the eastern and western margins of the basin developed delta front subfacies, which transitioned into delta plain and alluvial plain subfacies towards the outer areas. During this period, due to significant topographic variations, the sediment primarily consists of gray, gray-green, and flesh-red thick bedded sandstones, as well as gray, dark gray, and gray-black sandy mudstones, mudstones,

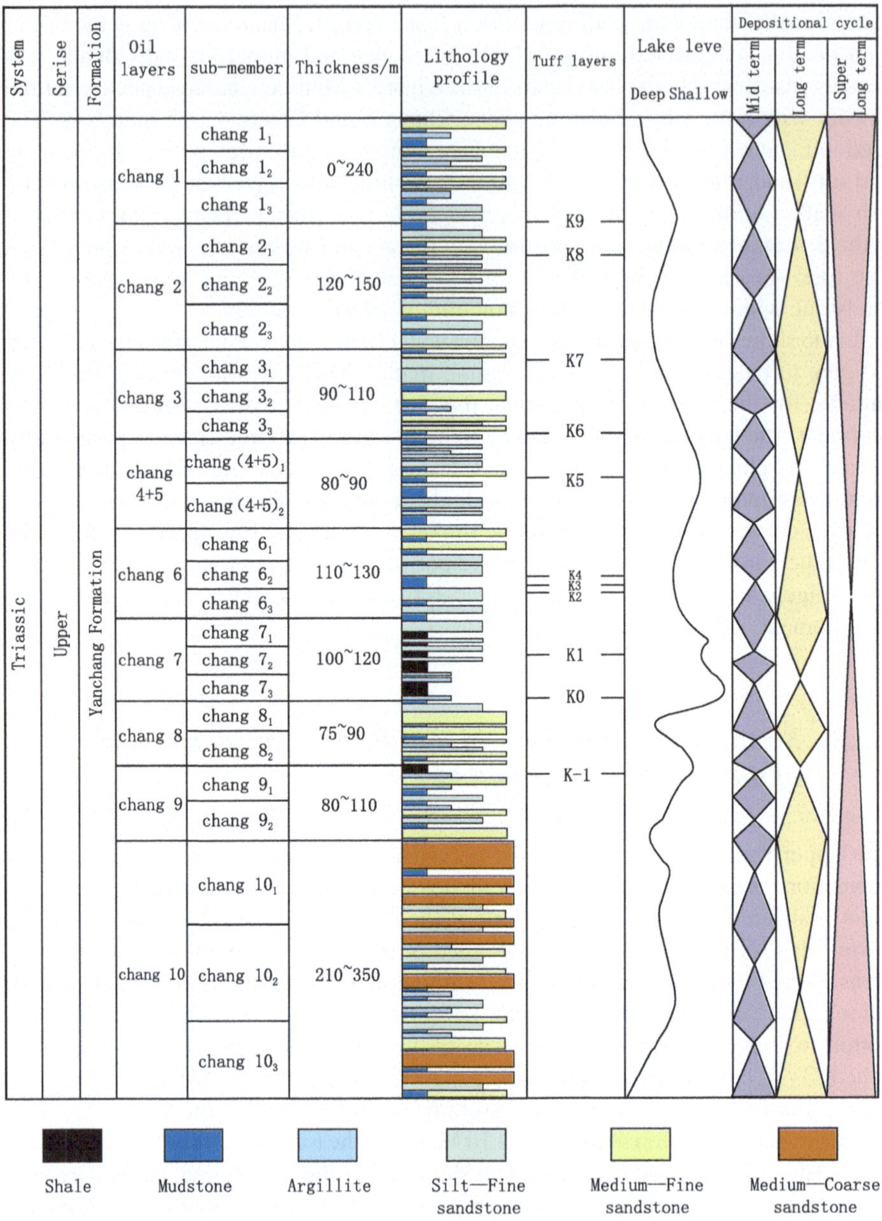

Fig. 6.2 Comprehensive stratigraphic column of the Yanchang Formation in Ordos Basin (after Jingli et al. 2018)

Table 6.4 Summary of the Yanchang Formation stratigraphy in Ordos Basin

System	Series	Formation	Member	Oil layers	Sub-member	Thickness/m	Lithologic characteristics
Jurassic	Lower	Fuxian Formation				0–150	Thick blocky sandy conglomerates interbedded with purple-red mud stones
Triassic	Upper	Yanchang Formation	Fifth member T_3y_5	Chang 1		70–90	Gray-green mudstones interbedded with fine-grained sandstones, carbonaceous shales, and the Wayaobao coal scams
			Fourth mem be T_3y_4	Chang 2	Chang 2_1	40–45	Gray-green blocky medium to fine-grained sandstones interbedded with gray mud stones
					Chang 2_2	40–45	Light gray medium lo fine-grained sandstones interbedded with gray mudstones
					Chang 2_3	45–50	Gray and light gray medium to fine-grained sandstones interbedded with dark-colored mudstones
				Chang 3	Chang 3_1	120–135	Light gray and grayish-brown fine-grained
					Chang 3_2		
					Chang 3_3		
			Third member T_3y_1	Chang 4 + 5	Upper	45–50	Dark-colored mudstones, carbonaceous mudstones, and coal seams interbedded with thin layers of fine-grained sandstone

(continued)

Table 6.4 (continued)

System	Series	Formation	Member	Oil layers	Sub-member	Thickness/m	Lithologic characteristics
					Lower	45–50	Interbedded light gray, fine-grained sandstones and dark-colored mudstones
				Chang 6	Chang 6_1	35–45	Dark greenish-gray sandstones interbedded with dark-colored mudstones
					Chang 6_2	20–30	Light grayish-green, fine-grained sandstones interbedded with dark-colored mudstones
					Chang 6_3	25–35	Interbedded gray-black mudstones, silty sandstones, and fine-grained sandstones with thin layers of tuffaceous shale
				Chang 7		80–100	Dark-colored mudstones interbedded with oil shales and thin layers of silt- to fine-grained sandstone
			Second member T_3y_2	Chang 8		70–85	Dark-colored mudstones and sandy mudstones interbedded with gray silt- to fine-grained sandstones

(continued)

Table 6.4 (continued)

System	Series	Formation	Member	Oil layers	Sub-member	Thickness/m	Lithologic characteristics
				Chang 9		90–120	Dark-colored mudstones and shales interbedded with gray silt- to fine-grained sandstones
			First member T_3y_1	Chang 10		280	Pinkish-red and grayish-green feldspathic sandstones interbedded with sandy mudstones, exhibiting mottled structure
	Middle	Zhifang Formation				300–350	Upper pail composed of grayish-green and brownish-purple muddy rocks interbedded with sandstones, lower part composed of grayish-green sandstones and rates

and carbonaceous mudstones, with thickness ranging from 100 to 300 m. The sandstone content is typically above 40%. The predominant color of the mudstones was dark with occasional purple-red hues along with carbonaceous mudstones. These sedimentary characteristics indicate a semi-arid to humid climate during this period.

During the deposition of the Chang 9 Member, the southwestern part of the basin experienced accelerated subsidence influenced by fault activity, resulting in an asymmetrical basin with steep southwestern margins and gentle northeastern margins (Fig. 6.3b). Against the backdrop of extensive lake transgression during the late Triassic, the basin area significantly expanded, with the development of both deep and shallow lake facies. The deltas on both sides of the basin retreated and became isolated from each other. Based on field outcrop observations, drilling core analysis, and the characteristics of sand body distribution, the northeastern part of the basin is dominated by meandering river delta environment, while braided river deltas prevail in the western and southwestern parts. The thickness of the Chang 9 Member

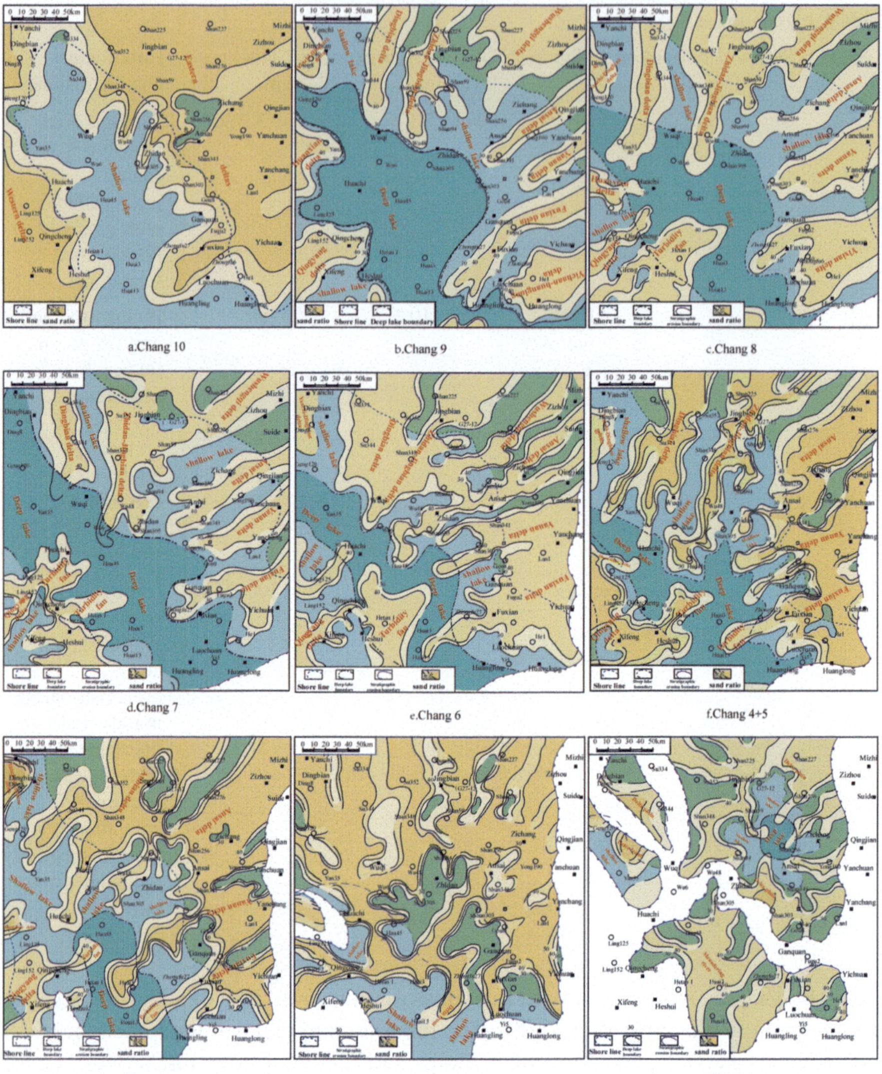

a.Chang 10 b.Chang 9 c.Chang 8

d.Chang 7 e.Chang 6 f.Chang 4+5

g.Chang 3 h.Chang 2 i.Chang 1

Fig. 6.3 Sedimentary facies map of the Yanchang Formation in the southern Yishan Slope, Ordos Basin (after Linsen et al. 2011)

is approximately 100 m, characterized by gray, dark gray, and gray-black sandy mudstone interbedded with grayish-green, medium to thick-bedded fine sandstones. The top of the layers is occupied by the laterally continuous "Lijiapan Shale" within the basin. The dark mudstone and shale interbedded with gray silt and fine sandstones form significant reservoirs of the Chang 9 Member. The presence of dark mudstone and shale serves as a significant foundation for the generation of oil. Additionally, the interlayering of silt and fine sandstones with the mudstone creates favorable conditions for the transformation of organic matter, thereby facilitating both the generation and storage of oil.

During the deposition of the Chang 8 Member, the basin was still in a development stage. Based on the sedimentary pattern established during the Chang 9 Member deposition, the deltaic systems slightly prograde with increased hydrodynamic forces, resulting in coarser sediment grain sizes. The thickness of the sedimentary package ranges from 45 to 150 m and is characterized by thick layers and lenses of gray-green and gray-yellow fine sandstones interbedded with gray-green, dark gray, and grayish siltstones and fine sandstones in its lower part. Meanwhile, its upper part consists of grey and dark grey mudstones, siltstones, and fine sandstones interbedded with thick layers of medium to fine sandstone that are colored either green or grey (Fig. 6.3c). The sand bodies within the Chang 8 Member generally exhibit favorable physical properties, and these well-permeable gray silt and fine sandstones, in conjunction with the underlying and overlying dark mudstones, constitute an effective oil source system within the Yanchang Formation. During the deposition of the Chang 7 Member, it represented the flourishing period of the paleolake, with a significant expansion of the basin area and increment of water depth, providing a favorable environment for the proliferation of aquatic organisms and planktonic organisms (Fig. 6.3d). The basin still exhibited a westward steep-eastward gentle configuration and the lithofacies along the margin of the basin show a zonal distribution. Various types of deltas developed near the shoreline, transitioning into shallow lake subfacies and deep lake subfacies towards the center of the basin. A package of dark mudstones and oil shales, approximately 80–150 m thick, was deposited, commonly referred to as "Zhangjiatan shale". The mudstones contain abundant fossils, including ostracods, bivalves, and conodonts, indicating a high organic content. The mudstones of the Chang 7 Member are interbedded with delta-front sand bodies or deep-water turbidites in the vertical direction, forming a significant suite of source rocks within the Yanchang Formation. These mudstones are rich in fossils such as ostracods, bivalves, and conodonts which suggest the presence of high organic content. In the Chang 7 Member, these mudstones are interbedded with delta-front sand bodies or deep-water turbidites vertically, creating an important set of source rocks within the Yanchang Formation.

During the deposition of the Chang 6 Member, there was an excess of sediment supply compared to subsidence in the basin. As a result, sediment from various directions comprehensively filled the basin (Fig. 6.3e). The hydrodynamic conditions of the delta during this period intensified significantly, leading to favorable reservoir properties. In contrast to the Chang 7 Member, the sediment found in the Chang 6 Member had relatively larger grain sizes. The Chang 7 Member mainly consisted of mudstones composed of fine to medium-grained sandstone layers. In contrast, the Chang 6 interval comprised alternating layers of mudstone and fine sandstone, with a sand content ranging from 30 to 45%. The thickness of this interval varied between 80 and 110 m. The mudstones contained fragments of plant stems, abundant ostracods, as well as occasional fish scales and fish fossils, indicating significant variations in sedimentary environment and facies distribution. Within the Yanchang

Formation, the Chang 6 Member is characterized by interbedded grayish-green and dark gray mudstones, along with light gray and light greenish-gray fine sandstones. This combination forms an effective self-sourcing and self-reservoir system.

During the deposition of the Chang 4 + 5 Member, there was another period of lake transgression in the basin, resulting in further expansion of the lake surface. The construction process of the deltas slowed down, leading to extensive delta plainification and marshland development in the western and northwestern parts of the basin. Within the delta front sand bodies, which developed during the deposition of the Chang 6 period, widespread mudstone deposits of floodplain swamp facies were observed (Fig. 6.3f).

During the deposition of the Chang 3 Member, the basin gradually shrank and all deltas significantly prograde towards the center of the paleolake. The water body in the basin experienced a large-scale shrink, and there was notable erosion of the strata in the southwestern part of the basin (Fig. 6.3g). The strata had a thickness ranging from 35 to 240 m. They were primarily composed of thick-bedded, gray and gray-green fine sandstone that was interbedded with dark gray mudstone, siltstone, and variable-thickness sandstone. The sandstone had a relatively coarse grain size and frequently appeared as thick, massive beds. The ratio of sand to mud ranged from 30 to 50%. The humic type was the primary organic matter found in the mudstone, which had some potential for oil generation. During the deposition of the Chang 2 Member, regional tectonic uplift caused intensified lake water to shrink. As a result, the northern part of the Yishan Slope became completely alluvial and plains-like (Fig. 6.3h). Throughout the basin, there were no deep lake subfacies present and only localized remnants of shallow lake subfacies persisted. The deltas around the basin are interconnected, forming a unified landmass. Along the Anbian-Ansai-Yanchang line, sedimentation consisted predominantly of river and floodplain deposits, characterized by thick-bedded, medium-grained arkosic sandstone. The sandstone had an average thickness of 50–70 m and a sand-to-mud ratio ranging from 45 to 60%. Large-scale crossbedding was commonly observed, while the development of source rocks was relatively limited.

During the deposition of the Chang 1 Member, significant uplift occurred in most parts of the basin, leading to the exposure of the strata at the surface (Fig. 6.3i). The humid climate during this period, coupled with abundant vegetation and gentle topography, resulted in insufficient sediment supply. As a result, extensive plain and swamp environment developed throughout the region, forming the well-known Wayaobao coal series. This series is characterized by deep gray and greenish-gray mudstone, accompanied by gray-black carbonaceous mudstone interbedded with shale and coal seams. It contains a significant amount of plant fossils and serves as a regional cap rock within the Yangchang Formation.

6.2.3 *Main Characteristics of Source Rocks in the Yanchang Formation*

The Yanchang Formation, predominantly found in the Ordos Basin, serves as the primary source rock. It is distributed mainly within the Chang 9 to Chang 4 + 5 members. These rocks are rich in organic matter, particularly in the Chang 7 Member, where the average TOC content can reach up to 5.17%. In other intervals, the TOC content ranges from 0.92 to 1.99% on average, indicating a generally moderate to good level of richness (Table 6.5). The high abundance of organic matter in the source rocks of the Yanchang Formation can be attributed to favorable conditions for organic matter enrichment. These conditions are created by the substantial supply of nutrients resulting from intense volcanic and hydrothermal activities, which promote the proliferation of algae and planktonic organisms within the formation.

The organic matter in the Yanchang Formation is primarily composed of a mixture of sapropelic-humic kerogen. The dark mudstones of the semi-deep lacustrine facies in the Lingyuan-Wuqi-Zhidan-Fuxian area, located at the northeastern front of a large delta, are primarily characterized by a predominance of a mixed type of kerogen. They exhibit relatively higher hydrocarbon transformation potential. The black oil shales of the shallow lake to semi-deep lacustrine facies in the Huanxian-Huachi-Qingyang area, located in the southwestern submarine fan, are primarily characterized by a predominance of sapropel kerogen. They exhibit relatively lower hydrocarbon transformation potential. The source rocks in the Yanchang Formation of the Ordos Basin have R_o values ranging from 0.73 to 1.06%, as indicated by vitrinite reflectance measurements. These values suggest that the source rocks are at a mature to highly mature stage, and there is a slight increase in R_o with increasing burial depth, indicating an increase in thermal evolution with depth.

During the Yanchang period, the depocenter of the Ordos Basin was in the Jiyuan-Huachi-Huangling area in the southwestern part of the basin. The majority of the Shanbei slope represents a transitional zone between lakes and deltas, where various

Table 6.5 Organic matter types and maturation the Yanchang Formation in the southern Ordos Basin

Strata			Facies	Kerogen type	R_o/%		TOC/%	
Triassic	Yanchang Formation	Chang 4 + 5	Shallow lake	Humic-sapropelic	0.73	0.73	0.7–1.11	0.92
		Chang 6			0.76	0.76	0.47–17.84	1.54
		Chang 7	Semi-deep lake	Sapropelic-humic	0.76–1.16	1.00	0.5–29.91	5.17
		Chang 8	Shallow lake	Humic-sapropelic	0.95–1.26	1.06	0.37–12.68	1.99
		Chang 9		Humic	0.96–1.05	1.01	0.08–18.11	1.27

types of delta fronts, prodeltas, and shallow lake subfacies developed. The sedimentary deposits are primarily composed of dark mudstones and oil shales interbedded with thin layers of silt and fine sandstones. Based on the bulk-rock and X-ray diffraction analysis of selected dark mudstone samples, the mineral composition of the dark mudstone is predominantly clay minerals and quartz, with minor amounts of carbonate and pyrite. Clay minerals are the most abundant, comprising an average of 49.0% of the content. Quartz is the second most prevalent mineral, with an average content of 25.9%. In addition, there are smaller amounts of pyrite (9.0%), plagioclase (9.5%), and calcite (3.6%)

6.2.4 Hydrocarbon Accumulation Conditions of the Yanchang Formation

The Yanchang Formation in the Ordos Basin is an oil-bearing stratum with low to ultra-low permeability, and it holds significant petroleum resources. Significant oilfields with reserves in the billions and tens of billions of tons have been discovered within the Yanchang Formation, thanks to advancements in inland petroleum exploration theories and reservoir development technologies. These include the Ansai Oilfield, Jingan Oilfield, Jiyuan Oilfield, Huaqing Oilfield, Longdong Oilfield, and Qingyang Oilfield, which are characterized by low porosity and low permeability reservoirs. By the end of 2020, the petroleum geological reserves within Mesozoic formations of the Ordos Basin were identified to be $592{,}919 \times 10^4$ t. The annual crude oil production in this area reached $24{,}670 \times 10^4$ t, solidifying its position as a major energy production base in China.

When reviewing the exploration history of the Ordos Basin, it becomes evident that there has been a shift in exploration strategies for the Yanchang Formation over the past 50 years. Initially, efforts were focused on exploring oil in peripheral structural zones of the basin. This was followed by a shift towards targeting oil in large river-delta sand bodies. Finally, attention turned to the "sweet spot" region within the central part of the basin. The understanding of oil accumulation models in the Ordos Basin has evolved from ancient landform-controlled and deltaic multilayer accumulation models to the current in-source reservoir-forming model. The development of these petroleum geological theories and the innovation of exploration techniques have effectively guided the exploration and exploitation of the basin, leading to significant oil exploration breakthroughs. The previous petroleum geological research approach in the Ordos Basin treated hydrocarbon generation and reservoir formation as two separate processes. It evaluated source rocks and reservoirs as distinct stratigraphic units, considering mudstones and sandstones as separate targets. During the evaluation process, the main reservoir units in the Yanchang Formation are typically considered to be the Chang 8, Chang 6, and Chang 2 oil-bearing members. These members are predominantly composed of feldspar-quartz sandstones and feldspar sandstones. The thick and widespread Chang 7 Member

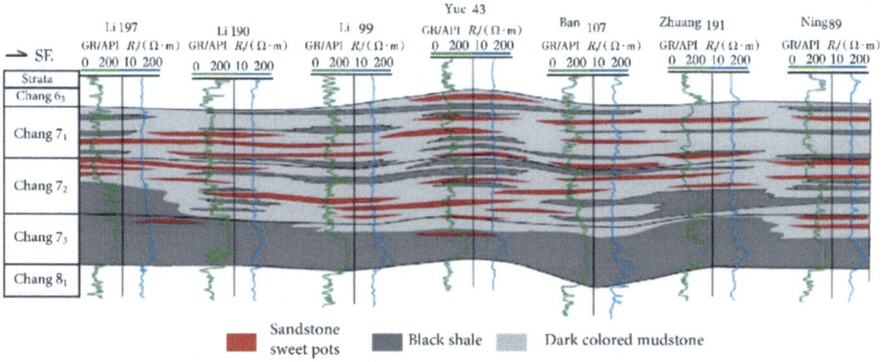

Fig. 6.4 Reservoir cross-section of Chang 7 oil formation in Qingcheng oilfield, Ordos Basin (after Fangzheng et al. 2020)

is recognized as the source rock, consisting of black shales and dark mudstones. Within this member, the most favorable source rock is the high organic-rich Chang 7_3 submember (Fig. 6.4), commonly referred to as the "Zhangjiatan Shale." It is characterized by thick black shales and extensive distribution. The secondary importance is given to the Chang 7_2 and Chang 7_1 submembers which are primarily composed of dark mudstones. The petroleum generated from the high-quality source rocks of the Chang 7 Member can migrate upward and downward through interbedded, relatively high-permeability sand bodies, microfractures, and pre-Jurassic paleoriver systems, leading to the formation of extensive lithological oil accumulations in the Chang 4 + 5, Chang 6, Chang 8, and Chang 9 members.

During the Chang 9 to Chang 4 + 5 members deposition period, the Ordos Basin underwent continuous subsidence, and the depocenter was in the Jinan-Huachi-Huangling area. The climate was warm and humid, which supported a wide distribution of abundant aquatic life within the basin. The basin experienced a transition from marginal shallow lakes to semi-deep lakes and then back to shallow lakes. Multiple sedimentary systems, including meandering river deltas, braided river deltas, and fan deltas associated with different sediment sources, developed around the central basin, resulting in a complex interlayered sequence of sandstones and mudstones in river-delta-lake depositional settings. In this study, using an organic geochemical lens, we regard the generation and accumulation of petroleum in the basin as a whole petroleum geological process. Characteristics of hydrocarbon accumulation in the Yanchang Formation were summarized.

(1) The continuous subsidence stage of the basin is a critical period for source rocks formation and the primary migration and accumulation of petroleum.

The continuous subsidence stage of the basin is characterized by the accumulation of sediment and the conversion and accumulation of energy through physical and chemical processes within these sediments. Throughout this stage, sediments experience continuous loading and an increase in pressure, while the temperature of the layers gradually rises with increasing burial depth. In the presence of appropriate temperature, pressure, and catalytic conditions, organic matter within the source rocks undergoes a gradual process of maturation and transformation, resulting in the formation of petroleum. As hydrocarbon expulsion pressure builds up, the generated petroleum starts migrating from the shale into the overlying and underlying sandstone layers, forming self-generated and self-stored shale oil reservoirs. During this continuous subsidence period, the basin remains a closed system, and large-scale migration of petroleum has not yet occurred.

During the late Middle Jurassic, as the basin continued to subside, sediments within the Yanchang Formation underwent increased loading and pressure. This led to the maturation of organic matter within the mudstones and oil shales. During this period, the Chang 7 oil-bearing formation was buried at a depth of approximately 1400 m, with a geothermal gradient of 4–4.5 °C/100 m.

The R_o value exceeded 0.6%, indicating that the organic matter had reached low maturity stage. Additionally, the paleogeothermal temperature was estimated to be around 70 °C. Furthermore, the basin was affected by the Yanshanian Orogeny, which resulted in increased faulting and volcanic activity. This led to the accumulation of heat energy within rock layers, triggering thermal catalysis that initiated the generation and conversion of oil from organic matter. Over time, this process created a favorable pressure differential between the mudstone and the sandstone layers overlying and underlying within the Yanchang Formation. The pressure difference between the mudstone and sandstone layers created favorable conditions for primary oil migration. The oil generated in the mudstone is stored in the nearby sandstone layers, creating shale oil reservoirs. The Chang 7 Member in the Qingcheng Oilfield provides a representative example of this case.

(2) The development stage of uplift and erosion is a critical period for oil secondary migration and accumulation within the Yanchang Formation.

During the uplift and erosion stage, the basin experiences a physical environment of unloading and decreased effective stress, leading to the erosion and thinning of overlying strata above the source rocks. When sandstone is subjected to external forces within its elastic deformation limit, it exhibits an elastic rebound after the removal of these forces. This causes the sandstone to return to its original state. As a result of this rebound phenomenon, negative pressure is generated within the pores of the sandstone. The amount of rebound is directly related to the negative pressure generated from sandstone rebound, which in turn is correlated with the erosion thickness. Typically, maximum erosion thickness occurs at peripheral slopes due to tectonic uplift. It gradually decreases towards the central part of the basin. As a result, sandstone layers in peripheral slopes exhibit larger rebound compared to those in source rock development areas where smaller rebound is observed. Consequently, there is a greater pressure drop in sandstone layers of peripheral slopes than in those of source rock development areas. Under this pressure differential, oil generated in source rocks migrates within connected sand bodies, namely secondary migration, towards regions with decreasing pressure. Eventually, it accumulated and preserved in well-sealed traps forming conventional oil reservoirs.

During the late Jurassic to early Cretaceous, the Ordos Basin experienced uplift and erosion influenced by the Yanshanian Orogeny, causing compression along all margins: east, west, south, and north. Specifically, the western margin underwent eastward thrusting, forming the Tianhuan syncline in the late Jurassic period. Concurrently, an eastern uplift belt advanced westward, uplifting the Shanxi block and creating a west-dipping monocline within the Ordos Basin. During this uplift and erosion stage, the physical conditions in the basin changed from continuous subsidence and pressure caused by loading to uplift and reduction of pressure due to unloading. This transition significantly influenced the oil primary and secondary migration and oil accumulation within the basin, especially during the peak period of hydrocarbon generation and expulsion within Yanchang Formation. The Chang 2 Member in the Jingan Oilfield serves as a typical example of this case.

(3) The fluvial-deltaic depositional systems serve as favorable targets for hydrocarbon generation and accumulation in the Yanchang Formation.

Numerous fluvial-deltaic depositional systems developed in the lacustrine basin gentle slope zone, which represent integrated hydrocarbon systems that are favorable for hydrocarbon generation, storage, migration, and accumulation, making it a promising area for oil generation and accumulation. The fluvial-deltaic depositional systems consist of organic-rich mudstones (or shales) and permeable sandstones. The mudstones serve as both source rock for hydrocarbon generation and cap rocks, while the sandstones act as reservoirs and conduits for petroleum migration. Within this deltaic system, sedimentary facies variations occur due to progradation and retrogradation, resulting in interbedding layers of mudstone and sand in the vertical profile. Mudstone-dominated intervals are typically found closer to the lake center, while sandstone-dominated intervals prevail towards the lake margin. The interbedding of sandstone and mudstone serves as both the source rock within the oil-bearing depression and as the self-generated and self-stored oil reservoir. If the sandstone reservoirs within the shale oil accumulation are connected to the sandstone bodies developed within fluvial-deltaic depositional system of the gentle slope zone, petroleum can be accumulated and stored within various traps in the slope zone through secondary migration, forming several conventional oil reservoirs.

The deposition of Yanchang Formation in the Ordos Basin involved a transition from shallow lakes to semi-deep lakes and eventually back to shallow lakes. This process gave rise to diverse sedimentary systems developed around the central basin, such as meandering river deltas, braided river deltas, and fan deltas. Consequently, a complex sequence of interbedded sandstones and mudstones were formed in river-delta-lake settings around the central basin. The Ordos Basin, with a focus on the river-delta depositional systems, has witnessed the successive discovery of billion-ton-class major oil fields. These include the Shenbei oilfield in the northeast, the Jiyuan oilfield in the northwest, the Longdong oilfield in the southwest, and the Huaqing oilfield in the central region.

(4) Interbedding of organic-rich mudstones and sandstones play crucial roles in the generation, migration, and accumulation of hydrocarbons.

This study focuses on the interbedded sequence of sandstones and mudstones that have the potential to generate, migrate, and accumulate hydrocarbons under specific geological conditions. These rocks are considered as effective source rocks for commercial oil reservoirs. The significance of this finding can be summarized in two key implications: Firstly, it involves assessing the organic-rich mudstones and sandstones within each reservoir unit as a whole. This means considering their porosity and permeability characteristics together. Secondly, it involves evaluating the processes of hydrocarbon generation and accumulation in source rocks as a complete petroleum geological process. The presence of sandstones within the source rocks is particularly important because they play crucial roles in the transformation, migration, and accumulation of hydrocarbons. In source rocks composed solely of mudstones, the generated oil exists as dispersed droplets within the pores. However, without sufficient space and pressure relief zones, these droplets struggle to gather and flow, hindering primary migration. The presence of interbedded sandstones solves this issue by providing storage space for the oil and facilitating continuous oil generation. Moreover, the pressure difference between sandstones and mudstones allows the dispersed droplets in the mudstone to coalesce and flow, enabling primary migration driven by pressure differentials. Therefore, the lithological combination of sandstones and mudstones not only act as source rocks but also serves as shale oil reservoirs.

The semi-deep to deep lacustrine shales, which serve as the source rocks for the Chang 9 to Chang 4 + 5 oil-bearing members. The shales are commonly found interbedded with delta-front sand bodies or deep-water turbidites vertically, with a sand content ranging from 25 to 45%. The mudstone contains abundant fossils such as ostracods, brachiopods, conchostracans, bivalves, and coelacanths, indicating a high presence of organic matter. The delta-front sand bodies mainly consist of blocky feldspathic sandstone and lithic sandstone with generally favorable physical properties. The biota in the lake basin, which thrives in brackish to saline water environments, creates ideal conditions for organic matter formation. The interlayered sequence of sandstones and mudstones deposited in the river-delta system provides favorable spatial conditions for hydrocarbon generation. Furthermore, the fluid pressure differences between the sandstones and mudstones aid in the migration and accumulation of petroleum.

The petroleum geological evolution of the Ordos Basin can be summarized into three stages: continuous subsidence, overall uplift, and comprehensive shrinkage adjustment. During the continuous subsidence stage, energy accumulation occurred in the basin, playing a crucial role in hydrocarbon generation and primary migration and accumulation. The stage of overall uplift witnessed energy release and was crucial for hydrocarbons large-scale migration and accumulation. In the comprehensive shrinkage stage, energy adjustment took place, resulting in the formation of new oil reservoirs and further enrichment and improvement of existing ones. Therefore, hydrocarbon generation and accumulation in the basin encompassed the entire life cycle of basin evolution. Throughout this process, deposition from river-delta systems was closely linked to the formation of source rocks as well as processes such as petroleum generation, expulsion, migration, and accumulation.

References

Caili Z, Xinshe L, Yajuan Y et al (2021) Oil and gas exploration history and enlightenment of Changqing Oilfield in Ordos Basin. Xinjiang Pet Geol 42(3):253–263

Dengfa H et al (2021) Periods and characteristics of key tectonic changes in Ordos Basin and its adjacent areas. Acta Petrolei Sinica 42(10):1255–1269

Fangzheng J, Ccaineng Z, Zhi Y (2020) Theoretical understanding and exploration and development practice of petroleum accumulation in continental source. Pet Explor Dev 47(06):1067–1078

Hua Y et al (2016) Biofacies division and geological significance of ancient lake basin in the late Triassic extension period of Ordos Basin. Acta Sedimentol Sin 34(4):688–693

Jingli Y et al (2018) Multi-source reservoir-forming model of Chang 9 member of Triassic in Ordos Basin. Pet Explor Dev 45(3):373–384

Linsen Z (2011) Mesozoic petroleum geological characteristics and efficient exploration in Yanchang Oilfield. Petroleum Industry Press, Beijing

Zhanli R, Qiang Y, Junping C et al (2017) Thermal evolution history of Ordos Basin and its control on oil and gas. Earth Sci Front 24(3):137–148

Chapter 7
Characteristics of Jurassic Source Rocks in the Sichuan Basin

Abstract The Sichuan Basin is a composite oil-and gas-bearing basin characterized bay Paleozoic marine deposits and Mesozoic-Cenozoic lacustrine deposits. Following the Late Triassic, the basin transitioned into an inland continental sedimentation stage, resulting in the development of lacustrine strata. The Upper Triassic Xujiahe Formation is the primary target for lacustrine gas reservoirs in the western Sichuan low-relief area, while the Da'anzhai Member of the Lower Jurassic Ziliujing Formation serves as the main target for lacustrine oilfields in the central Sichuan gentle slope area. This chapter focuses on the Da'anzhai Member, specifically in the Langzhong to southern Sichuan region, as a case study. It examines the depositional environment, biogeographic zoning, stratigraphy, sedimentary facies, primary source rock characteristics, and petroleum accumulation conditions. These insights provide valuable guidance for the exploration and development of lacustrine petroleum resources in the Sichuan Basin.

The Sichuan Basin, covering approximately 20×10^4 km^2, is in the Sichuan-Chongqing region. This area encompasses the eastern part of Sichuan Province, most parts of Chongqing Municipality, and minor parts of Hubei, Guizhou, and Yunnan provinces. The basin is diamond-shaped, surrounded by high mountains; Longmen Mountain to the northwest, Daba Mountain to the northeast, Daliang Mountain to the southwest, and Dalou Mountain to the southeast. The interior of the Sichuan Basin can be divided into six first-order structural units (Pan Changlin 2011). These units are: the east Sichuan high and steep fault folding belt, south Sichuan low and steep curved belt, north Sichuan low and gentle belt, central Sichuan gentle belt, southwest Sichuan low and gentle folding belt, and west Sichuan low and steep belt (Fig. 7.1).

The Sichuan Basin is a compound petroliferous basin dominated by Paleozoic marine sediments and Meso-Cenozoic continental sediments. The Sinian System-Lower Paleozoic was a stable platform-type deposit. However, the Caledonian movement at the end of Silurian caused the basin to uplift and suffer denudation, essentially erasing the Devonian system from the basin's records. From the Carboniferous period to the Middle Triassic, a set of platform marine strata were deposited, but the Indosinian movement in the Middle Triassic withdrew all the sea water from

D. Guan et al., *Theory and Simulation Experimental Technology of Oil Generation From Nonmarine Source Rocks in China*, https://doi.org/10.1007/978-981-96-3145-2_7

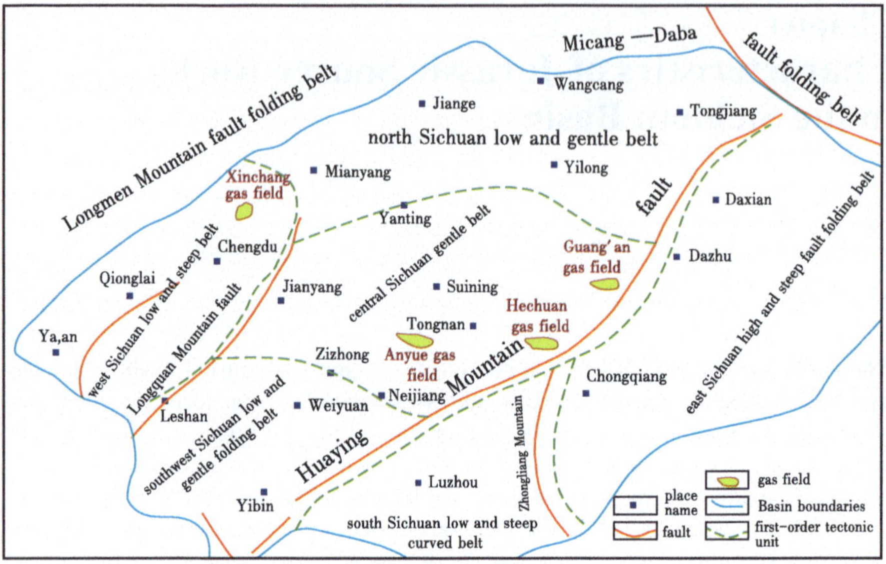

Fig. 7.1 Schematic map of Sichuan Basin structural division (modified from Pan Changlin 2011)

the basin. These platform marine strata contain abundant oil and gas resources, but the high degree of thermal evolution of organic matter has generally transformed the oil in many marine reservoirs into natural gas. Thus, the marine carbonates of the entire Sichuan Basin in the Sinian, Carboniferous, Triassic, and other systems are gas-bearing strata, mainly located in the east Sichuan high and steep fold belt, south Sichuan low and steep curved belt, and southwest Sichuan low and gentle fold belt. In the Late Triassic, the Sichuan Basin entered the sedimentary stage of a continental lacustrine basin and developed a set of continental strata from the Late Triassic to the Quaternary (Fig. 7.2). Notably, the Upper Triassic Xujiahe Formation consists of sandstone and shale interbedded with coal seams, making it the main target for searching continental gas reservoirs in the west Sichuan low and steep belt. On the other hand, the Da'anzhai Member of Lower Jurassic Ziliujing Formation, composed mainly of dark mudstone intercalated with ostracod shell limestone, is the primary target for searching continental oil fields in the central Sichuan gentle belt. The present study analyzes the characteristics of Jurassic source rocks in the Sichuan Basin using data from the Da'anzhai Member, located in Langzhong to the south of the Sichuan Basin.

The Sichuan Basin is a compound petroliferous basin dominated by Paleozoic marine sedimentary and Meso-Cenozoic continental sedimentary rocks. During the Sinian System-Lower Paleozoic era, the basin had stable platform-type deposits. However, the Caledonian movement at the end of Silurian caused uplift and denudation, resulting in the absence of the Devonian system in the basin. From the Carboniferous to Middle Triassic periods, the basin had a deposition of platform marine strata which were later withdrawn due to the Indosinian movement in the Middle Triassic.

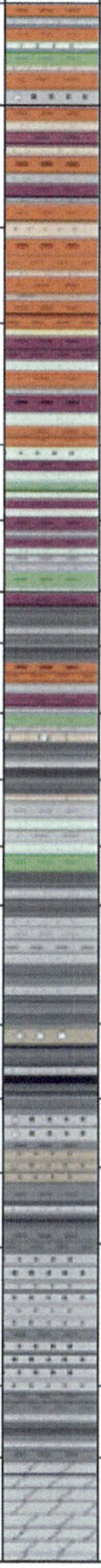

Strata						Lithologic section	Regional thickness/m	Comprehensive lithological description	Tectonic movement	Sedimentary facies
Erathem	System	Series	Formation	Member	Code					
Mesozoic	Cretaceous	Lower			(K1)		0~700	Dark red and gray mudstone interbedded with yellow gray, brown, gray, white sandstone and siltstone with unequal thickness. It is in parallel unconformity with the underlying strata.	Middle Yanshannian	Fluvial-lacustrine facies
	Jurassic	Upper	Penglaizhen Formation		(J3p)		1000~1400	The upper section consists of brown-gray, light gray, and light-yellow-gray siltstone and fine sandstone interbedded with dark brown and brown mudstone of equal thickness. The lower section comprises light gray, light-yellow-gray high-medium sandstone interbedded with brown and dark-brown mudstone of equal thickness. Mudstone in general contains green-gray spot group calcareous masses. They also conformably contact the underlying strata.		Fluvial-lacustrine facies
			Suining formation		(J3sn)		300~350	Purplish-red, dark purplish-red, and brown-purple mudstone and silty mudstone are interbedded with light gray and gray-green fine sandstone of uneven thickness.		Lacustrine facies
		Middle	SUpper Shaximiao Formation		(J2s)		1300~1600	The upper section consists of interbedded brown-gray, light gray, and light yellow-gray siltstone and fine sandstone, with dark brown and brown mudstone of equal thickness. The lower section comprises interbedded light gray, light yellow-gray high-medium sandstone and brown and dark-brown mudstone of equal thickness. Mudstone in general contains green-gray spot group calcareous masses. These sections conformably contact the underlying strata.		Fluvial-lacustrine facies
			Lower Shaximiao Formation		(J3x)		300~350	The strata comprises interbedded purplish-red, dark purplish-red, brown-purple mudstone and silty mudstone with light gray and gray-green fine sandstone of varying thickness.		Fluvial-lacustrine facies
			Qianfoya Formation	3th	(J2q3)		50~100	Brown and gray mudstone are interbedded with silty mudstone, light gray and gray-green lithic feldspar sandstone (referred to as upper Variegated Interval).		
				2th	(J2q2)		100~150	The middle black section consists of interbedded deep gray, black-gray shale, and sandstone with varying thickness		Lacustrine facies
				1th	(J2q1)		20~60	The lower Variegated Interval comprises interbedded brown and gray mudstone with silty mudstone, as well as light gray and gray-green lithic feldspar sandstone.	Early Yanshannian	Fluvial-lacustrine facies
		Lower	Ziliujing Formation	4th	(J1z1)		70~100	Brown ash, gray mesoclastic limestone, marl interbedded with black shale, mudstone and greenish-gray calcareous siltstone		Lacustrine facies
				3th	(J1z3)		60~150	Gray, dark gray mudstone, silty mudstone interbedded with greenish-gray siltstone, argillaceous siltstone		
				2th	(J1x2)		30~60	Gray black, dark gray shale, mudstone, silty mudstone, mixed gray siltstone		
				1th	(J1s1)		200~300	Interbedded with varying thickness, the black and gray-black shales are mixed with gray and black-gray silty mudstone, as well as light-gray fine-grained quartz sandstone and dark-gray argillaceous siltstone. At the bottom, the variegated sandstone and glutenite are interbedded with uneven dark-gray and gray shale. These strata exhibit parallel unconformity with the underlying rock layers.	Late Indosinian	
	Triassic	Upper	Xujiahe Formation	5th	(J3x5)		60~100	The strata consists of interbedded gray-black shale and silty shale with gray siltstone and siltstone of varying thickness. Additionally, there is a mixed black coal seam (coal section).		Lacustrine facies and delta facies
				3th	(J3x4)		50~100	The main composition is gray-pebbled sandstone and gray-calcareous medium-fine sandstone with gray-black mudstone in the middle.		
				3th	(J3x3)		60~120	Interbedded with varying thickness, the strata comprises gray-black mudstone and silty mudstone mixed with gray siltstone and fine sandstone. Additionally, there is a mixed black coal seam (coal section).	Middle Indosinian	
				2th	(J3x2)		150~300	The upper section comprises gray, light gray, medium, and fine-grained lithic quartz sandstone, gray fine-grained lithic sandstone, and gray calcareous sandstone with gray-black shale. In the middle section, gray-black shales and silty shales are interbedded with gray fine-grained lithic sandstone and siltstone of relatively uniform thickness. The lower section consists of gray fine-grained lithic sandstone and calcareous sandstone with gray-black shale.		
				1th	(J3x1)		30~100	The strata is composed of gray-black shale interbedded with dark gray siltstone, fine sandstone, and other layers that have relatively uniform thickness.		
		Middle	Leikoupo Formation	4th	(J2l4)		80~150	Dark gray, gray micrite dolomite	Early Indosinian	Evaporative platform

Fig. 7.2 Typical Mesozoic member and Jurassic stratigraphic characteristics in the Sichuan Basin (after Shuai 2014)

The organic matter within these marine reservoirs has undergone a high degree of thermal evolution leading to the cracking of oil into natural gas. This transformed the marine carbonates of the whole Sichuan Basin from the Sinian to other systems into gas-bearing strata. These natural gas resources are mainly found in east Sichuan's high and steep fold belt, south Sichuan's low and steep curved belt, and southwest Sichuan's low and gentle fold belt.

During the Late Triassic period, the Sichuan Basin entered a sedimentary stage of a continental lacustrine basin and developed a set of continental strata from the Late Triassic to Quaternary (Fig. 7.2). Of these strata, the Upper Triassic Xujiahe Formation is a sandstone-shale interbedded with coal seams and is the primary area of interest for searching continental gas accumulations in the west Sichuan low and steep belt. Additionally, the Da'anzhai Member of Lower Jurassic Ziliujing Formation is mainly composed of dark mudstone intercalated with ostracod shell limestone and is the main target for searching continental oil accumulations in the central Sichuan gentle belt. Analyzing data from the Da'anzhai Member, a study conducted in Langzhong to the south of Sichuan Basin analyzed the characteristics of Jurassic source rocks in the basin.

7.1 Evolution Characteristics of Continental Sedimentary Stage in the Sichuan Basin

According to the data of Exploration and Development Research Institute of Southwest Oil and Gas Company, Sinopec, the stratigraphic division of Jurassic in Sichuan Basin is shown in Table 7.1.

The Jurassic system is well-preserved overall, and its various series, formations, and members can be described as follows:

Lower Series: The Ziliujing formation is divided into four members from bottom to top: Zhenzhuchong Member, Dongyuemiao Member, Ma'anshan Member, and Da'anzhai Member. The Zhenzhuchong Member consists of variegated sandstone and mudstone interbeddings. The sediments in the Dongyuemiao Member are primarily dark mudstone, shale, and Ostracod shell limestone. The Ma'anshan Member is dominated by interbedded variegated and red sandstone and mudstone. As the lake water deepens toward the basin boundary to the north, the sandstone and mudstone gradually become predominantly gray and gray-black. The Da'anzhai Member is the primary source rock of the Jurassic in the Sichuan Basin.

Middle Series: From bottom to top, the strata consist of the Qianfoya Formation, lower Shaximiao Formation, and upper Shaximiao Formation. The Qianfoya Formation is a set of meandering fluvial delta deposits primarily composed of variegated shale, mudstone, and sandstone. The lower Shaximiao Formation is composed of purple mudstone and gray-green sandstone interbeddings of varying thickness. The upper Shaximiao Formation consists of light-gray sandstone and brown mudstone interbeddings.

Table 7.1 Stratigraphic division of Jurassic in Sichuan Basin

Cretaceous		Jianmenguan formation/Cangxi formation		
Jurassic	Upper	Penglaizhen formation	莲花口组	
		Suining formation		
	Middle	Upper Shaximiao formation		
		Lower Shaximiao formation		
		Qianfoya formation	Xintiangou formation	Lianggaoshan formation
	Lower	Ziliujing formation	Baitianba formation	Da'anzhai member
				Ma'anshan member
				Dongyuemiao member
				Zhenzhuchong member
Upper triassic		Xujiahe formation		
		Xiaotangzi formation		
		Kuahongdong formation		

Upper Series: From bottom to top, the strata consist of the Suining Formation and Penglaizhen Formation. The Suining Formation is dominated by purple mudstone and shale mixed with sandstone. The Penglaizhen Formation is mainly composed of sandstones and has undergone denudation except for its central part in western Sichuan.

Cretaceous and Cenozoic sediments in the Sichuan Basin are primarily located in the northwestern region, as the eastern part of Sichuan has been experiencing continuous uplift while other areas only have scattered sediments.

The evolution of the continental sedimentary stage in the Sichuan Basin can be divided into three developmental processes.

7.1.1 Continuous Subsidence Development Stage—Energy Accumulation Process

During this developmental stage, from the Late Triassic sedimentary period of the Xujiahe Formation to the end of the Jurassic, a set of inland fluvial-delta facies and lacustrine facies sand-mud (shale) rock interbeddings were primarily developed. The Upper Triassic Xujiahe Formation, which is intertwined with black shale and coal seam, is the main gas source rock in western Sichuan. The Lower Jurassic Ziliujing Formation, especially the mudstone (shale) mixed with Ostracod shell limestone of Da'anzhai Member, is the source rock of the central Sichuan area and the primary source rocks of continental oil fields in central Sichuan. The thickness of the strata during this settling period ranges from 3780 to 5330 m, with the Upper Triassic

Xujiahe Formation being 350 to 720 m thick, and the Jurassic being 3430 to 4610 m thick. By the end of the Jurassic, the gas source rocks of the Upper Triassic Xujiahe Formation had entered the gas generation stage, while the main source rocks of the Da'anzhai Member of the Lower Jurassic Ziliujing Formation had entered the oil generation stage.

7.1.2 Uplift to Denudation Development Stage—Energy Release Process

At the end of the Jurassic period, the entire Sichuan Basin underwent uplift and denudation. The Penglaizhen Formation located to the north, east, and south of the basin was less well-preserved and only relatively intact in the central region of western Sichuan. During this stage, the basin underwent denudation for over 10 million years. As the basin experienced unloading and decompression for an extended period, various energies accumulated in the source rocks and reservoirs within the basin were fully released. The source rocks had entered the massive oil generation stage followed by extensive primary and secondary migration. Therefore, this developmental phase is the primary period for the formation of oil and gas accumulations in the basin.

7.1.3 Overall Atrophy Stage—Energy Adjustment Process

This developmental period spans from the Cretaceous to Quaternary era and marks the total shrinkage and extinction of the Sichuan Basin. The primary characteristic of this stage is the basin's frequent uplift and subsidence, multiple deposition and denudation events, all aimed at achieving material balance with surrounding geological formations and balancing the basin's physical and chemical fields. Consequently, this developmental stage represents the final completion and finalization of all oil accumulations in the basin.

7.2 Characteristics of the Jurassic Main Source Rocks in the Sichuan Basin

The Ziliujing Formation of the Lower Jurassic in the Sichuan Basin is the primary layer of interest for discovering continental oil and gas reserves. Within this formation, both the Zhenzhuchong and Ma'anshan Members are unsuitable for hydrocarbon generation, while the kerogen of the Dongyuemiao Member belongs to type III and serves as the main source of gas. Only the Da'anzhai Member possesses significant oil generation capacity and is thus the main source rock for oil reserves in the Sichuan Basin. This study focuses on analyzing the characteristics of the Jurassic main source rocks in the Sichuan Basin, taking the Da'anzhai Member at Langzhong-Southern Sichuan as an example. By examining the source rock properties of this formation, we gain a deeper understanding of the characteristics and potential of oil reserves in the region.

7.2.1 Sedimentary Environment and Biogeographic Zoning

The publication "Paleoenvironment and Oil and Gas," which is the second volume of Jurassic in Northern China and was released by the Petroleum Industry Press in 2003, outlines the biogeographic zones of the early, middle, and late Jurassic periods in northern China. The authors constructed these zones based on paleontological fossils and sedimentary characteristics observed in Jurassic sedimentary basins throughout northern China, in addition to reference to biofacies models. The Jurassic sedimentary basin in China is segregated into three biogeographical zones, namely the Circumtethys, Coastal Pacific, and Ussuri Bay biogeographical zones, running in an east-to-west direction (Figs. 7.3, 7.4 and 7.5). The Sichuan Basin is classified within the Circumtethyan biogeographic zone, which suggests that the Sichuan Basin and several other Jurassic basins located in the northwest region of China, such as Xinjiang, Qinghai, Gansu, western Inner Mongolia, Ordos, and western Henan, share the same tectonic settings of sedimentary environments and warm-humid climates. These factors allowed for a diverse array of organisms, notably abundant aquatic animals and plants like bivalves, Phyllostachys, Charophyta in the brackish- saline water environment, to flourish in the early Jurassic period of the Sichuan Basin. This flourishing ecosystem provided the high-quality material basis for source rocks found in the Da'anzhai Member.

7.2.2 Stratigraphic and Sedimentary Facies Characteristics of Da'anzhai Member

During the deposition of the Da'anzhai Member, the inland lacustrine basin of Sichuan was in a stable stage of structural development without any noticeable external provenance system. The main sedimentary types were shallow and semi-deep lacustrine facies. The thickness of the Da'anzhai Member is approximately 80–90 m and is divided into three sub-members. The first sub-member consists.

mainly of gray-brown micrite ostracod shell limestone, clastic ostracod shell limestone with a thin layer of gray-black shale. The ostracod shells are mostly bivalves, accounting for up to 85–95% of the content, with fragments of different sizes arranged in parallel and a thickness of about 38 m. The second sub-member is primarily

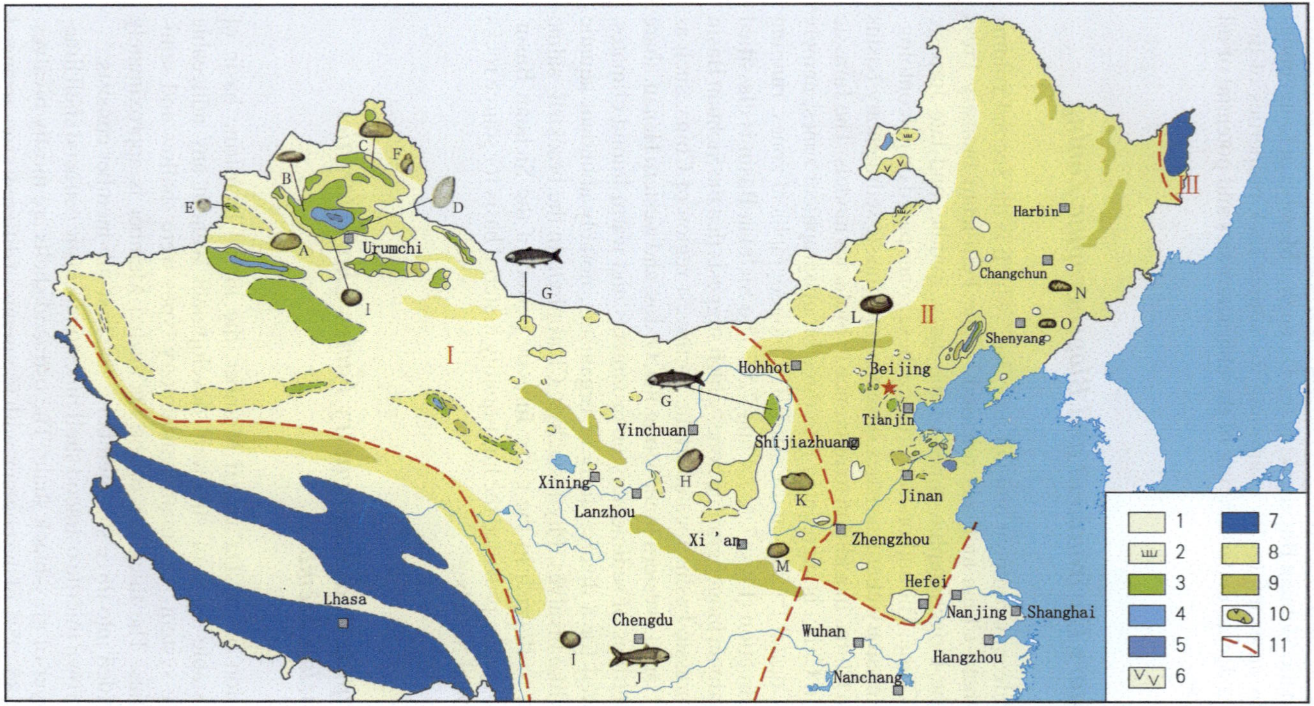

Fig. 7.3 Early Jurassic biogeographic zones in Northern China (after Xiaochun et al. 2003) I—Circumtethys biogeographical zone; II—Coastal Pacific biogeographical zone; III—Ussuri Bay biogeographical zone; 1—Fluvial facies; 2—Swamp facies; 3—Coastal and shallow lacustrine facies; 4—Shallow lacustrine facies; 5—Semi-deep lacustrine facies; 6—Volcanic rock distribution area; 7—Marine facies; 8—paleo-highlands; 9—paleo-mountain ranges; 10—paleo-uplift; 11—Biogeographic zoning lines A—*Unio manasensis*; B—*Unio zhungarica*; C—*Margaritifera delunshanensis*; D—*Waagenoperna mytiloides*; E—*Illiestheria nikaensis*; F—*viviarus shanbeiensis*; G—*Yuchoulepis gansuensis*; H—*Eosolimnadiopsis shanbeiensis*; I—*Palaeolimnadia sichuanensis*; J—*Yuchoulepis szechuanensis*; K—*Unio* cf. *lucaogouensis*; L—*Yananoconcha bella*

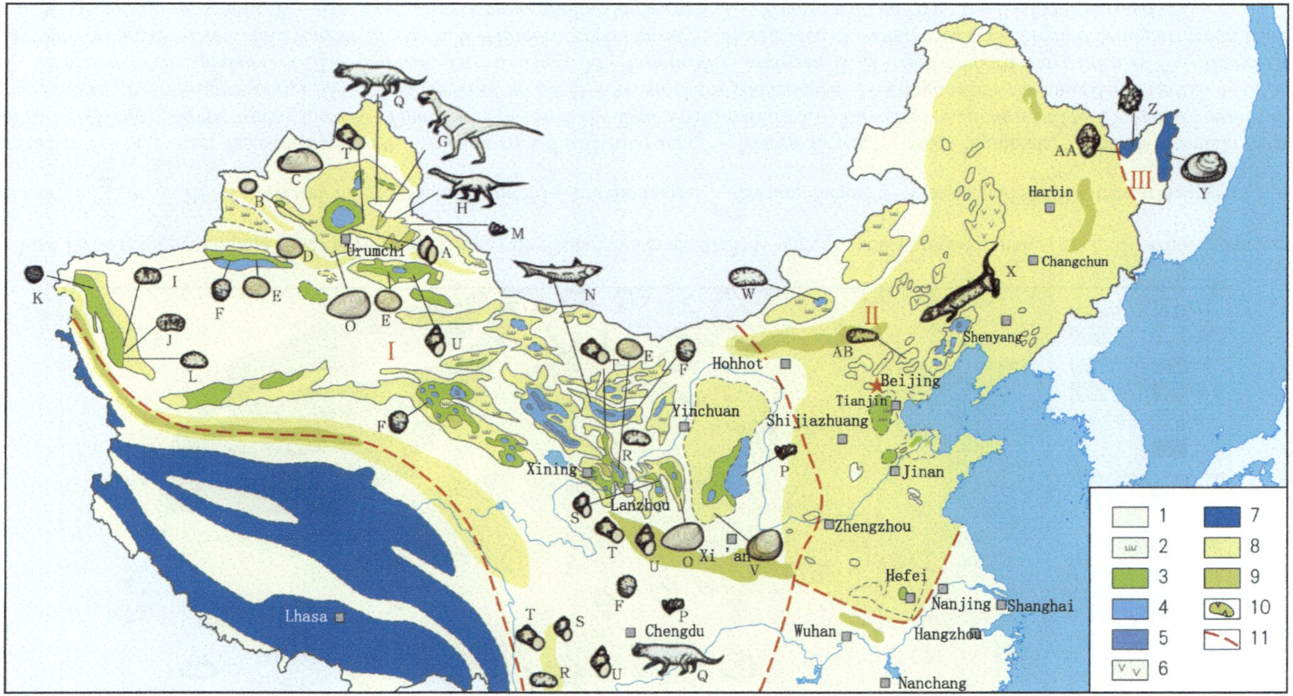

Fig. 7.4 Middle Jurassic biogeographic zones in Northern China (after Xiaochun et al. 2003) I—Circumtethys biogeographical zone; II—Coastal Pacific biogeographical zone; III—Ussuri Bay biogeographical zone; 1—Fluvial facies; 2—Swamp facies; 3—Coastal and shallow lacustrine facies; 4—Shallow lacustrine facies; 5—Semi-deep lacustrine facies; 6—Volcanic rock distribution area; 7—Marine facies; 8—paleo-highlands; 9—paleo-mountain ranges; 10— paleo-uplift; 11—Biogeographic zoning lines A—*Lioplacodes orientalis*; B—*Triglvpta manasensis*; C—*Unio karamaica*; D—*Psilunio manasensis*; E—*Sinokontikia lianmuqinensis*; F—*Aclistochara jiangyouensis*; G—*Monolophosaurus jiangi*; H—*Bellusaurus sui*; I—*Bisulcocypris nodifera*; J—*Mandelstamia* sp; K—*Xinjiangichara wuqiaensis*; L—*Timiriasevia incongnata*; M—*Ceratodus*; N—*Hybodus*; O—*Psilunio jinyuanensis*; P—*Ceratodus szechuanensis*; Q—*Bienotheroides zigonensis*; R—*Darwinula yunlongensis*; S—*Viviparus jingguensis*; T—*Amplovalvata jingguessis*; U—*Lioplacodes yunnanensis*; V—*Psilunio trigonus*; W—*Darwinula erenhotensis*; X—*Yabeinosaurus tenuis*; Y—*Palaeonucula makitoensis*; Z—*Pareodinia ceratophora*

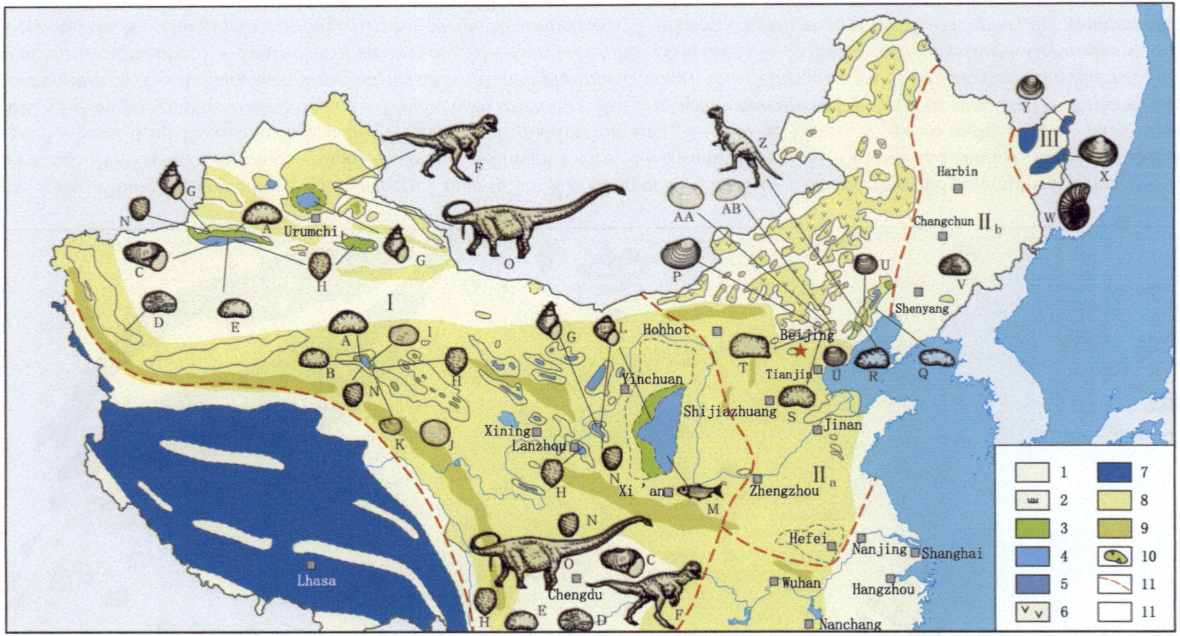

Fig. 7.5 Late Jurassic biogeographic zones in Northern China (after Xiaochun et al. 2003) I—Circumtethys biogeographical zone; II—Coastal Pacific biogeographical zone; a. 内陆小区; b. 近海小区; III—Ussuri Bay biogeographical zone; 1—Fluvial facies; 2—Swamp facies; 3—Coastal and shallow lacustrine facies; 4—Shallow lacustrine facies; 5—Semi-deep lacustrine facies; 6—Volcanic rock distribution area; 7—Marine facies; 8—paleo-highlands; 9—paleo-mountain ranges; 10—paleo-uplift; 11—Biogeographic zoning lines A—*Djungarica procera*; B—*Djungarica tumida*; C—*Valvata zhongjiangenis*; D—*Cetacella inermis*; E—*Djungarica yunnanensis*; F—*Sinraptor dongi*; G—*Psendamnicola acuta*; H—*Aclistochara yunnanensis*; I—*Qinghaiestheria hungshuigouensis*; J—*Mangyalimnadia*; K—*Danlengiconcha qaidamensis*; L—*Amnicola kushixiaensis*; M—*Baleiichthys antingensis*; N—*Aclistochara jiangyouensis*; O—*Mamenchisaurus*; P-*Pseudograpta orbita*; Q—*Darwinula vangshulingensis*; R—*Eoparacypris obesa*; S—*Djungarica* cf. *yunnanensis*; T—*Luanpingella postacuta*; U—*Nestoria reticularia*; V—*Cetacella substriata*; W—*Arctocephalites* (*Canocephalites*)*hulinensis*; X—*Mesosaoella morrisi*; Y—*Buchia fisheriana*

interbedded with gray-black shale and gray clastic shell limestone of varying thicknesses. The thickness of a single-layer shale ranges from 2.5–8.5 m, and the cumulative thickness of shale accounts for approximately 90% of the submember's thickness, making it the primary source rock of the Da'anzhai Member. The submember has a thickness of about 35 m. The third sub-member mainly comprises gray micritic ostracod shell limestone and gray-black shale. The ostracod shells are mostly bivalves, with a content of 60–80%. Similar to the first sub-member, most of the ostracod shells are fragments arranged in parallel, but the thickness is only about 12 m. During the sedimentary period of the Da'anzhai Member, the sedimentary environment consisted mostly of clast beach of shallow lake and semi-deep lake facies and mudstone of shallow lake and semi-deep lake facies. Vertically, the sedimentary facies showed characteristics of lake transgression-maximum lake transgression-lake regression, with a change in sedimentary facies from shallow lake to semi-deep lake to shallow lake from the third sub-member to the first sub-member. The division and characteristics of sedimentary facies can be found in Tables 7.2 and 7.3.

Figure 7.6 illustrates the paleogeographic division of sedimentary facies, while Figs. 7.7, 7.8, and 7.9 show the distribution characteristics of sedimentary facies in each sub-member of the Da'anzhai Member. The classification of sedimentary facies is primarily based on core observation data, well logging facies and sedimentary facies analysis, seismic facies analysis, and sedimentary facies analysis, ultimately determining the sedimentary facies types in each sub-member (Figs. 7.10, 7.11 and 7.12).

The Da'anzhai Member's source rock is relatively uncomplicated, consisting primarily of ostracod shell limestone and black shale. The sedimentary environment is stable, comprising shallow lacustrine facies and semi-deep lacustrine facies, and is largely influenced by factors such as variations in lake depth and the elevation of lake level. These characteristics are evident in the fifth-order sequence characteristics of each sub-member of the Da'anzhai Member:

During the sedimentary period of the third sub-member, there was a lacustrine transgression, with alternating ostracod shell beach and lacustrine mudstone sequences. The sedimentary cycle was from ostracod shell limestone to black shale, followed by argillaceous ostracod shell limestone to black shale. The logging characteristics show a transition from high to low natural gamma changes and low to high resistivity changes.

Table 7.2 Brief table of stratigraphic facies division of Da'anzhai Member in Langzhong-Southern Sichuan area

Sedimentary facies	Subfacies	Microfacies	Developed strata
Lake	Shallow lake	Beach nucleus	1st to 3rd sub-member
		Beach margin	1st to 3rd sub-member
		Shallow lake mud	1st to 3rd sub-member
	Semi-deep lake	Beach front lake slope	
		Semi-deep lake mud	2nd sub-member

Table 7.3 Sedimentary facies characteristics of Da'anzhai Member in Langzhong-southern Sichuan area

Sedimentary facies	Lake				
Subfacies	Shallow lake			Semi-deep lake	
Microfacies	Beach nucleus	Beach margin	Shallow lake mud	Beach front lake slope	Semi-deep lake mud
Depth of water	0–10	0–10	5–10	10–20	10–20
Lithologic type	Mesoclastic limestone	Argillaceous mesoclasti c limestone	Shale, mudstone	Argillaceous mesoclastic limestone, mesoclastic marl	Shale, mudstone
Color	Gray—Dark gray	Gray—Dark gray	Dark gray—gray black	Gray-black—black	Gray-black—black
Bedding structure	Massive bedding	Wavy bedding	Horizontal bedding, drainage structure, coiling bedding	Banded bedding	Horizontal bedding, graded bedding, biodisturbance structure
Log feature	20API<GR<55API 45 μs/m<AC <65 μs/m	55API<GR<75API 65 μs/m<AC <85 μs/m	75API<GR<105API 85 μs/m<AC <110 μs/m	55API<GR<105API 65 μs/m<AV <110 μs/m	85API<GR<140API 90 μs/m<AV <105 μs/m
Depositional environment	Highest energy	Medium energy	Weak reduction—reduction, low energy	Reduction—Strong reduction, minimum energy	

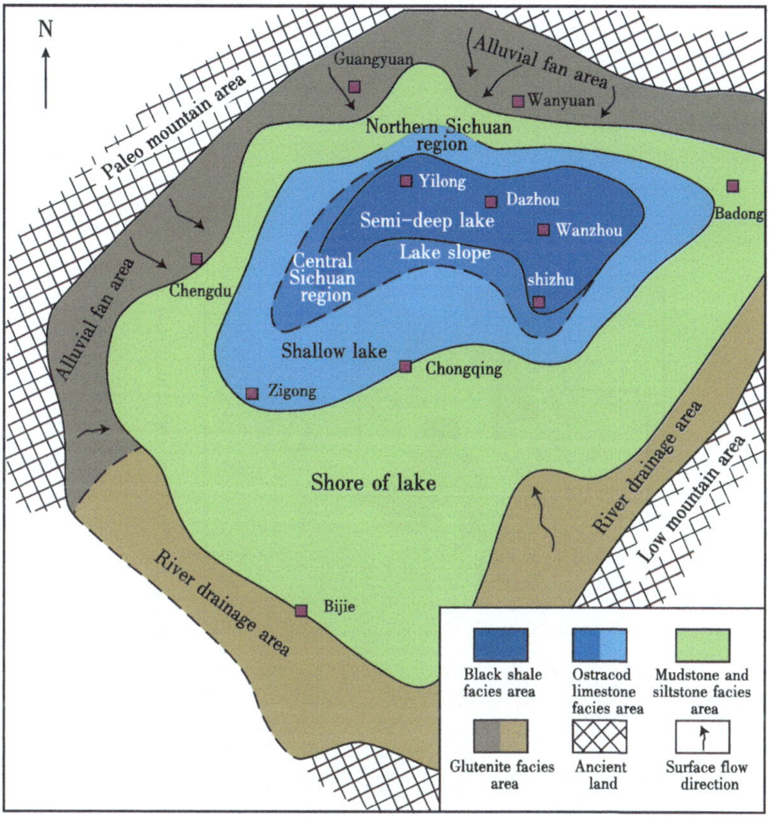

Fig. 7.6 Paleogeography map of the Da'anzhai member in the Sichuan Basin during deposition period (after Bingxiong et al. 2014)

The sedimentary period of the second sub-member was the maximum lake transgression period, characterized by black shale and argillaceous ostracod shell limestone. The sequence in this period was an argillaceous ostracod shell limestone-black shale cycle. The natural gamma value of the logging curve transitions from high to low, while the resistivity value changes from low to high area (after Xiaoming et al. 2014)

The deposition of the first sub-member has entered the lacustrine retreat phase, with lithology and sequence consisting of argillaceous mesoclastic limestone to black shale and argillaceous mesoclastic limestone to argillaceous ostracod shell cycles. The logging curve's gamma ray exhibits a shift from high to low, while the resistivity changes from low to high (Fig. 7.13).

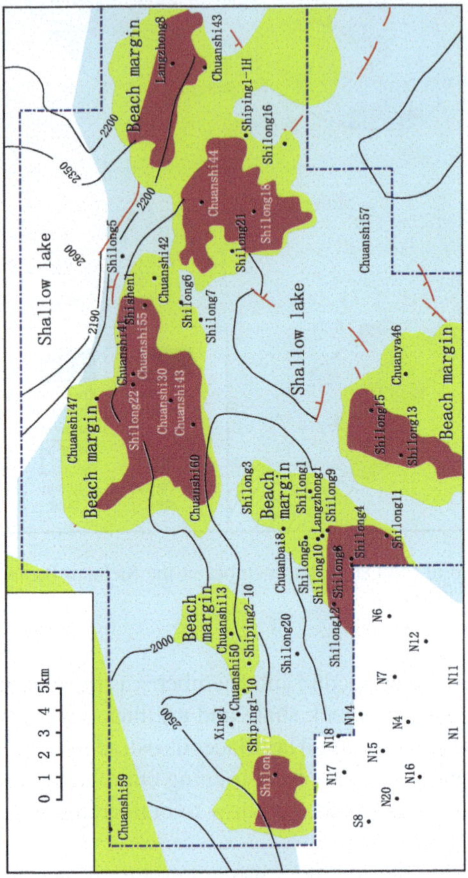

Fig. 7.7 Sedimentary facies map of the third sub-member in Langzhong-Southern Sichuan area (after Xiaoming et al. 2014)

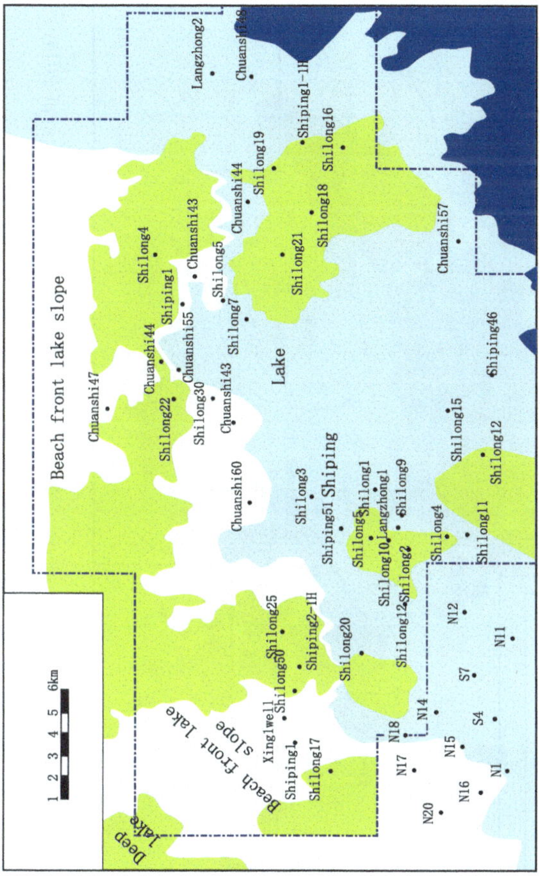

Fig. 7.8 Sedimentary facies map of the second sub-member in Langzhong-Southern Sichuan area (after Xiaoming et al. 2014)

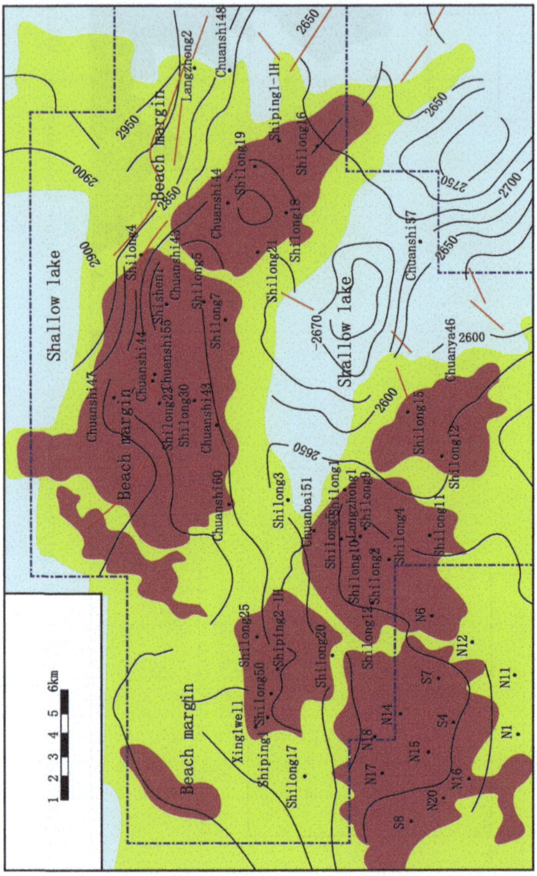

Fig. 7.9 Sedimentary facies map of the first sub-member in Langzhong-Southern Sichuan area (after Xiaoming et al. 2014)

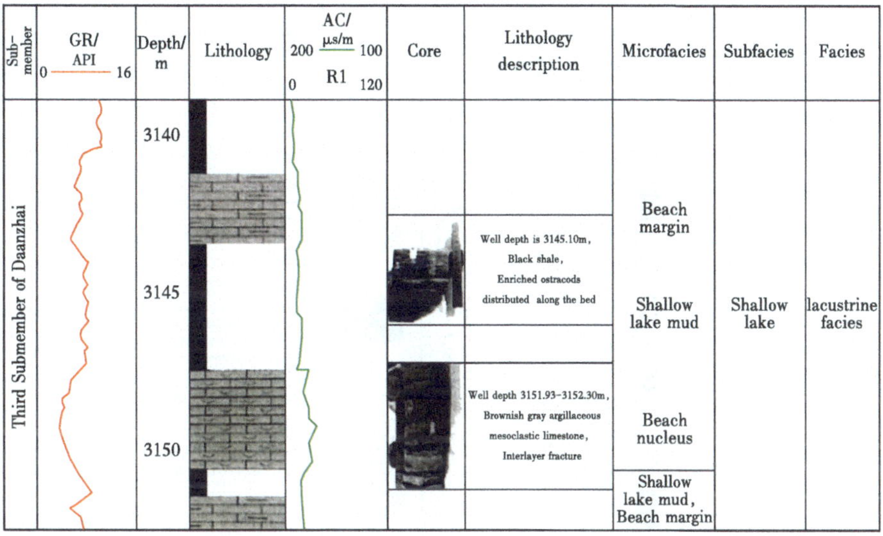

Fig. 7.10 Lithology-well log-facies model of shallow lacustrine carbonate in the third sub-member (well Chuanshi 57) (after Xiaoming et al. 2014)

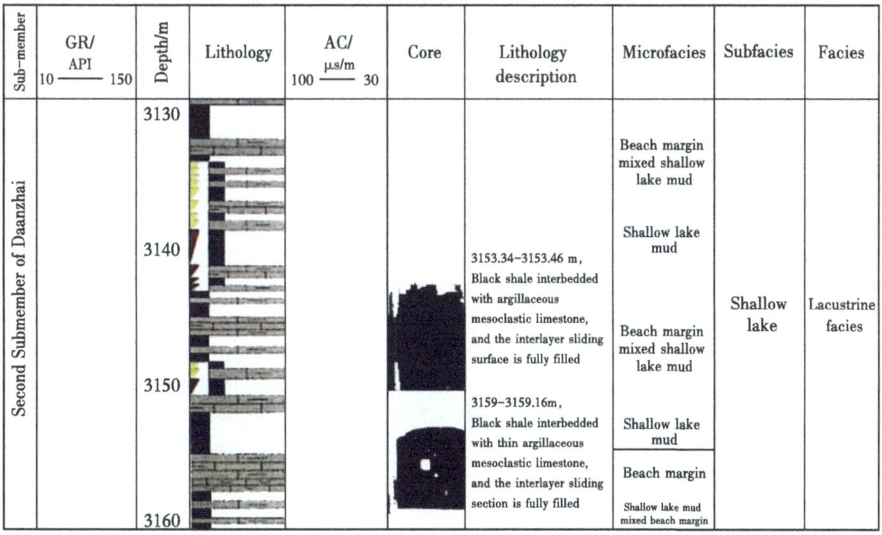

Fig. 7.11 Lithology-well log-facies model of shallow lacustrine carbonate in the second sub-member (well Shilong 8) (after Xiaoming et al. 2014)

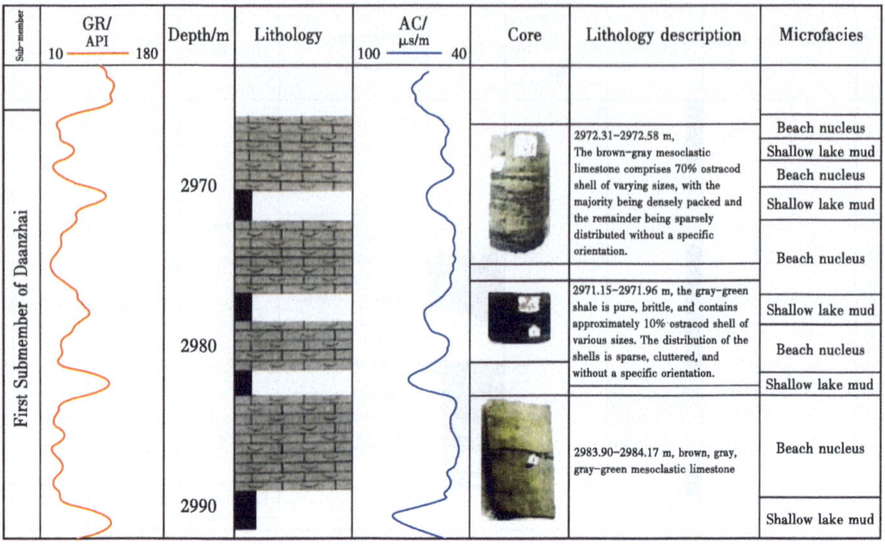

Fig. 7.12 Lithology-well log-facies model of shallow lacustrine carbonate in the first sub-Member (well Shilong 17) (after Xiaoming et al. 2014)

7.2.3 Main Characteristics of Source Rocks in the Da'anzhai Member

The Da'anzhai Member's source rock has an approximate thickness of 86 m, with the black shale making up about 30 m and the shell limestone about 56 m. Typically, only black shale serves as the source rock, while ostracod shell limestone functions as the reservoir rock. However, according to authors'new line of thought, the Da'anzhai Member source rock should comprise both black shale and ostracod shell limestone. Black shale serves as the parent rock for petroleum generation, and its organic content and kerogen type directly impact its ability to generate petroleum. The combination characteristics of ostracod limestone and black shale, along with the physical conditions of ostracod limestone and the size of the pore pressure difference between them, directly control black shale's oil generation. Therefore, merely evaluating the black shale's organic matter content and kerogen type isn't sufficient to assess the oil generation potential of the Da'anzhai Member's source rocks. Other factors such as porosity and permeability of ostracod shell limestone and other physical conditions are equally important. The higher the pore pressure difference between black shales and ostracod limestone, the more oil may migrate from black shales to ostracod shell limestone.

Based on the biogeographic subdivision of the Early Jurassic period in northern China, the Sichuan Basin's sedimentary environment during that period was comparable to those of other basins such as Junggar, Turha, and Qaidam. The Sichuan Basin's environment featured brackish-saline shallow to semi-deep lacustrine facies,

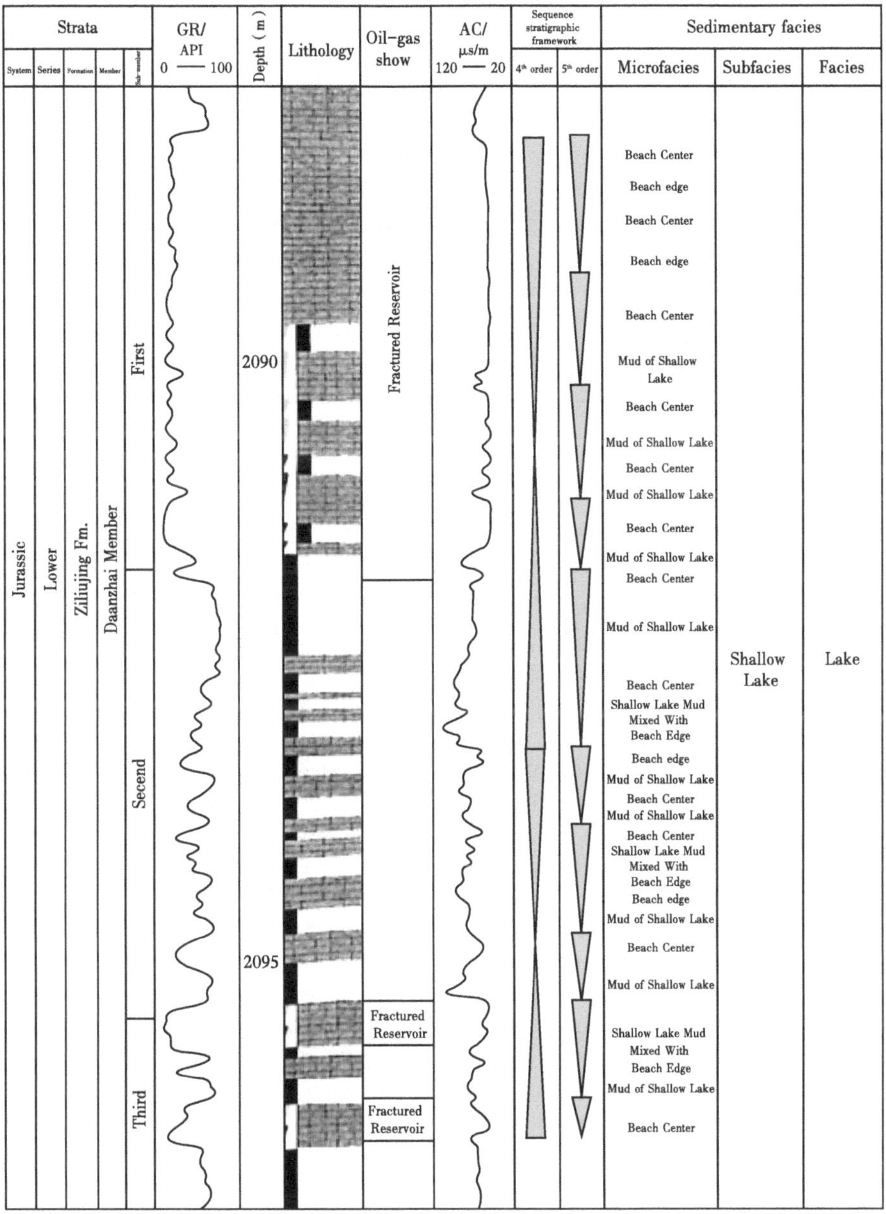

Fig. 7.13 Comprehensive column of sedimentary facies and sequence stratigraphy of Da'anzhai member of the Lower Jurassic Ziliujing Formation, Langzhong (well Shilong 12) (after Xiaoming et al. 2014)

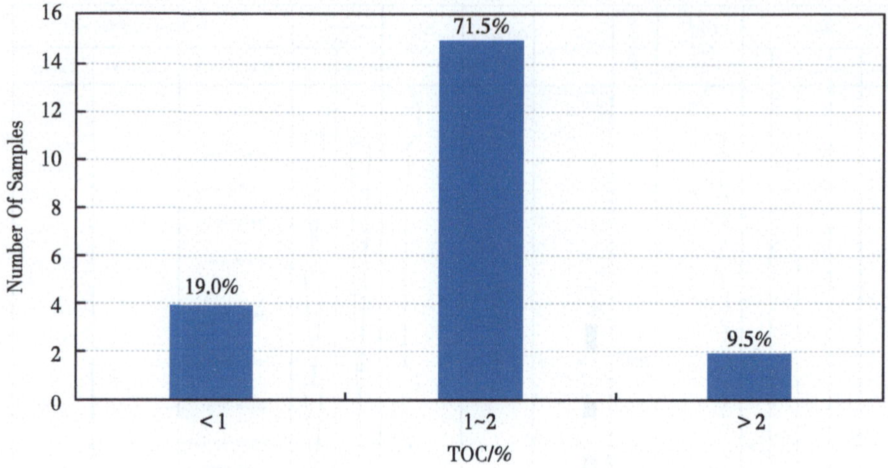

Fig. 7.14 Frequency distribution histogram of organic matter abundance in shale core of the Da'anzhai Member (after Dehua et al. 2020)

and was home to Granodiscus, an algae similar to Botryococcus, that is abundant in organic matter and oil generating capacity.

Da'anzhai Member's black shale has an organic matter content of >1.0%, accounting for over 80% (as illustrated in Fig. 7.14), with an average content of 1.33% (Dehua et al. 2020). The dominant kerogen type was II, with an average sapropelic substance content of 30.28% and an average liptinite content of 45.28%. The black shale's primary mineral composition comprises quartz and clay minerals, with an average content of 44.6% and 41.3% respectively. This mineral composition favors petroleum formation and primary migration. Moreover, the black shale has a high rare earth element content, which plays an essential catalytic role in the organic matter catagenesis process during oil generation.

The Da'anzhai Member source rocks possess promising oil generation potential based on an analysis of their organic geochemical conditions. However, their evaluation of pore pressure difference conditions regarding primary oil migration indicates significant insufficiencies. This is primarily due to the compact nature of both black shale and ostracod shell limestone, resulting in low porosity and permeability, with minimal differences in pore pressure between them. Oil droplets generated through thermal cracking of dispersed organic matter in black shale are unable to accumulate easily in the pores or fractures of ostracod shell limestone. Concentration of the dispersed oil droplets from black shale into flow towards adjacent ostracod shell limestone can only occur when the pore space and fracture aperture in the ostracod shell limestone near the contact plane between the two rocks are sufficiently extensive. Only under these conditions, driven by the pore pressure difference, the primary migration of oil from the black shale to the adjacent ostracod shell limestone can be realized. The accumulation of oil is particularly favorable when the material composition of black shale contains a higher content of quartz particles.

7.2.4 Oil Accumulation Conditions from the Lower Jurassic Da'anzhai Member Source Rocks in the Sichuan Basin

Two types of oil accumulation exist in continental lacustrine source rocks in China. The first is comprised of shale and conventional reservoirs formed by lacustrine and fluvial delta sedimentary environments. If oil droplets generated from the source rock achieve primary migration in the lacustrine sedimentary environment, then the source rock itself functions as a shale reservoir. The black mudstone (shale) in the source kitchen serves as both the source rock and cap rock, while the sandstone (or reef limestone) in contact with the black mudstone (shale) facies in the source kitchen serves as the reservoir. Therefore, once source rocks in the continental lacustrine basin reach primary migration, shale reservoir formation begins. If the sandstone in the shale system connects with the sandstone body of the fluvial delta sedimentary facies, then the oil generated in the source rock undergoes secondary migration, forming a conventional reservoir in the suitable trap located within the fluvial delta sandstone body. This is the most common type of oil accumulation in the continental lacustrine sedimentary environment in China.

The second type of oil accumulation is pure lacustrine sedimentary environments without a foreign provenance system, meaning no fluvial-delta sedimentary facies are present. Instead, they are composed simply of black shale or black mudstone mixed with biological limestone or reef limestone. The Lower Jurassic Da'anzhai Member of the Sichuan Basin belongs to this type of source rock-oil accumulation assemblage, which is also the only type of source rock and shale reservoir united system in continental lacustrine basins in China. However, the formation of this type of oil accumulation must adhere to certain petroleum geological conditions.

(1) For oil to be expelled to ostracod shell limestone, the organic matter content of black shale (or mudstone) must exceed 1%, with dispersed organic matter being relatively concentrated near the contact surface of black shale and ostracod shell limestone. Da'anzhai Member source rocks have an average organic matter content of 1.33%. Thin interbedded layers of black shale and ostracod shell limestone promote both oil generation and expulsion.
(2) The matrix porosity of ostracod shell limestone is similar to that of black shale. To achieve primary migration of oil from black shale (mudstone), physical conditions such as the development of secondary pores, high density of karst caverns, good connectivity of pore size, and high density of fractures must exist in ostracod shell limestone, creating a large pore pressure difference between the two.
(3) For oil accumulation to take place, the cumulative thickness of the ostracod shell limestone must exceed 10 m, and the monolayer must be well-connected. This can be observed through isopach maps of ostracod shell limestone from the Da'anzhai Member (Figs. 7.15, 7.16, 7.17, 7.18, 7.19 and 7.20). By comprehensively evaluating black shales and ostracod shell limestone in each sub-member of the Da'anzhai Member, the reserve potential of source rocks and shale reservoirs can be predicted (Figs. 7.21,7.22, 7.23 and 7.24).

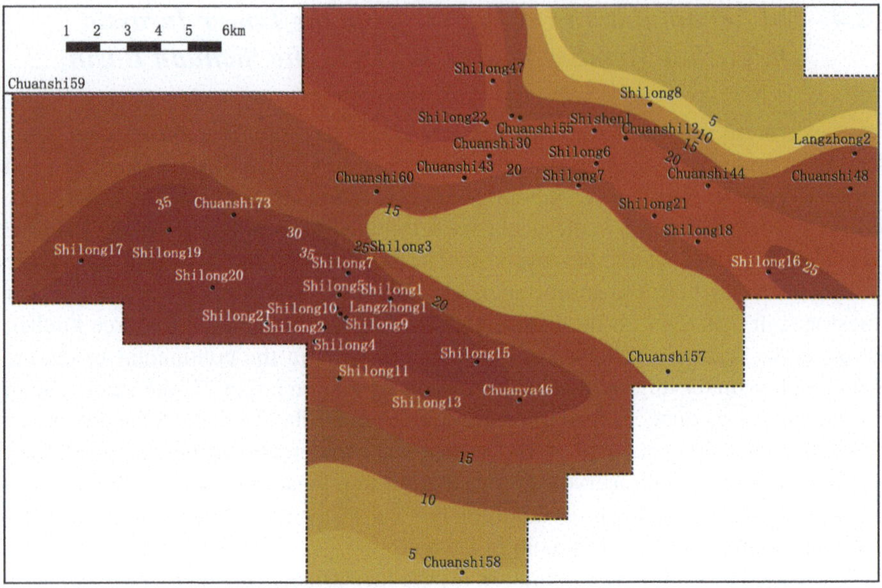

Fig. 7.15 Mesoclastic limestone thickness isopach (m) map of the first sub-member of the Da'anzhai Member, Langzhong-Southern Sichuan area (after Xiaoming et al. 2014)

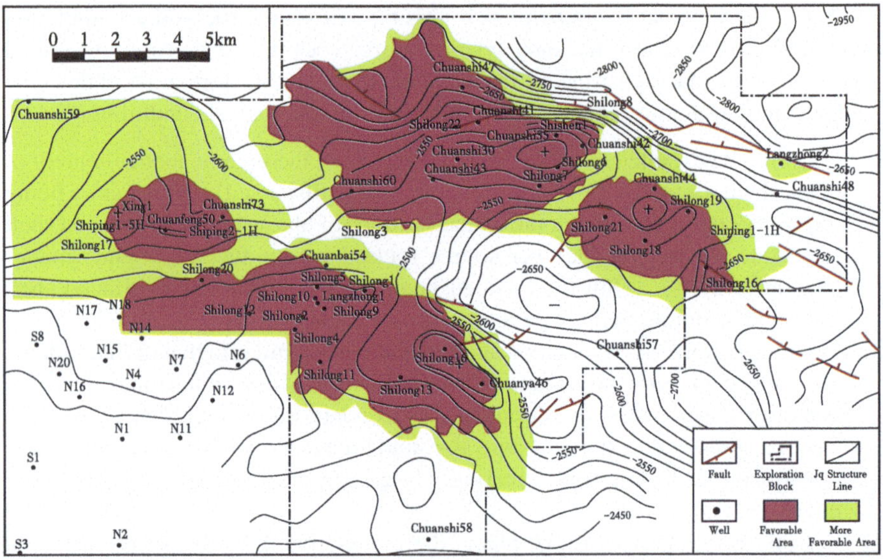

Fig. 7.16 Comprehensive reservoir evaluation map of the first sub-member of Da'anzhai member in Langzhong-Southern Sichuan area (after Xiaoming et al. 2014)

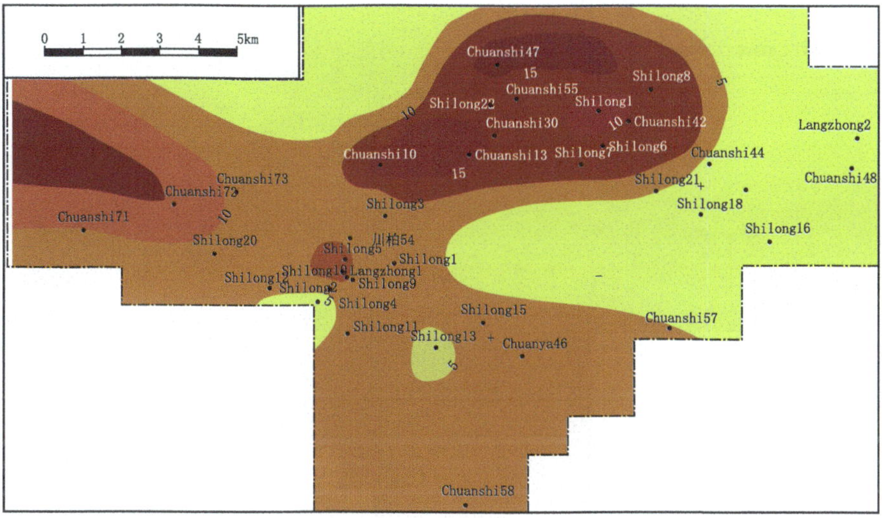

Fig. 7.17 Mesoclastic limestone thickness isopach (m) map of the second sub-member of the Da'anzhai member, Langzhong-Southern Sichuan area (after Xiaoming et al. 2014)

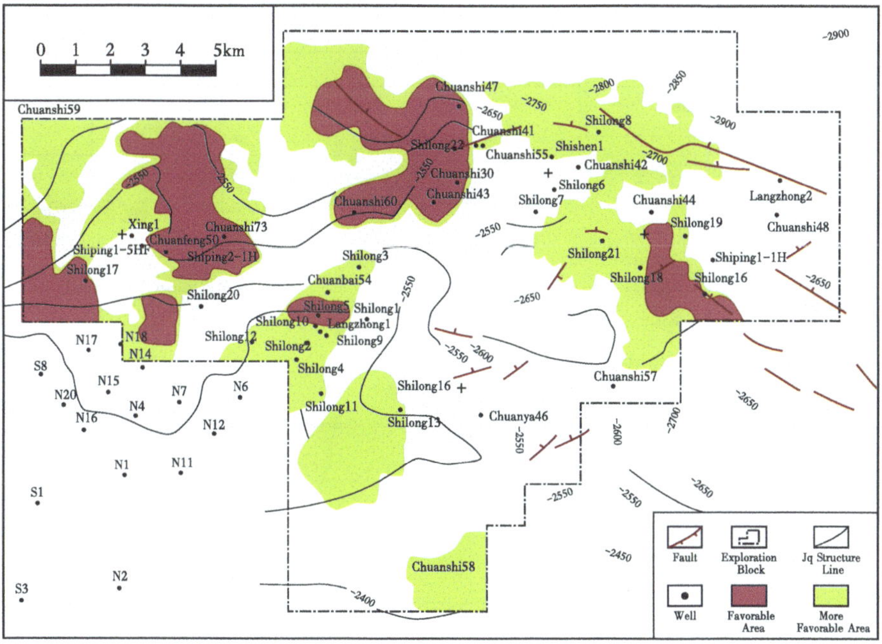

Fig. 7.18 Comprehensive reservoir evaluation map of the second sub-member of the Da'anzhai member in Langzhong-southern area (after Xiaoming et al. 2014)

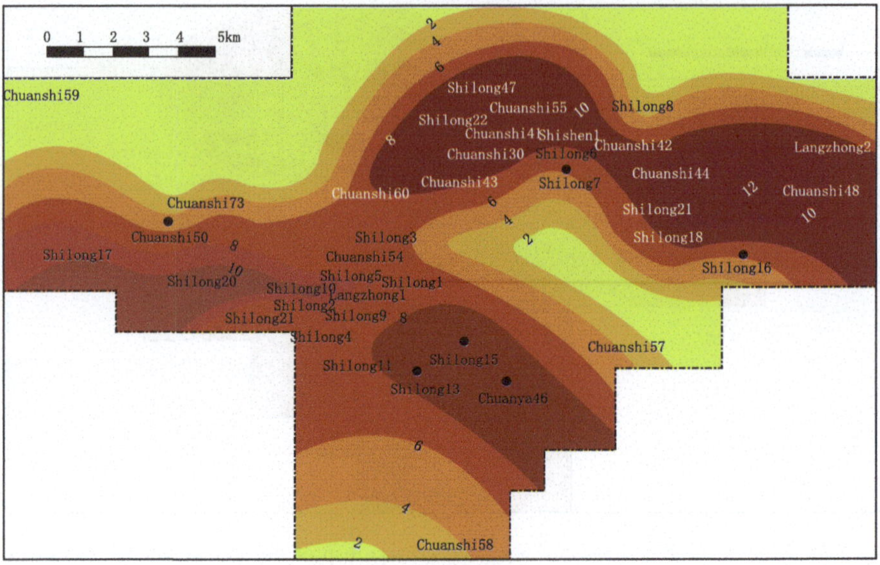

Fig. 7.19 Mesoclastic limestone thickness isopach (m) map of the third sub-member of Da'anzhai member, Langzhong-Southern Sichuan area (after Xiaoming et al. 2014)

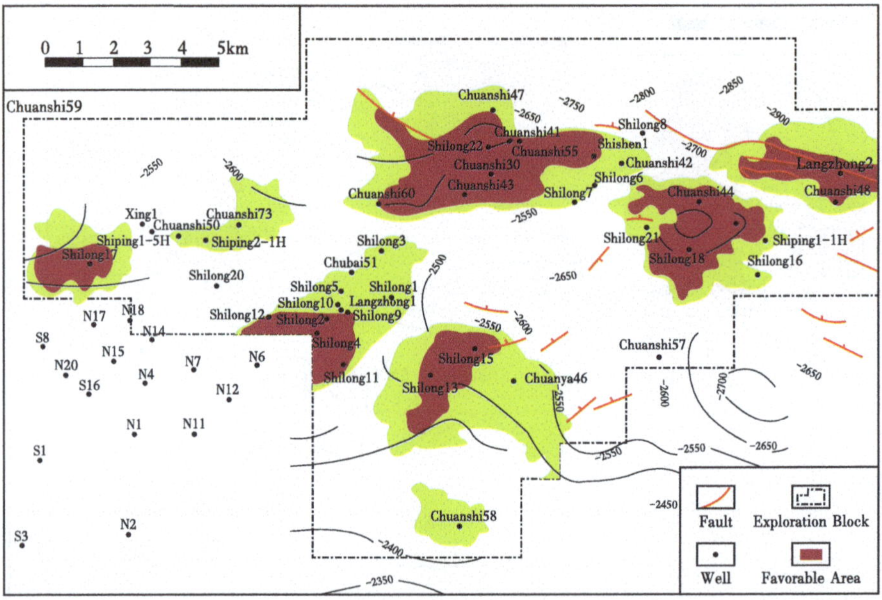

Fig. 7.20 Comprehensive reservoir evaluation map of the third sub-member of the Da'anzhai member in Langzhong-Southern Sichuan area (after Xiaoming et al. 2014)

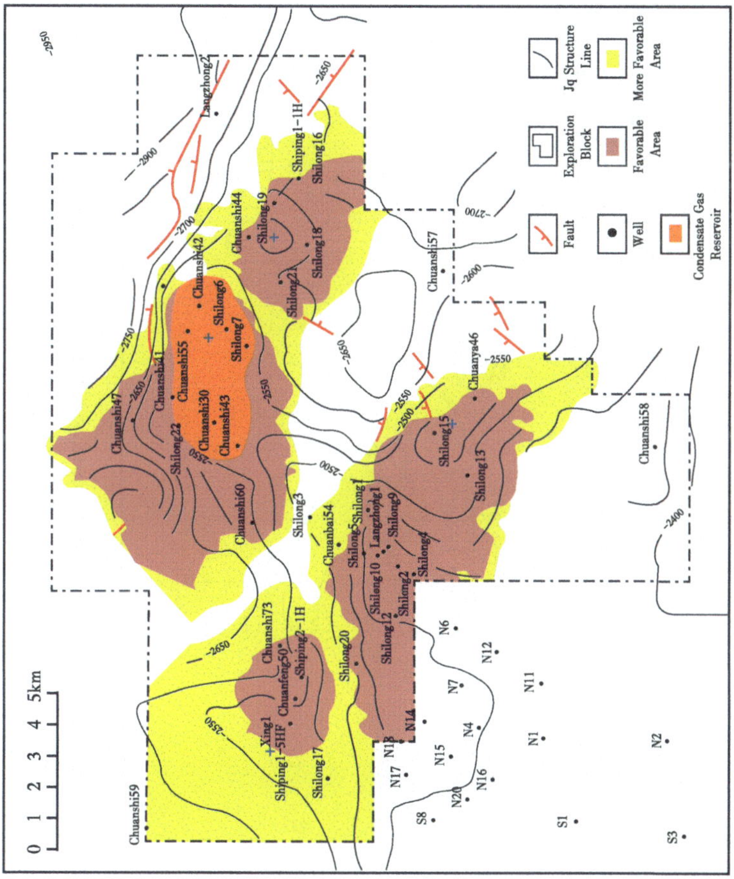

Fig. 7.21 Predicted reserve distribution of the potential area within the first sub-member of the Da'anzhai member in Langzhong-Southern Area (after Xiaoming et al. 2014)

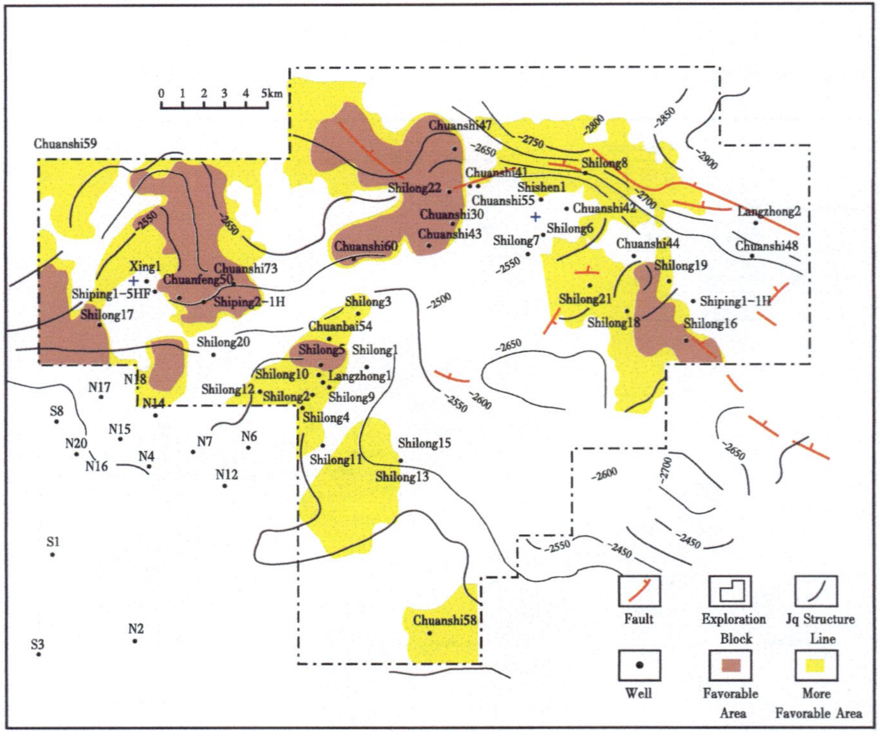

Fig. 7.22 Predicted reserve distribution of the potential area within the second sub-member of the Da'anzhai member in Langzhong-Southern Sichuan Area (after Xiaoming et al. 2014)

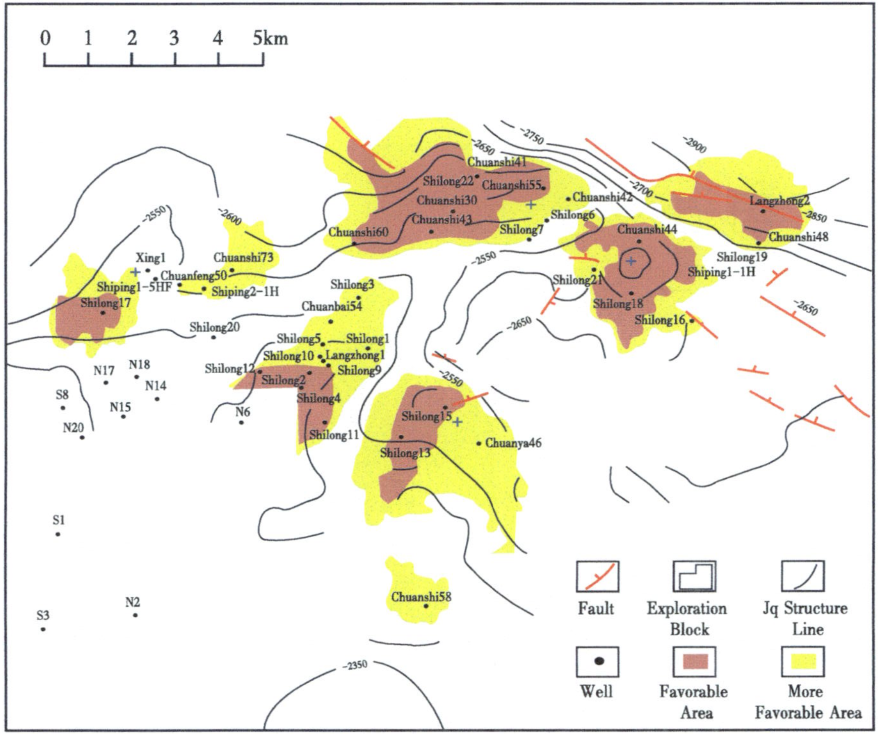

Fig. 7.23 Predicted reserve distribution of the potential area within the third sub-member of the Da'anzhai member in Langzhong-Southern Sichuan area (after Xiaoming et al. 2014)

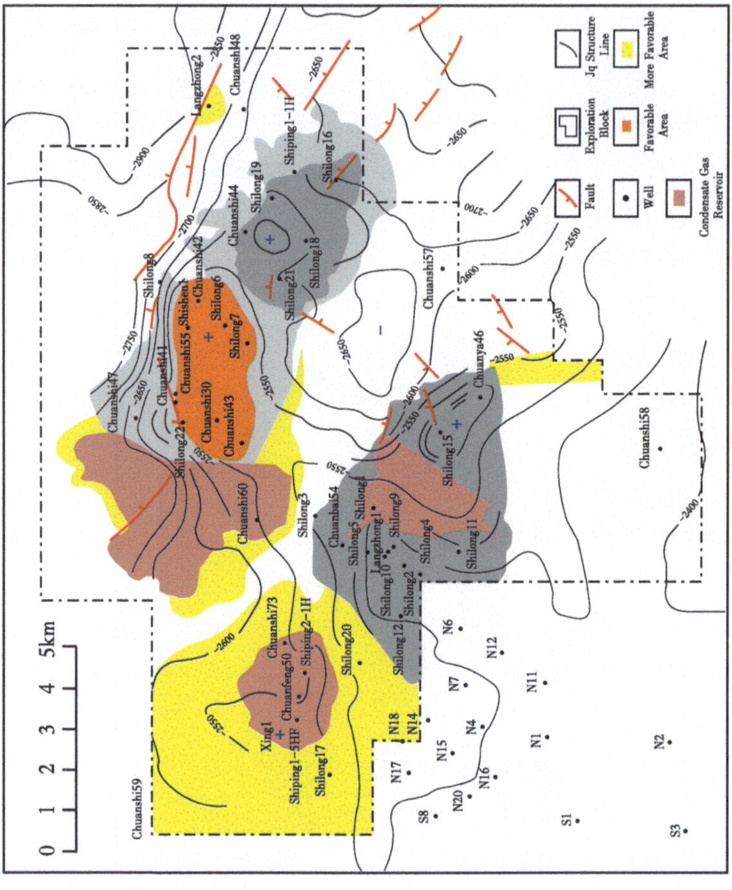

Fig. 7.24 Superposition map of predicted reserves in the potential area of the Da'anzhai member in Langzhong-Southern Sichuan area (after Xiaoming et al. 2014)

References

Dehua Z, Chuanxiang S, Zhongbao L, Nie H (2020) Geological characteristics of continental shale gas reservoir in Daanzhai section of Northeast Sichuan. Pet Explor China 25(5):32–42

Feng X et al (2014) Optimization of rolling production target of tight reservoirs in Langzhong-Southern Sichuan Basin. Chengdu: Exploration and Development Research Institute of Southwest Oil and Gas Company of Sinopec

Lu B, Rongcai Z, Liang X (2014) Shale gas (oil) reservoir characteristics in Daanzhai section of eastern Sichuan Basin. Geol China 41(4):1387–1398

Pan C, Liu S, Yongsheng M et al (2011) Reservoir characteristics and controlling factors of Xujiahe formation in Northeastern Sichuan. J Southwest Pet Univ: Nat Sci Ed 33(3):27–34

Xiaochun Z, Chuanben Z, Shizhong Y et al (2003) Jurassic in Northern China (II) Paleoenvironment and Oil and Gas. Petroleum Industry Press, Beijing

Yang S (2014) Jurassic sedimentary evolution and facies-controlled reservoir prediction in Sichuan Basin. Doctoral dissertation, Chengdu University of Technology

Chapter 8
Characteristics of Cretaceous Source Rocks in the Songliao Basin

Abstract The Songliao Basin, located in northeastern China, features a basement composed of Paleozoic and Precambrian metamorphic rock series, overlain by sedimentary strata from the Jurassic, Cretaceous, Paleogene, Neogene, and Quaternary periods. Among these, the Cretaceous system is the most significant, serving as the primary oil-bearing sequence. It is characterized by extensive fine-grained clastic deposits from fluvial-lacustrine environments formed during the basin's overall subsidence phase. This chapter provides a comprehensive analysis of the Lower Cretaceous Qingshankou Formation source rocks. It examines their depositional environments, paleontological features, stratigraphy, sedimentary facies, primary source rock properties, and petroleum generation and accumulation conditions. These findings are vital for guiding precise exploration and efficient development of petroleum resources in the Songliao Basin.

The Songliao Basin, spanning an area of approximately 26,000 km^2, is situated in northeastern China and covers central Heilongjiang Province, western Jilin Province, northern Liaoning Province, and southeastern Inner Mongolia Autonomous Region. The basin stretches about 700 km north to south and 370 km east to west (refer to Fig. 8.1). It comprises seven primary structural units: the western slope belt, north dip out belt, central depression belt, northeast uplift belt, southeast uplift belt, southwest uplift belt, and Kailu Depression. The basin's basement is composed of Paleozoic and Pre-Paleozoic metamorphic rock series, while its sedimentary cover primarily consists of layers from the Jurassic, Cretaceous, Paleogene, Neogene, and Quaternary periods. Notably, the Cretaceous period serves as the basin's most crucial sedimentary and oil-bearing layer.

Formed during the late Jurassic period, the Songliao Basin is a continental sedimentary basin of Meso-Cenozoic age. The basin's most prominent developmental period occurred during the Cretaceous period, which resulted in the deposition of thick, widely distributed, and well-developed sediments. As such, the Songliao Basin is considered the location of the most complete development of continental sediments from the Cretaceous period in China. Within the Lower Cretaceous, several formations can be distinguished, namely the Huoshiling Formation, Shahezi

D. Guan et al., *Theory and Simulation Experimental Technology of Oil Generation From Nonmarine Source Rocks in China*, https://doi.org/10.1007/978-981-96-3145-2_8

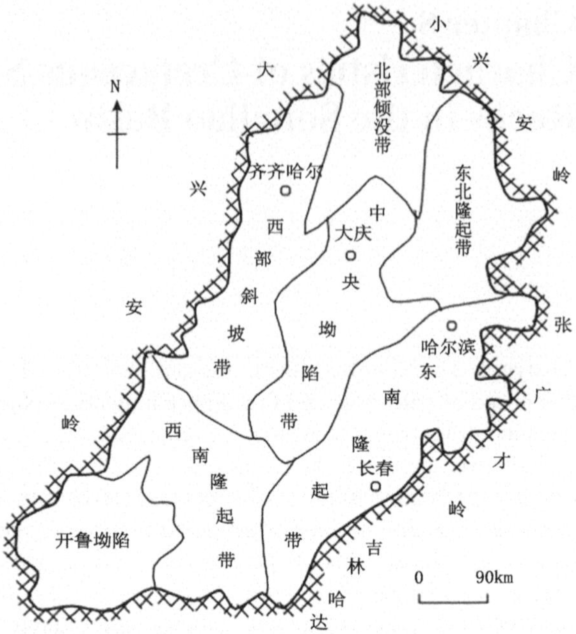

Fig. 8.1 Structural zoning map of Songliao Basin (after Zaiyi and Qingchun 1996)

Formation, Yingcheng Formation, Denglouku Formation, and Quantou Formation. The Huoshiling Formation, Shahezi Formation, and Yingcheng Formation consist primarily of pyroclastic rocks, volcanic rocks, and underwater deposits from terrigenous sources in faulted lacustrine basins. In contrast, the Denglouku Formation is composed predominantly of coarse-grained clastic rocks, marking a transition stage within the rift-depression process, with a thickness of up to 1700 m and an unconformable contact with the underneath Yingcheng Formation. The Quantou Formation is dominated by fluvial-lacustrine facies mixed with fine-grained clastic rocks at the beginning of the transition stage from rift to overall depression, with a thickness ranging from 700 to 1200 m and a maximum thickness exceeding 1900 m. The Quantou Formation contacts the underlying Denglouku Formation in pseudo-conformity and overlies the old strata of various ages. Notably, the third and fourth members of the Quantou Formation correspond to the Yangdachengzi and Fuyu Oil Layers, respectively, within the Songliao Basin.

The Upper Cretaceous in the Songliao Basin is comprised of several formations, including the Qingshankou Formation, Yaojia Formation, Nenjiang Formation, Sifangtai Formation, and Mingshui Formation. The basin was in the whole depression stage and was characterized by fine-grained clastic deposits with fluvial-lacustrine facies. The Qingshankou Formation consists of black, gray-green mudstone and sandstone, and its first member contains black and gray-black mudstone and shale with oil shale, measuring between 60 and 164 m thick and serving as the main source rock in the basin. The second and third members are gray-green and gray-black mudstone with calcareous siltstone, with a thickness of 200–400 m and a maximum thickness exceeding 500 m, and are the main source rock and reservoir rock (Gaotaizi Oil Layer) in the basin, conformably contacting the underlying Quantou Formation.

The Nenjiang Formation is composed of deep and shallow lacustrine facies fine-grained clastic rocks and conforms to the underlying Yaojia Formation. This formation can be divided into five members from the bottom up, with the first member containing a large gray-black mudstone mixed with siltstone, measuring between 60 and 200 m thick, serving as a minor source rock and important reservoir rock in the basin (along with the second and third members of Yaojia Formation, forming the Sartu Oil Layer). The second member is the main regional caprock in the basin, measuring between 150 and 250 m thick. The third member consists of light-colored silty mudstone and siltstone interbedded, measuring between 60 and 130 m thick, while the fourth and fifth members are gray-green sand-mudstone interbeds, measuring between 0 and 650 m thick. The third member of Nenjiang Formation, together with the fourth and fifth members of Nenjiang Formation, constitute the Hedimiao Oil Layer in the basin. At the end of the Nenjiang Formation deposition, the Songliao Basin rose and experienced denudation, indicating a change from a whole depression to a whole rise stage.

During the shrinkage period of the basin, the Sifangtai Formation and Mingshui Formation consisted of coarse-grained clastic rocks. Mingshui Formation was limited to the Qijia-Gulong sag and Sanzhao sag in the central basin. The thickness generally ranges around 300 m, with the thickest being about 600 m. At the end of the Mingshui Formation deposition, the Songliao Basin underwent another rise and denudation process, and the Mingshui Formation was unconformable with the overlying Paleogene. The Paleogene Yi'an Formation only developed in the northern part of the basin. At the end of the Paleogene, the Songliao Basin experienced another upward denudation process, leading to further shrinkage. From the late Paleogene to Quarternary, the Songliao Basin exhibited surface loose deposition and plain swamp landforms dominated by fluvial facies, entering an atrophy and extinction stage (Fig. 8.2).

8.1 Petroleum Geological Evolution at Different Development Stages in the Songliao Basin

The geological evolution of petroleum in the Songliao Basin can be divided into three main stages: continuous subsidence, overall rise, and overall shrinkage.

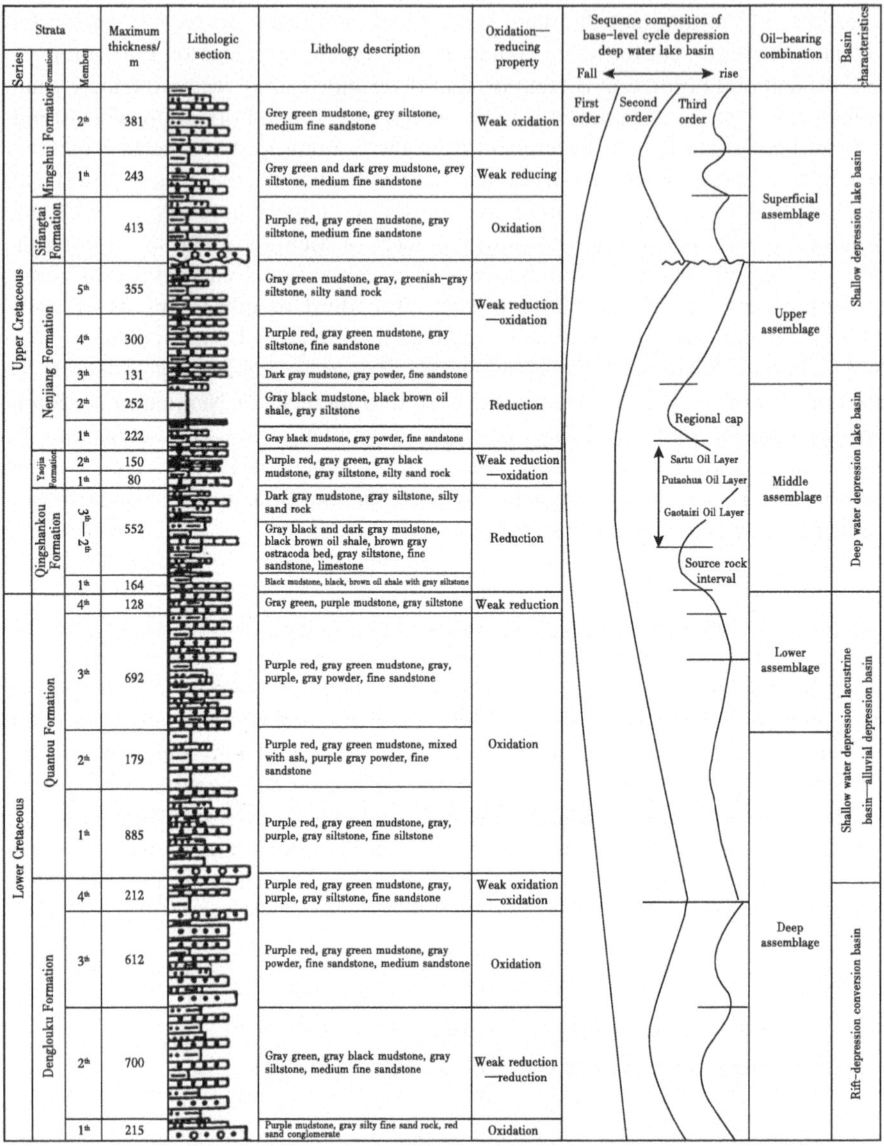

Fig. 8.2 Histogram of Cretaceous oil assemblage in Songliao Basin (according to Defan 2004)

8.1.1 *Basin Continuous Subsidence Stage—Energy Accumulation Process*

During the continuous subsidence stage, which occurred from the late Jurassic to the Nenjiang Formation deposition in the Cretaceous period, the basin experienced rapid subsidence and deposition. It was characterized by a fault-subsidence packing during the early period and continuous depression subsidence during the later period. Despite the different subsidence characteristics of the rift and depression, this stage was crucial for the accumulation of materials with petroleum geological significance, including organic-rich argillaceous sediments that formed the main source rocks. The dark mudstones of the Qingshankou Formation in particular, were the primary source rock in the basin, having accumulated a significant amount of organic matter. With the process of compaction and thermal evolution, the dark mudstone of the Qingshankou Formation matured into source rock, leading to significant oil generation. Along with the accumulation of organic and biological energy, it transformed into chemical and hydrocarbon energy. The primary pores in mudstones were filled with oil due to the large amount of oil generation, which not only restricted further compaction but also maintained the mudstone porosity at 10–15%. Additionally, major reservoir sandstones such as Sartu, Putaohua, and Gaotaizi, had increasing internal density due to the compaction of overlying strata, with rigidity gradually emerging. The sandstone showed a gradual increase in elastic energy when the pressure of overlying strata did not reach the fracture pressure of sandstones.

Thus, the continuous subsidence stage of the Songliao Basin was crucial for completing the accumulation of primary materials such as source rock, reservoir, caprock, and overburden rock. During this stage, the Qingshankou Formation's source rock underwent a series of physical and chemical processes, leading to the oil generation process and the accumulation of oil under abnormally high pressure. Subsequently, major oil layer sandstones like Sartu, Putaohua, and Gaotaizi accumulated significant elastic energy.

8.1.2 Whole Basin Uplift to Denudation Stage—Energy Release Process

This stage occurred from the end of Nenjiang Formation deposition to the pre-deposition of Sifangtai Formation, marking the transition of Songliao Basin from continuous subsidence to overall uplift and denudation. This period lasted for about 8–10 million years. During this time, the 5th and even the 4th member of Nenjiang Formation experienced significant denudation, with the erosion strata thickness in Daqing placanticline estimated to be around 200–300 m.

The overall uplift and denudation process at the end of Nenjiang Formation deposition signaled a change in the physical field properties of the basin; from loading and pressurization during continuous subsidence to unloading and decompression during overall uplift and denudation. Consequently, the development areas of primary materials such as source rock and reservoir rock underwent significant changes under the conditions of unloading and decompression physical fields.

8.1.2.1 Major Source Rock Development Area

The Qingshankou Formation is primarily found in the Qijia-Gulong and Sanzhao depressions of the Songliao Basin. By the end of Nenjiang Formation deposition, the source rocks within the Qingshankou Formation were matured, with a significant amount of oil generated and stored within mudstone pores under abnormally high pressure. The unloading and decompression process, along with the pore pressure difference, facilitated the expulsion of oil into nearby sand bodies, thereby enabling primary migration of the oil.

8.1.2.2 Major Reservoir Rock Development Area

The primary reservoir rocks in the Daqing Placanticline area of the Songliao Basin comprise fluvial delta sandy rocks, ranging from the Qingshankou Formation to the Nenjiang Formation. As the process of unloading and decompression takes place, a significant amount of elastic energy accumulates within the sandstones and must be released. This leads to the generation of a strong suction force, known as the "sandstone rebound" force, directed towards the surface. Due to this phenomenon, low pressure, and even negative pressure, areas develop within the sandstones across a large area. Therefore, under the overall upward unloading and decompression conditions of the Songliao Basin, the oil generated in the Qingshankou Formation within the Qijia-Gulong Sag and Sanzhao sag must undergo secondary migration to low-pressure reservoirs situated between the two depressions after experiencing primary migration. Once the migrating oil encounters a suitable trap, such as one with differential pressure or potential energy between the source rock and the reservoir rock, accumulation of oil will occur, forming a pool. The Lamadian, Sartu, and Xingshugang structures in the northern part of the Daqing placanticline have the most extensive thickness and distribution area of sandstones, resulting in the greatest suction force during unloading and decompression. Hence, these three structures exhibit the most abundant oil accumulation.

It is important to note that both primary and secondary migration of petroleum, as well as the process of oil and gas accumulation, occur concurrently with the unloading and decompression process during basin uplift. These phenomena are a result of energy release triggered by the alteration of physical field properties. Therefore, a comprehensive understanding of the mechanism underlying oil and gas accumulation within the Songliao Basin requires a careful examination of the change characteristics of the physical field induced by the overall upward unloading and decompression process in this developmental stage.

8.1.3 Basin Atrophy Stage—Energy Adjustment Process

The stage of development, spanning from the depositional period of the Sifangtai Formation to the Quaternary period, signifies the atrophy adjustment stage of the Songliao Basin following the late depositional period of the Nenjiang Formation. During this stage, the main characteristic of the basin is frequent energy adjustment. There were two short periods of uplift and denudation in the Songliao Basin, which occurred towards the end of the Mingshui Formation deposition and during the Early Paleogene period. Although the magnitude of these uplift and denudation events was small, they hold great significance to the formation of reservoirs within the Songliao Basin, and are demonstrated in the following three aspects.

(1) The Qingshankou Formation source rocks in the Qijia-Gulong and Sanzhao sags was matured towards the end of the Nenjiang Formation deposition. As the basin began to break and undergo unloading and decompression, oil expulsion occurred, and the source rock underwent further compaction and discharge due to overlying strata. With the basin entering a depression stage, source rocks reaching maturity thresholds could efficiently expel oil due to overlying strata compaction.
(2) During the later sedimentary stage of the Nenjiang Formation, oil reservoirs developed within the Daqing placanticline and other areas along with the overall basin elevation. These oil reservoirs were further enhanced by the continuous tectonic amplitude and area expansion under the suction force produced by two "sandstone rebounds" during the basin's uplift and decline into its shrinking stage.
(3) Within the Songliao Basin's shrinking stage, certain new oil reservoirs like the Longhubao oilfield developed due to the suction force induced by the "sandstone rebound" during two instances of uplift and denudation.

In summary, the petroleum geology development stage of the Songliao Basin is closely linked to the hydrocarbon accumulation process. The deposition of muddy and sandy sediments from the Quantou Formation to the Nenjiang Formation during the basin's continuous subsidence stage established the material foundation for the primary oil reservoirs in the Songliao Basin. Due to overlying strata compaction and thermal evolution, this development stage facilitated both the organic matter's oil generation process and provided energy for the primary and secondary migration and accumulation of petroleum. The uplift and denudation of the whole basin towards the end of the Nenjiang Formation deposition not only supplied the necessary physical field conditions for oil accumulation in the Songliao Basin but also marked the complete onset of the oil accumulation period, which continued into the overall shrinking stage of the basin's development.

During the continuous sedimentation stage within the oil-generating depression, the Songliao Basin underwent the accumulation of muddy and sandy sediments, as well as organic matter. Sedimentary and diagenetic evolution of sediments and catagenesis of organic matter were completed during this time. At the end of this

Table 8.1 Petroleum geological evolution stages and characteristics of Songliao Basin

Era	Basin stage of development	Physical nature of the field	Petroleum geology characteristics	Geothermal field characteristics
Quaternary period	Basin overall atrophy adjustment development stage	Temporary subsidence (loading pressure buildup) temporary uplift (unloading pressure release) balance adjustment	Old oil and gas reservoirs continue to be filled and finalized; new oil and gas reservoir formed	
Paleogene—Neogene period				
Sedimentary period of Mingshui Formation				
Sifangtai Formation depositional period				
Sedimentary period of Nenjiang Formation—before Sifangtai formation deposition	The basin as a whole rose to the stage of denudation	Unloading decompression	Accumulation process	
Sedimentary period of Nenjiang Formation	Basin continuous subsidence stage	Load pressure buildup	Oil-accumulation process	On average, 4.2 °C/100 m, Qingshankou Formation had the largest sedimentary period, reaching 5.6 °C/100 m
Yaojia Formation sedimentary period				
Qingshankou Formation sedimentary period				

stage, the oil-generating depression reached a high-pressure and high-temperature state, initiating a gradual release, unloading, and decompression of accumulated energy within the basin, ultimately leading to a variety of oil accumulations. The basin's transition to the stage of comprehensive atrophy adjustment marked the final formation of diverse oil accumulations (Table 8.1).

8.2 Source Rock Characteristics of Lower Cretaceous Qingshankou Formation in the Songliao Basin

8.2.1 Sedimentary Environment and Paleontological Characteristics

During the Cretaceous period, the Songliao Basin was a vast continental lake basin, with the Qingshankou Formation, Yaojia Formation, and Nenjiang Formation marking the most significant periods of its development. The sizeable lake area

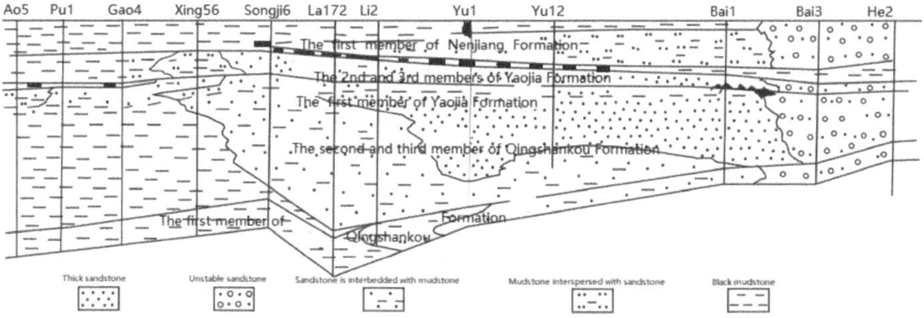

Fig. 8.3 Lithologic profile of oil-bearing assembly foliated delta complex in the central Songliao Basin (after Wanli 1985)

reached up to 20×10^4 km^2. The convergence of four primary source water systems from the outskirts of the basin created an alluvial fan, alluvial plain, delta (shore-shallow lake facies), semi-deep lake, and deep lake depositional system, extending from the basin's edge to its center, known as the river delta depositional system. The largest sedimentary system, the Daqing river delta, also referred to as the Black fish Bubble, covers nearly 3×10^4 km^2, spanning 220 km in length and 150 km in width, while the volume of sandy rock within this system is 2900 km^3 (Fig. 8.3).

During the sedimentary period of the Qingshankou-Nenjiang Formation, the climate was humid, and the lake water was extensive, yet shallow. The coastal shallow lake and delta front area had a water depth of less than 10 m, while the semi-deep lake-deep lake area typically had a depth under 30 m. Due to the basin's flat paleo-topography along its long axis, with a slope only ranging from $0°4'$ to $0°57'$, there was no noticeable slope break. As a result, the shallow lake area was vast, and the shoreline range was wide, reaching up to 20–30 km. During this sedimentary period, the lake basin water body consisted of alternating environments of fresh water, brackish water, and saline water. The sedimentary period of the Qingshankou Formation primarily took place in a brackish-saline water environment, evident from the distribution of various paleontological species. Currently, over twenty fossil species have been discovered, including not only organisms found in freshwater environments, but also marine organisms in the brackish-saline water environment, such as the megalostoma-pinefish, marine bivalves, dinoflagellates, among others. Dinoflagellates, in particular, possess strong oil-generating capacity, often contributing to over 60% of algae content, making them the primary oil source for the Qingshankou Formation (Fig. 8.4).

The oil-bearing areas of northern China, including the Hailar Basin, Erlian Basin, Yanji Basin, and Tarim Basin, contain Cretaceous source rocks rich in fossils of dinoflagellates and coccolithophores (Fig. 8.5). Dinoflagellates are a type of algae that can survive in both freshwater and saline water environments. However, the practice of petroleum exploration at home and abroad has demonstrated that only dinoflagellates living in saline water environments can produce a substantial amount of oil

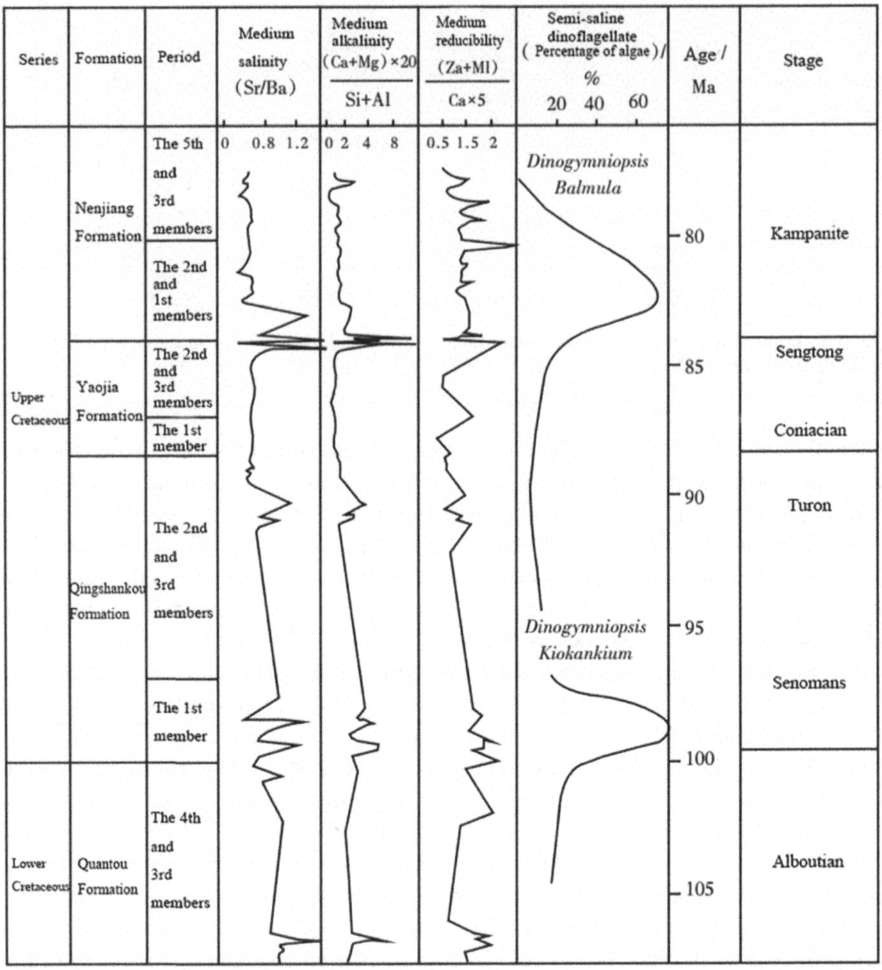

Fig. 8.4 Vertical variation of Cretaceous lake water conditions and content of brackish water dinoflagellate in Songliao Basin (after Ruiqi 1997)

source material and create good source rocks. Dinoflagellates thriving in freshwater environments typically fail to form good source rocks.

8.2.2 Stratigraphic and Sedimentary Facies Characteristics of the Qingshankou Formation

The Qingshankou Formation is primarily composed of a series of grayish-black mudstone, shale, and sandstone. The formation contains a wealth of biological fossils,

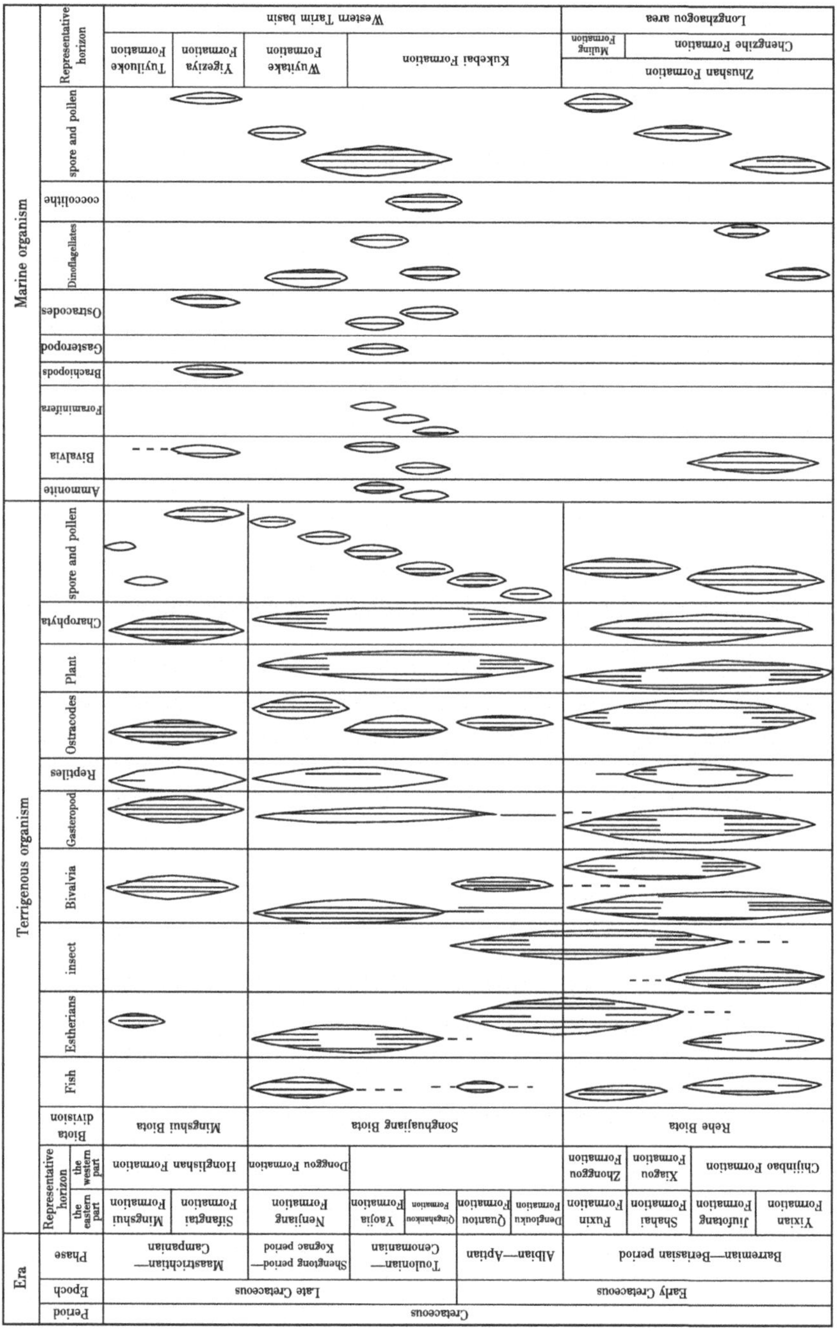

Fig. 8.5 Development map of Cretaceous biota in northern China. Simplified from Ye Dequan (1990)

consisting of over 200 species from almost 20 genera, including ostracods, phyllometes, and trichodinospora. Typically, the formation's thickness ranges from 300 to 500 m, with a maximum thickness of 639.5 m (in well Hei 2). Based on lithologic characteristics, the formation can be divided into three sections.

8.2.2.1 The First Member of Qingshankou Formation

The first member of the Qingshankou Formation features a mixture of black and gray-black mudstone and shale, along with lower quality oil shale. Thin siderite bands or lenses and dispersed pyrite are also present in this layer. The formation is rich in ostracoda, dinoflagellates, and fish fossils, with ostracoda fossils distributed among specific layers. In the western part of the basin, the formation varies to gray-black and gray-green mudstone and interbedded sandstone. In the southern and southwestern parts of the basin, red mudstone and sandstone are prevalent, with a 10–30 m layer of black mudstone featuring inferior oil shale serving as the primary marker for stratigraphic correlation in the basin. The thickness of the formation ranges from 60 to 164 m.

8.2.2.2 The Second and Third Member of Qingshankou Formation

The second and third members of the Qingshankou Formation are composed of gray-black and gray-green mudstone, with intermittent layers of gray calcareous siltstone, fine sandstone, and ostracoda. The lithology undergoes significant changes. In the central basin, gray-black mudstone predominates, while in the southern, northern, and western areas, sandstone is prevalent, which constitutes the Gaotaizi oil-bearing layer. Thin layers of red mudstone and siltstone can be found in the eastern region, whereas purple mudstone is present in the southeastern area. The formation contains abundant phyllometa, ostracoda, bivalve, and fish fossils, with ostracoda fossils notably distributed in thin bands within either mudstone or sandstone layers. Typically, the stratum ranges from 200 to 400 m thick, with a maximum thickness of 503 m. Figures 8.6 and 8.7 illustrate the stratigraphic changes between the first and second/third members of the Qingshankou Formation.

Over the deposition period for the first member of the Qingshankou Formation, the lake area continuously expanded, with depositional centers primarily located in Qijia-Gulong and Sanzhao areas, lending a thickness range of 80–100 m. Lithofacies differentiation was pronounced, ranging up to 46,000 km^2 in the deep lacustrine facies area. In the basin's eastern part, a single deep lacustrine facies zone was present, while in the west, a river-delta-shore-shallow lake-deep lacustrine facies developed along the basin's long axis from north to south. The sedimentary period witnessed broad-ranging silt deposits occurring through multiple provenance directions, with numerous fluvial-lacustrine or alluvial fan sand bodies developing, such as Keshan-Xingshugang in the north, Baokang in the south, and Qiqihar, and Yingtai in the west (Fig. 8.8).

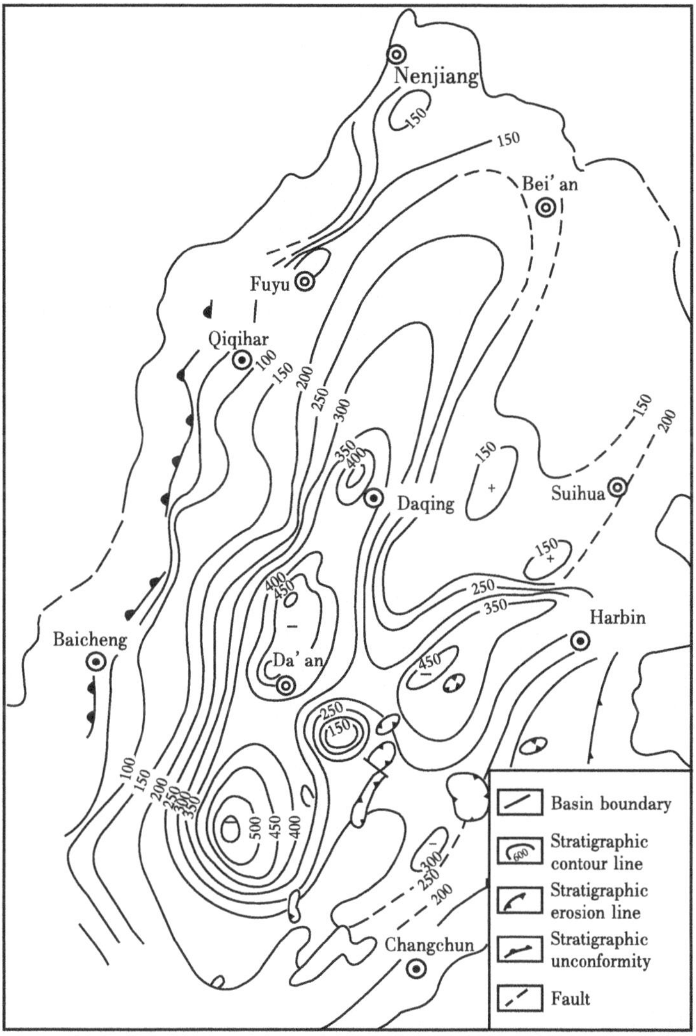

Fig. 8.6 Stratigraphic thickness map of the first member of the Cretaceous Qingshankou Formation in the Songliao Basin (after Ruiqi 1997)

The depositional range of the second and third members of the Qingshankou Formation is comparable to that of the first member. In the eastern basin, a shallow lake silty facies area lacking sandstone deposits can be found. Meanwhile, in the western basin, the deep lake area lies in the Sanzhao-Putaohua-Qijiagulong region, covering an area of around 16,000 km^2, and featuring a thickness of black mudstone at approximately 250–300 m. Towards the north of the deep lake lies a large fluvial delta depositional system, ranging from Bei'an to Daqing, with an area of roughly 10,000 km^2 and sandstone thickness of 150–250 m (Fig. 8.9).

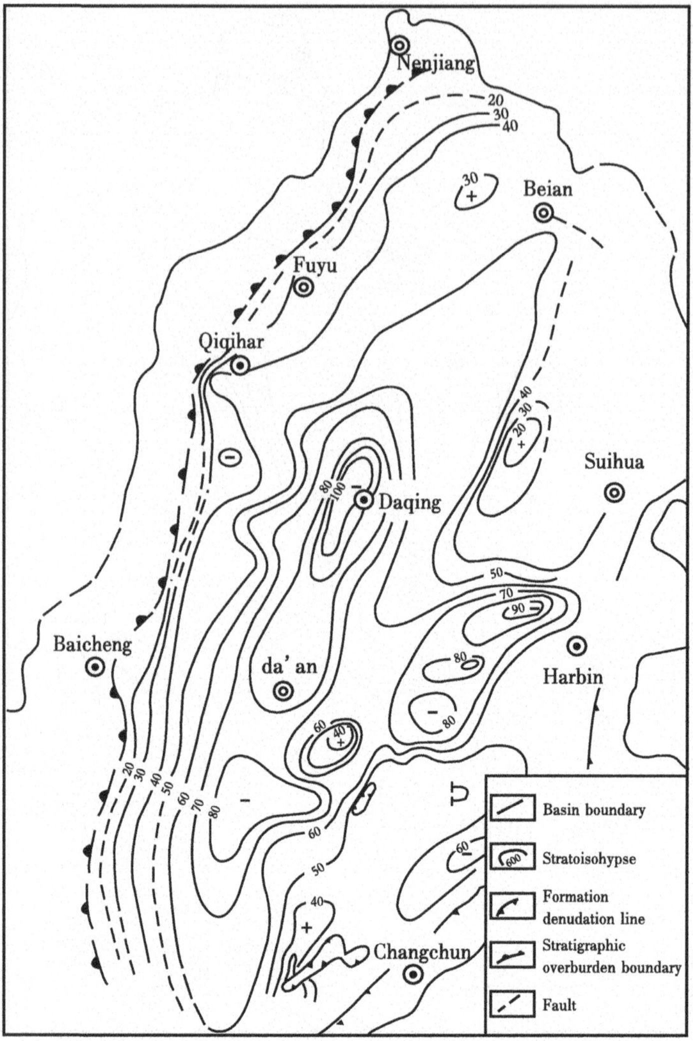

Fig. 8.7 Thickness map of the second and third member of the Cretaceous Qingshankou Formation in the Songliao Basin (after Ruiqi 1997)

8.2.3 Main Characteristics of Source Rocks in Qingshankou Formation

The Qingshankou Formation has a thickness ranging from 350 to 550 m. The first member is approximately 100 m thick, while the second and third members are around 350 m thick. The lithology consists of interbedded mudstone, shale, and sandstone. Mudstone and shale are more dominant in the first member, while numerous

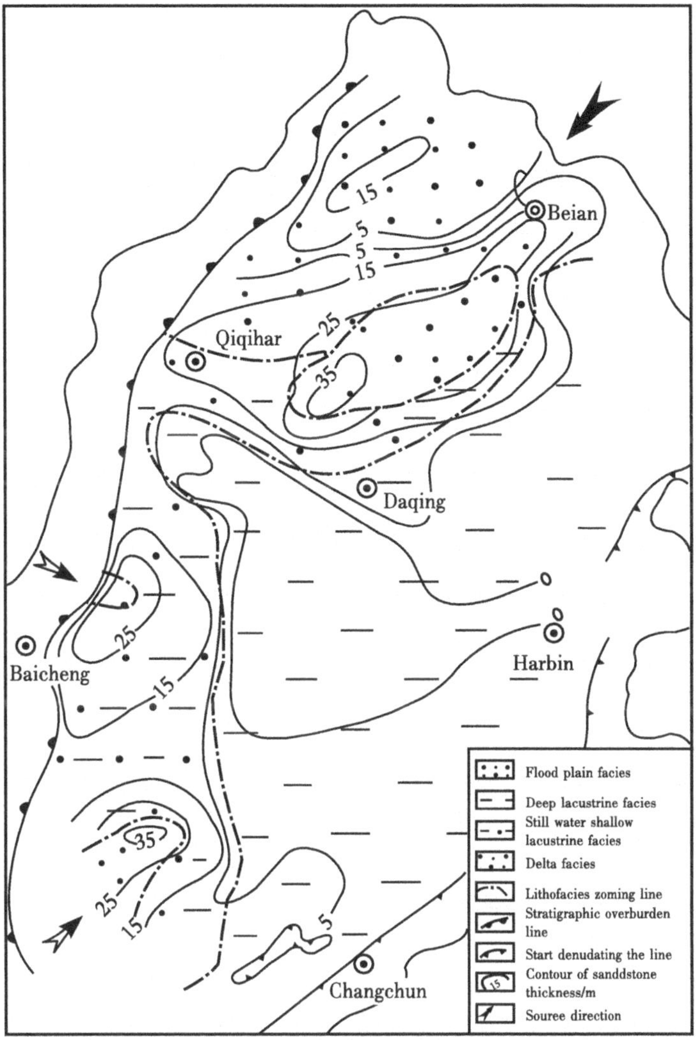

Fig. 8.8 Lithofacies map of the first member of the Cretaceous Qingshankou Formation in the Songliao Basin (after Ruiqi 1997)

interbedded sandstones are present in the second and third members (Table 8.2). As indicated in Figs. 8.6 and 8.7, the individual layer sandstone thickness in the Qingshankou Formation is not significant. The average thickness of single-layer sandstone ranges over 1 m in the Keshan and Heiyubao regions in the north, while the average thickness of sandstone in the Lamadian, Xingshugang, and Putaohua oil fields is roughly 1 m. The sedimentary characteristics of sandstone interbedded with mudstone and shale are highly conducive to the primary migration of oil in mudstone, shale, and other source rocks.

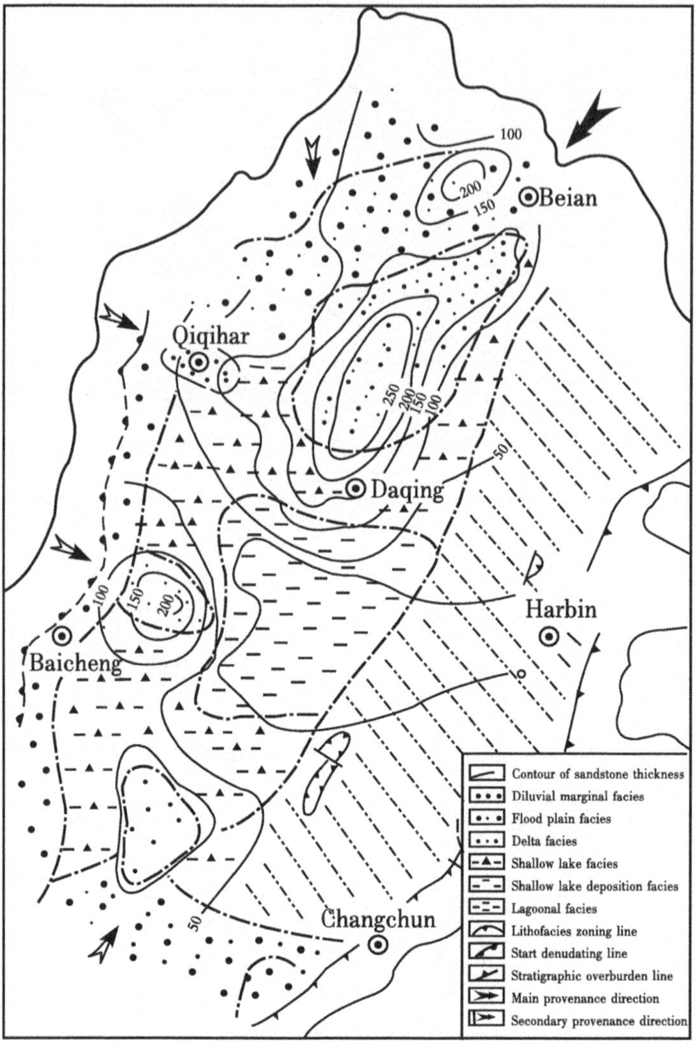

Fig. 8.9 Lithofacies map of the second and third Member of the Cretaceous Qingshankou Formation in the Songliao Basin (after Ruiqi 1997)

Table 8.2 Distribution of thickness and layers of sandstones in the Qingshankou Formation and northern Yaojia Formation

Region	Keshan	Heiyupao	Lamadian	Xingshugang	Putaohua
Sandstone cumulative thickness/m	141	368	293	88	13
Layer number	61	194	281	91	12
Location	Near source ← → Near depocenter				

The black mudstone in the Qingshankou Formation can accumulate to a thickness of up to 300 m. The first member is 80 m thick, has a 2.21% TOC content, and contains type I kerogen. The second and third members are approximately 200–250 m thick, have type II_1 kerogen, and an average TOC content of 0.71%. With an average geothermal gradient of 4.2 °C/100 m in the oil-generating depression coupled with the catalytic effect of rare and metal elements like strontium, nickel, gallium, vanadium, iron, copper, and manganese from magmatic hydrothermal solutions, the source rock of the Qingshankou Formation is buried at 1900 m, and Ro has reached 1.0%, entering the mature stage. This stage corresponds to the early middle diagenesis stage, where the mudstone porosity of Qingshankou Formation is around 8–10% (Fig. 8.10). The sandstone porosity of the Qingshankou Formation and Yaojia Formation ranges between 12 and 20%. At this stage, the porosity difference between mudstone and sandstone ranges from %, which significantly increases the pore pressure difference, making it highly favorable for primary migration of oil generated in mudstone. In summary, the source rocks of the Qingshankou Formation in the Songliao Basin possesses the following characteristics:

(1) The oil source material of this rock is dinoflagellate, which exists in the brackish-saline water depositional environment where it was developed.
(2) This sapropelic source rock contains type I and type II_1 kerogen with high organic matter abundance.
(3) It contains various rare and metal elements, and its organic matter undergoes low temperature catalytic cracking, which facilitates oil generation.
(4) This rock comprises mainly of interbedded mudstone, shale, and sandstone, with a sandstone–mudstone ratio of 20–60%. The combination of these layers is favorable for both oil generation and primary migration.

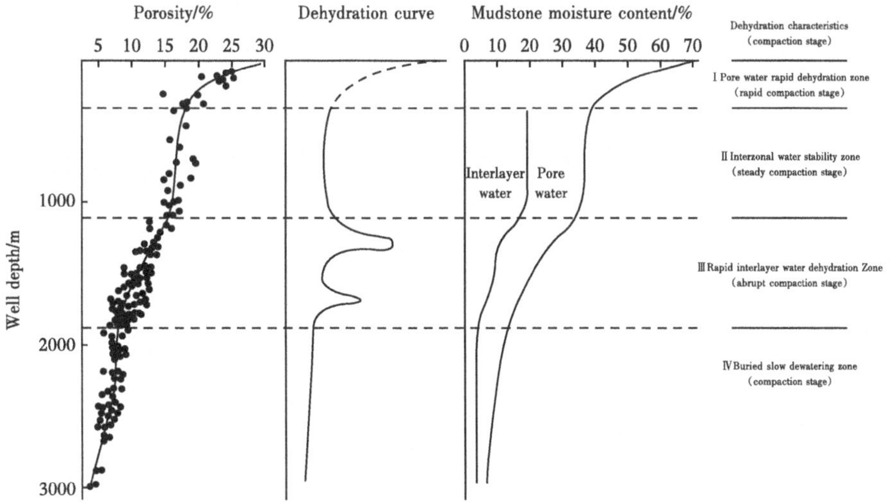

Fig. 8.10 Mudstone dehydration characteristics in the Songliao Basin (Ruiqi 1997)

(5) The distribution area of the source rock is vast, and oil generation is rapid due to shallow burial. The effective pore space available for oil generation is also significant, making the combination conditions for oil generation and accumulation unique.

8.2.4 The Qingshankou Formation as a Source Rock and the Conditions for Oil Accumulation in the Middle Oil-Bearing Assemblage

The Lower Cretaceous in the Songliao Basin contains three oil-bearing assemblages: upper, middle, and lower. The upper assemblage comprises of the third and fourth members of the Nenjiang Formation, known as the Heidimiao OIL Layer. The middle assemblage is composed of the Qingshankou Formation, Yaojia Formation, and the first member of the Nenjiang Formation. This includes the Gaotaizi Oil Layer, Putaohua Oil Layer, and Sartu Oil Layer from bottom to top. The lower assemblage consists of the third and fourth members of the Quantou Formation, which can be further divided into Yangdachengzi and Fuyu oil layers from the bottom up. The middle assemblage is not only the most crucial oil-producing target of the Cretaceous in the Songliao Basin but also represents a typical oil-forming system of a large river delta in China's continental oil-bearing basins. The river-delta oil-forming system is comprised of three parts.

The development of a superimposed large river delta depositional system—a large leafy delta complex.

The basin's initial deposition began with the first member of the Qingshankou Formation, and the entire basin rapidly settled. During this time, the climate was humid, and the lake expanded quickly, covering an area of 8.7×10^4 km^2 while the local shoreline was in the northern part of the basin. Five major depositional systems formed around the basin, with the northern depositional system being the largest. Five sedimentary facies zones developed from north to south: alluvial facies, fluvial facies, delta plain facies, delta front facies, and semi-deep to deep lacustrine facies. At the basin's edge, the alluvial facies comprised red and green mudstone as well as grayish-white conglomerate and breccia that overlaid the weathered crust. The fluvial facies were characterized by red mudstone interbedded with off-white pebbled sandstone and sandstone of varying thickness. Siltstone-fine sandstone was interbedded with green mudstone and gray-black mudstone, dominating the delta plain facies. Meanwhile, the delta front facies consisted of siltstone, silt-fine sandstone, gray-black mudstone, and several layers of shell limestone of varying thickness. Finally, the semi-deep to deep lacustrine facies was dominated by dark mudstone interspersed with thin siltstone (Fig. 8.11).

During the deposition of the second and third members of the Qingshankou Formation, the depositional system inherited from the first member of the formation remained largely unchanged, except that the lake area decreased from 6.8×10^4 km^2 in the early stage to 3.5×10^4 km^2 in the late stage (Fig. 8.12). This period

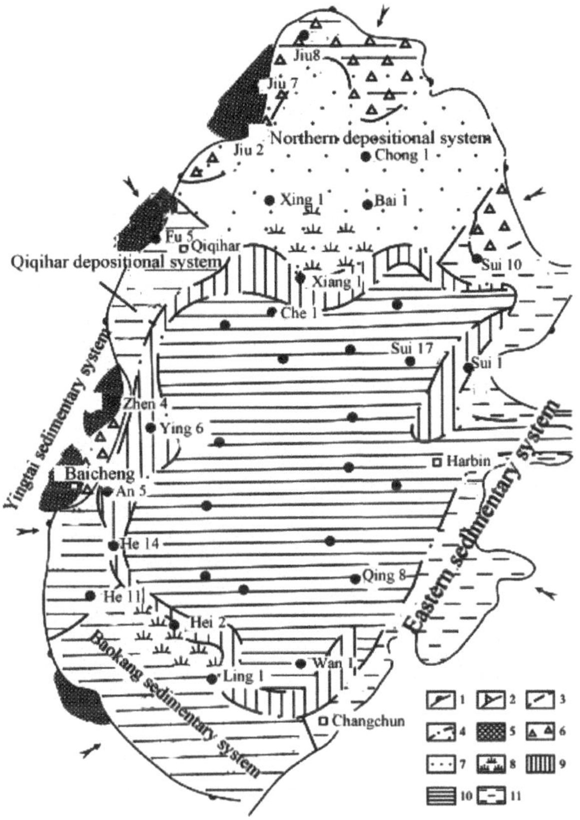

Fig. 8.11 Depositional facies of the first member of the Qingshankou Formation (after Ruiqi 1997). 1—depositional boundary line; 2—stratigraphic denudation boundary; 3—sedimentary system boundary line; 4—depositional (sub-)facies boundary; 5—denudation zone; 6—alluvial facies; 7—fluvial facies; 8—delta plain subfacies; 9—delta front and shore-shallow lake subfacies; 10—semi-deep to deep lacustrine subfacies; 11—plain depositional facies

marked the peak of delta development, with several lake transgressions occurring under the backdrop of a retreating lake. However, due to slow subsidence and a small paleo-gradient, the shore-shallow lacustrine facies belt was wide, spanning up to 60–70 km. The deltas that formed in the early stage were primarily leaf-shaped, whereas those in the late stage resembled bird-foot shapes (Figs. 8.13, 8.14 and 8.15).

During the deposition of the first member of the Yaojia Formation, the lake basin was uplifted due to a dry climate, and the lake area shrank to less than 1×10^4 km^2. With the significant reduction of the lake surface, river progradation occurred rapidly, and the delta distributary plain extended widely (Figs. 8.16 and 8.17).

Fig. 8.12 Depositional facies of the second and third members of the Qingshankou Formation (after Ruiqi 1997). 1—Depositional boundary line; 2—stratigraphic denudation line; 3—sedimentary system boundary line; 4—depositional (sub-)facies boundary; 5—denudation zone; 6—alluvial facies; 7—fluvial facies; 8—delta plain subfacies; 9—delta front and shore-shallow lake subfacies; 10—semi-deep to deep lacustrine subfacies; 11—plain depositional facies

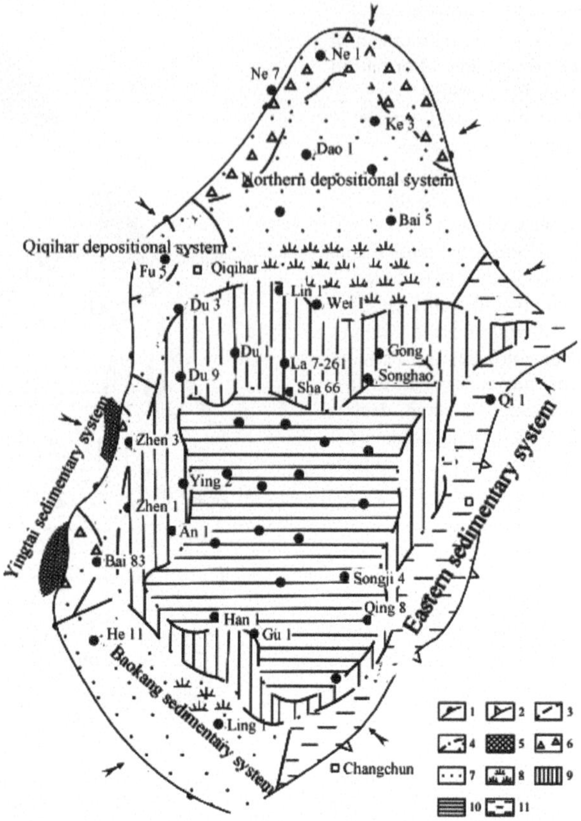

As the lake expanded during the deposition of the second and third members of the Yaojia Formation, covering an area of nearly 2×10^4 km^2, the lithologic characteristics of the coastal shallow lacustrine facies were similar to those observed during the depositional period of the second and third members of the Qingshankou Formation. They were mainly composed of gray-black mudstone, siltstone, and interbedded ostracods shell sandstone. In the semi-deep lacustrine and deep lacustrine sedimentary areas, only dark and black mudstone were present. In addition to foliated and bird-foot delta deposits, sheet delta sedimentary systems were also formed (Figs. 8.18, 8.19 and 8.20).

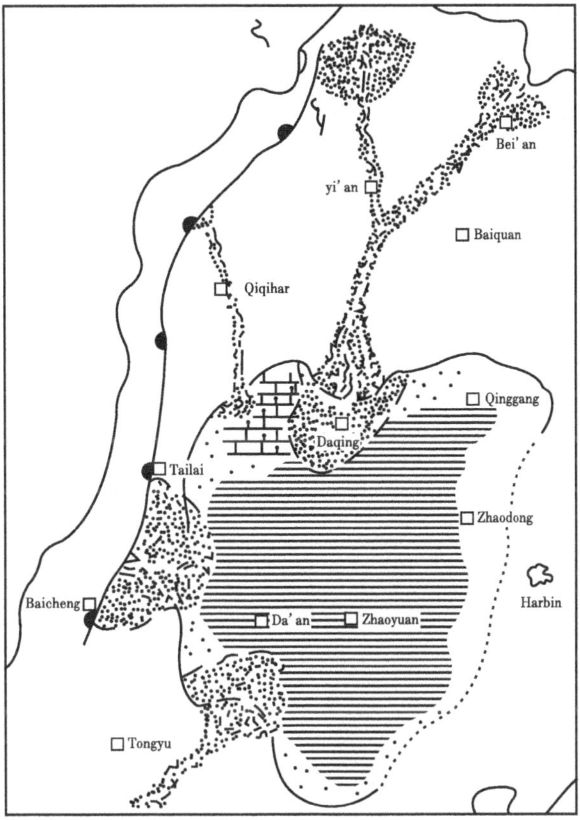

Fig. 8.13 Paleogeographic map of the Songliao Basin during the deposition period of the Qingshankou Formation (after Ruiqi 1997)

During the deposition period of the first member of the Nenjiang Formation, the basin underwent large-scale subsidence, resulting in a vast water area that extended beyond the deposition period of the Qingshankou and Yaojia Formations. By the time the second member of the Nenjiang Formation was deposited, the entire basin had been covered by the lake. The lake began to gradually shrink during the deposition of the third member of the Nenjiang Formation, and the lake area continued to decrease during the deposition of the fourth and fifth members until the end of the Nenjiang Formation when the crust uplifted and underwent denudation. The first member of the Nenjiang Formation is primarily composed of black mudstone mixed with sandstone and siltstone, representing a set of delta front deposits, as well as coastal shallow lake and semi-deep lacustrine facies deposits (Fig. 8.21).

Overall, the oil-bearing assemblage in the central Songliao Basin is primarily controlled by the large fluvial delta-leafy delta complex of the northern sedimentary system. This leafy delta complex, which developed along the long axis of the basin, has a vast scale of over 200 km and a shoreline and shallow lacustrine facies bandwidth of 60–70 km (Fig. 8.22).

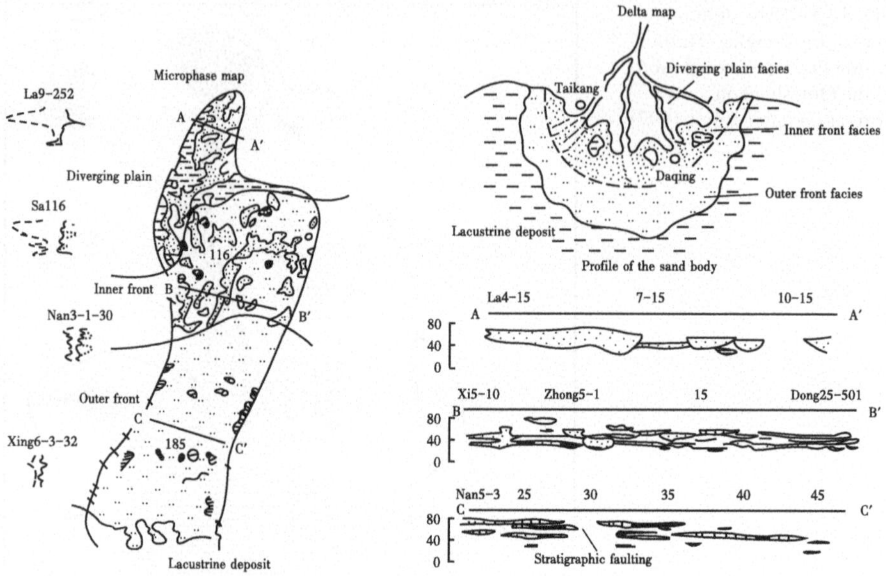

Fig. 8.14 Schematic diagram of the deposition model of the leaf delta (after Ruiqi 1997)

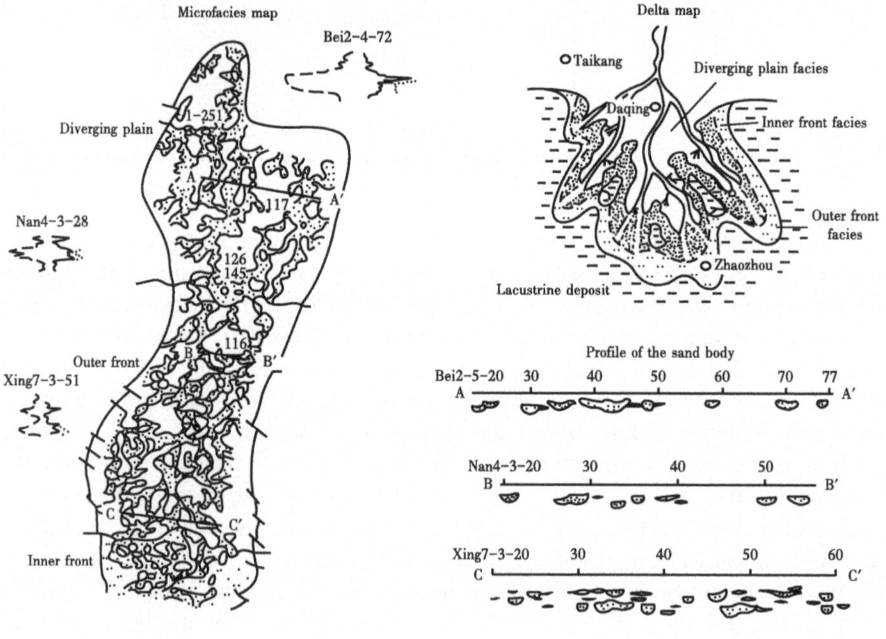

Fig. 8.15 Schematic diagram of the deposition model of birdfoot delta (after Ruiqi 1997)

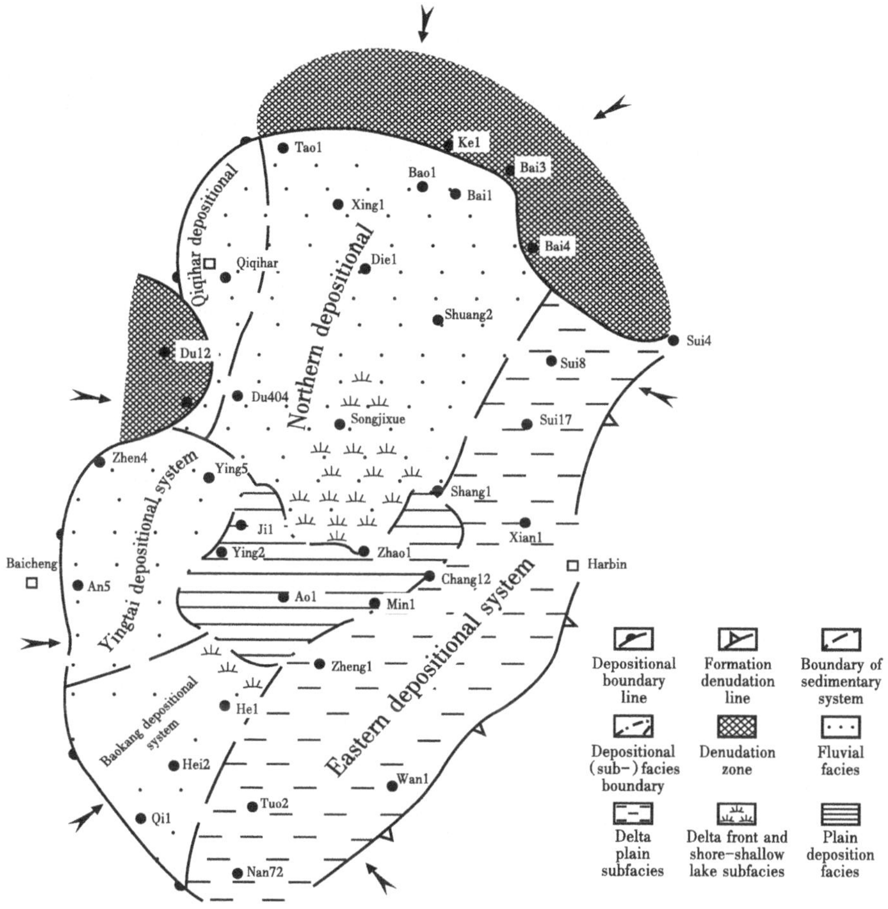

Fig. 8.16 Planar distribution of depositional facies of the first member of the Yaojia Formation (after Ruiqi 1997)

The ideal interbedded combination of dark mudstone and sandstone is developed.

In the Songliao Basin, large fluvial deltas developed during the deposition periods of the first member of the Upper Cretaceous, the second and third members of the Qingshankou Formation, the first member of the Yaojia Formation, the second and third members of the Yaojia Formation, and the first member of the Nenjiang Formation. Although the lake area varied across the different periods, the semi-deep lacustrine facies, shore-shallow lacustrine facies, delta facies, and fluvial facies all contained black mudstones and sandstones. Specifically, the thin interbedded formations of black mudstone and sandstone in the second and third members of the Qingshankou and the second and third members of the Yaojia Formation have higher oil-generating capacity than the large black mudstone in the first and Nenjiang members. The area of semi-deep lacustrine and shore-shallow lacustrine facies is

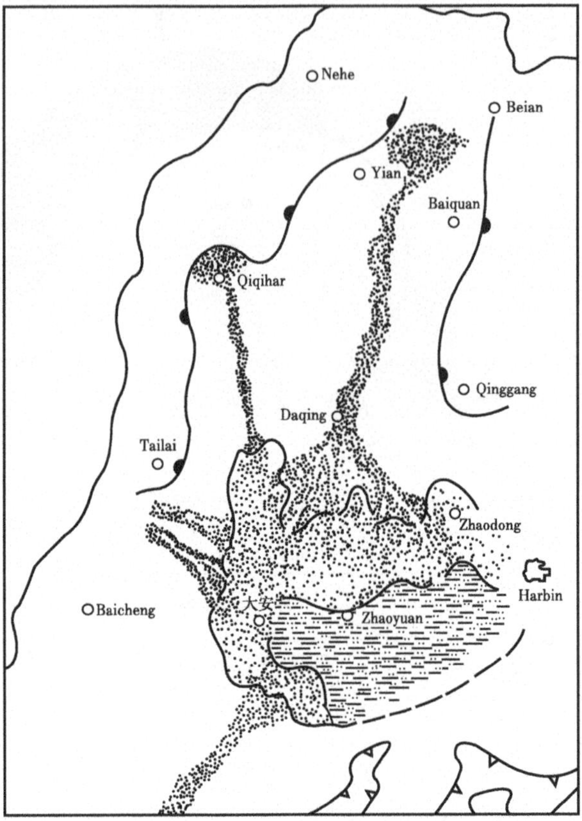

Fig. 8.17 Paleogeographic map of the Songliao Basin during the deposition period of the first member of the Yaojia Formation (after Ruiqi 1997)

considered to have the best oil-generating capacity among the various facies, making it the main source of oil in the region.

Various forms of sandstone bodies give rise to the development of synsedimentary structures, specifically differential compaction structures.

Due to the predominance of up-and-down movements and minimal fold activity in continental sedimentary basins in central and eastern China, synsedimentary structures are the primary trap structures. These synsedimentary structures are controlled by the distribution of sedimentary systems, specifically differential compaction structures. The Daqing placanticline structure is a classic example of a differential compaction structure, as evidenced by the paleostructural development history of the top surface of the Yaojia Formation in Daqing placanticline (Fig. 8.23).

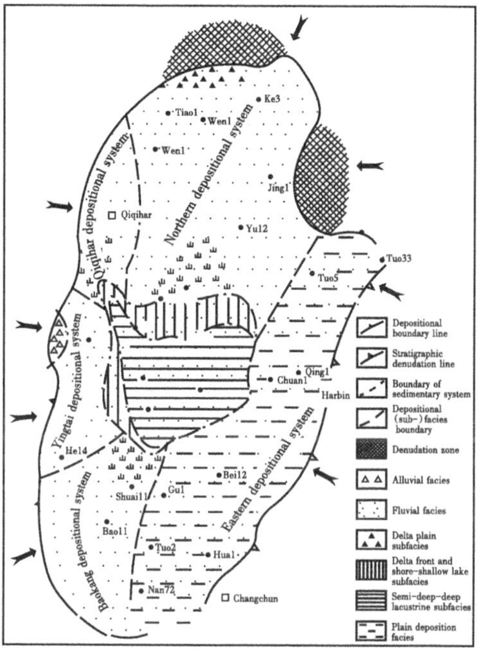

Fig. 8.18 Planar distribution diagram of depositional facies of the second-third members of the Yaojia Formation (after Ruiqi 1997)

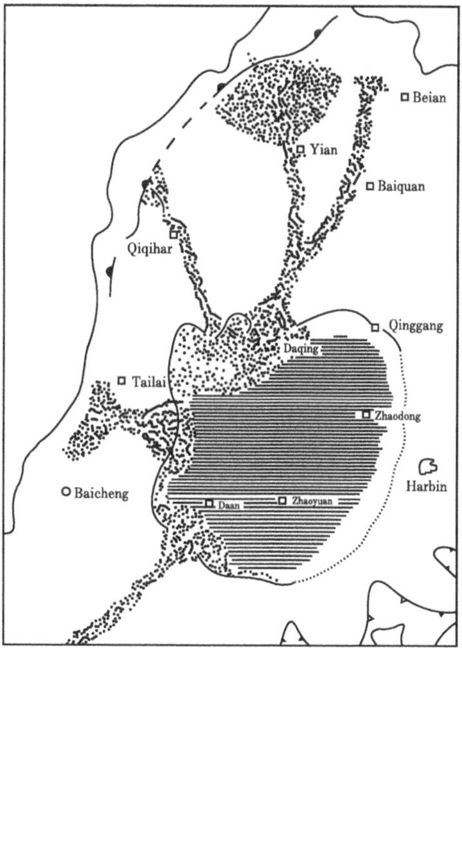

Fig. 8.19 Paleogeographic map of the second-third members of the Yaojia Formation in the Songliao Basin (after Ruiqi 1997)

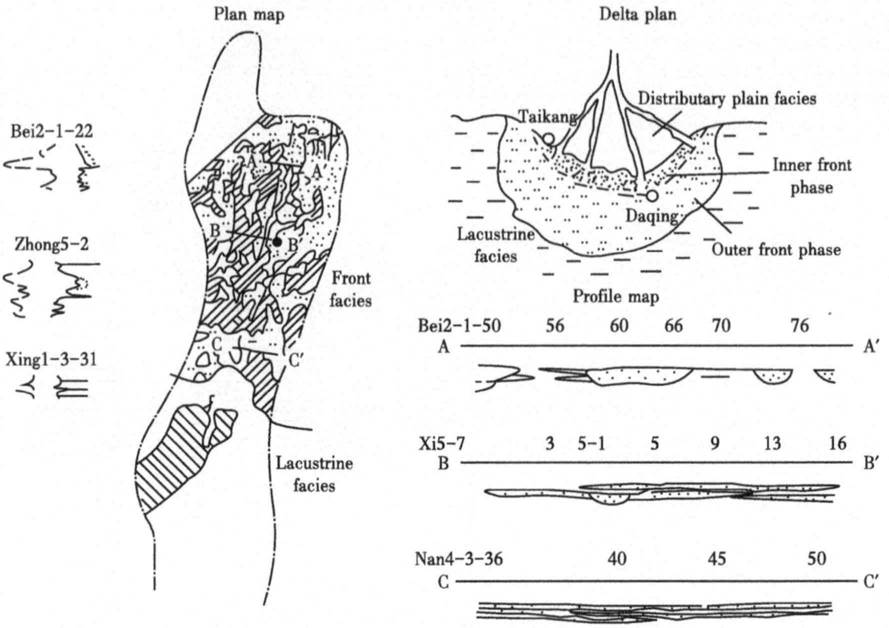

Fig. 8.20 Schematic diagram of the deposition model of sheet delta (after Ruiqi 1997)

Differential compaction structures mainly form in the transition zone of river and lake, particularly in the axial part of the river-delta complex where there is often a difference in the sand-mud ratio between two sides of the area. This differential settling can lead to a high value of sandstone thickness, which causes the formation of differential compaction structures. During the basin's whole uplift denudation stage, the structural morphology of areas with high sandstone thickness becomes more prominent under the influence of "sandstone rebound", as clearly depicted in Figs. 8.24 and 8.25.

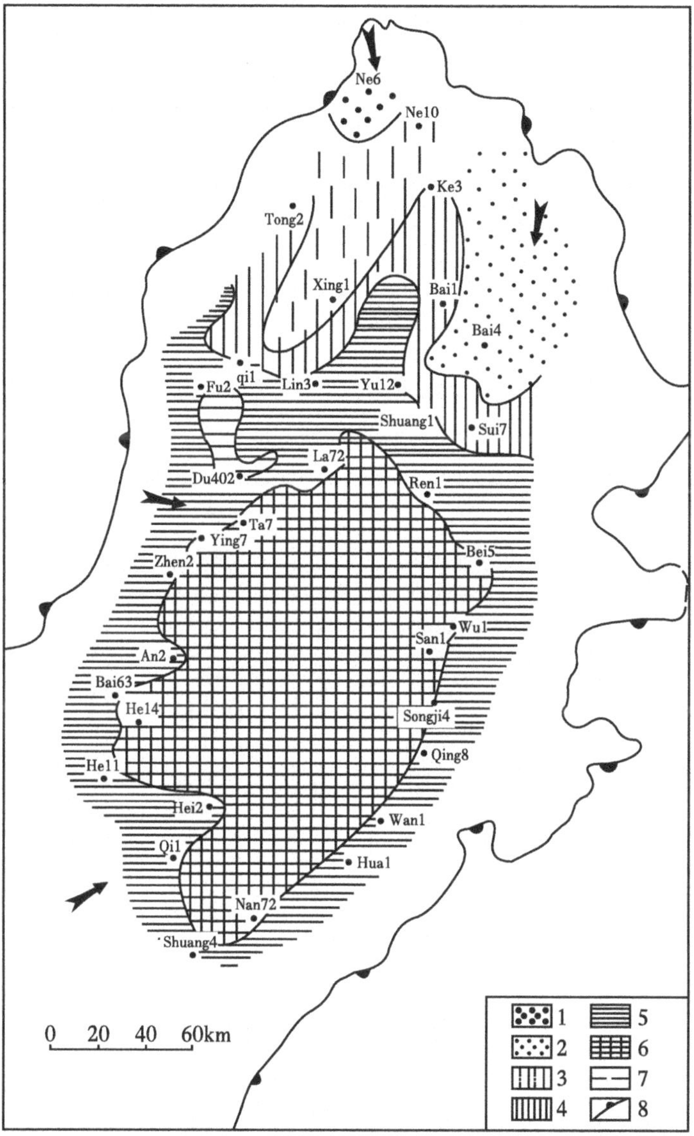

Fig. 8.21 The sedimentary facies plane distribution map of the first member of the Nenjiang Formation (number zero sandstone combination of Sartu oil layer) (after Ruiqi 1997). 1—Alluvial facies; 2—fluvial facies; 3—river and delta interaction deposition; 4—delta front subfacies; 5—shoreline shallow lake subfacies; 6—deep lacustrine subfacies; 7—boundary line of sedimentary system; and 8—sedimentary boundary line

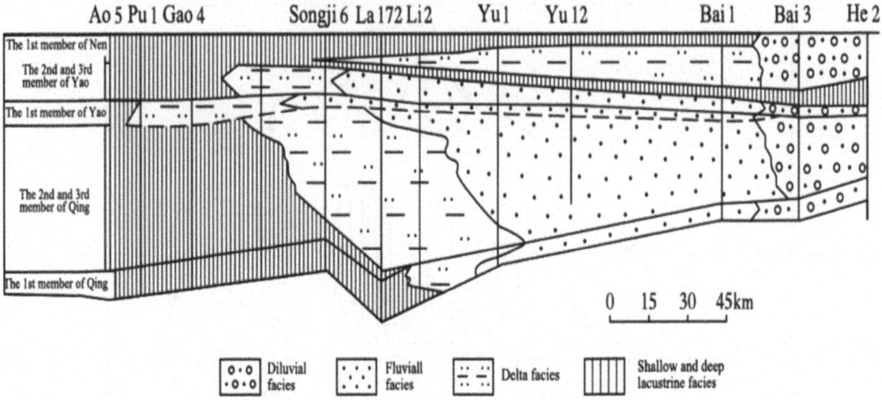

Fig. 8.22 Profile of foliated delta complex facies in Songliao Basin (after Chongjun 1992)

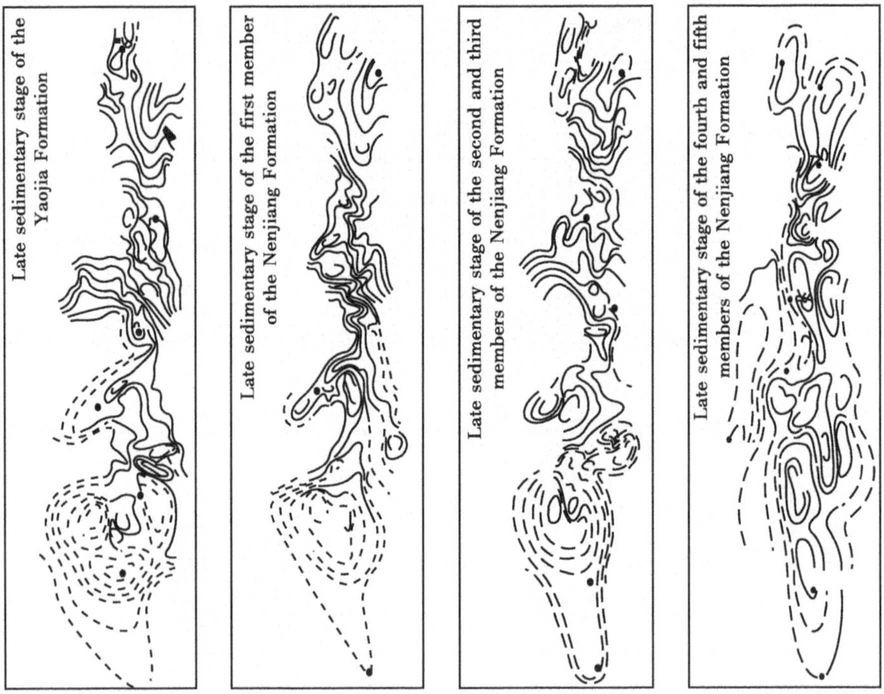

Fig. 8.23 The paleo-structural development diagram of the top surface of the Yaojia Formation on Daqing placanticline

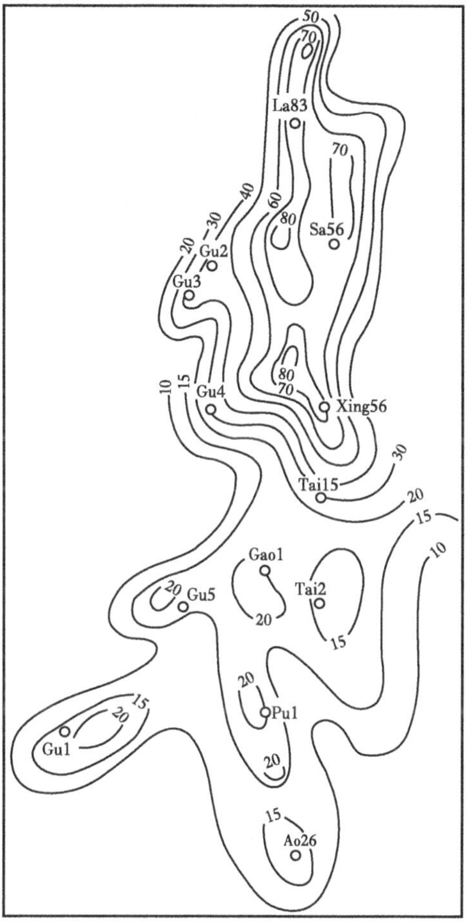

Fig. 8.24 Isopach map (m) of the Yaojia Formation sandstone in Daqing placanticline

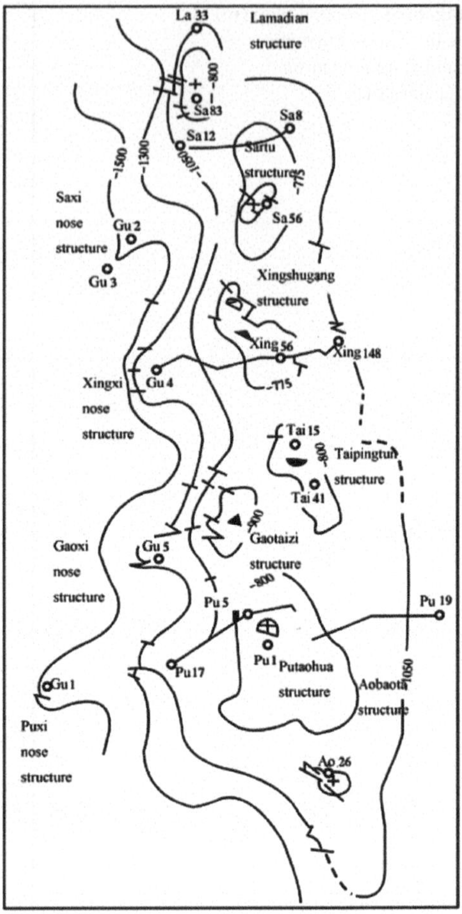

Fig. 8.25 Structure diagram of the top surface of the Yaojia Formation on Daqing placanticline

References

Chongyun W (1992) Sedimentology of China petroliferous basins. Petroleum Industry Press, Beijing
Defan G et al (2004) Theoretical thinking of basin-forming hydrocarbon accumulation: from basin to oil and gas reservoir. Petroleum Industry Press, Beijing
Dequan Y et al (1990) Cretaceous in the oil and gas bearing area of northern China. Petroleum Industry Press, Beijing
Ruiqi G et al (1997) Formation conditions and distribution of oil and gas fields in Songliao Basin. Petroleum Industry Press, Beijing
Wanli Y (1985) Petroleum geology in Songliao continental basin. Petroleum Industry Press, Beijing
Zaiyi T, Qingchun Z (1996) On oil-bearing sedimentary basins in China. Petroleum Industry Press, Beijing

Chapter 9
Characteristics of Paleogene Source Rocks in the Bohai Bay Basin

Abstract The Bohai Bay Basin, situated along the eastern lacustrine margin of China, primarily developed during the Paleogene period. Its petroleum geological evolution can be categorized into three stages: continuous subsidence (energy accumulation), overall uplift (energy release), and shrinkage adjustment (energy redistribution). This chapter takes the Jiyang Depression as a case study to provide a detailed analysis of the Paleogene Shahejie Formation. Key topics covered include its depositional environment, paleontological features, stratigraphy, sedimentary facies characteristics, and primary source rock properties. These insights are essential for understanding the petroleum generation conditions and assessing the resource potential of the Paleogene in the Bohai Bay Basin.

Located on the eastern China, the Bohai Bay petroliferous basin is a Mesozoic–Cenozoic fault-depression basin superimposed on the Paleozoic sedimentary cover of the North China Platform, with the basement consisting of the Archaeozoic Taishan Group crystalline bedrocks. This basin covers an area of approximately 20,000 km^2, with a land area of 12,700 km^2, including Hebei Province, northern and western Shandong Province, southern Liaoning Province, northern Henan Province, Beijing, and Tianjin. The Bohai Sea area has an area of 7,300 km^2, with the water depth up to 70 m (average 18 m).

The Bohai Bay Basin is bounded by the Tancheng-Lujiang Fracture and Taihang Mountain Fracture to the east and the west, and by the Luxi Uplift and the Yanshan Fold Belt to the south and the north, respectively. Within the basin, there are seven sub-basins including the Liaohe, Bozhong, Jiyang, Huanghua, Jizhong, Linqing, and Changwei, as well as two uplifts called the Chengning and Cangxian. Each sub-basin is further divided by secondary uplifts into more than 50 Palaeogene dustpan-shaped or graben fault depressions. Together, they form the fundamental tectonic framework of the basin (Fig. 9.1).

D. Guan et al., *Theory and Simulation Experimental Technology of Oil Generation From Nonmarine Source Rocks in China*, https://doi.org/10.1007/978-981-96-3145-2_9

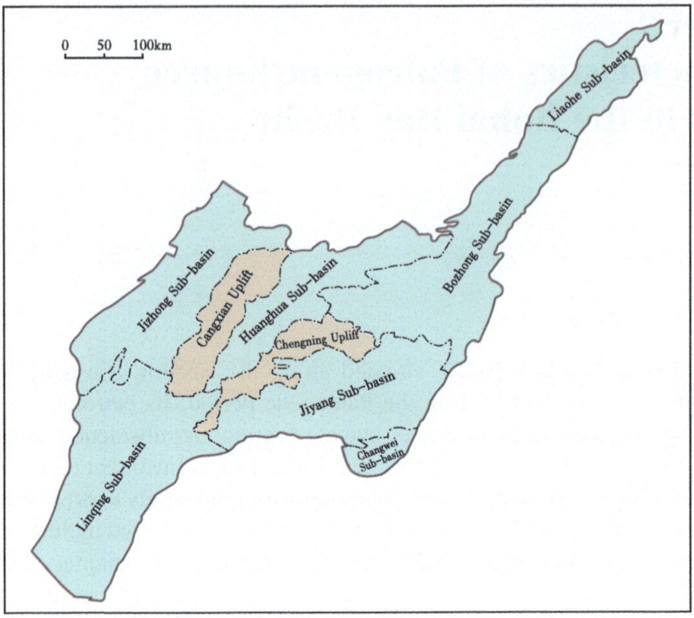

Fig. 9.1 A sketch map of tectonic division of Bohai Bay Basin. Modified from Guoyu and Minggang (2002)

9.1 Evolution of Petroleum Geology in Different Development Stages of the Bohai Bay Basin

The main developmental period of the Bohai Bay basin is the Paleogene. Its petroleum geology evolution can be divided into three stages: continuous subsidence, overall uplift, and contraction adjustment (see Table 9.1).

9.1.1 Continuous Subsidence and Development Stage of the Basin—Process of Energy Accumulation

During this period, which corresponds to the deposition of the Paleogene Kongdian Formation, Shahejie Formation, and Dongying Formation, dark lacustrine mudstones and deltaic sandstones are the main sediments in this basin. The sedimentation rate reached 0.42–0.64 mm/year. Among them, the Shahejie Formation is the main source rock in the basin, characterized by interbedded mudstones and sandstones of variable thickness. This formation has an average thickness of around 2500 m, with some areas exceeding 3000 m. It is in conformable contact with the underlying Kongdian Formation, with locally unconformable contact. At the end of the Dongying Formation, the main source rocks of the Shahejie Formation had become mature. Thermal

Table 9.1 Petroleum geological evolution stages and characteristics of Bohai Bay basin (taking Jiyang sub-basin as an example)

Epochs	Evolution stages	Physical field properties	Characteristics of petroleum geology	Characteristics of geothermal field
Quaternary	Comprehensive contraction stage of the basin	Short-term subsidence (pressure loading) short-term uplift (unloading-depressurization) adjustment of balance	Formation of new reservoirs, enrichment and adjustment of existed reservoirs	
Deposition period of Minghuazhen Formation				
Deposition period of Guantao Formation				
The end of Dongying Formation deposition period to before Guantao Formation deposition period	Overall uplift and erosion stage of the basin	Unloading-depressurization	Process of accumulation	
Deposition period of Dongying Formation	Continuous subsidence stage of the basin	Pressure loading	Process of oil generation	3.6 °C/100 m, average
Deposition period of Shahejie Formation				

decomposition of organic matters occurred gradually, resulting in an increase in total hydrocarbon content from 200–1000 to 1000–5000 μg/g. Abnormally high pressures within source rocks, which provide the driving force for primary oil migration, was formed due to significant amount of generated hydrocarbons. Therefore, during the continuous subsidence stage of the basin, the continuous accumulation of sedimentary materials also led to the gradual pressure loading and energy accumulation in the basin. Simultaneously, it facilitated the thermal decomposition of source rocks into oil.

9.1.2 Overall Uplift and Erosion Stage of the Basin—Process of Energy Release

During the late stage of sedimentation in the Dongying Formation until the deposition of the Guantao Formation, the basin underwent a notable uplift and erosion phase. This resulted in the erosion of the overlying strata above the source rocks, leading to a decrease in thickness and effective stress. Consequently, there was a release of energy within the stratigraphic layers, causing a shift in the basin's physical field from a loading-pressurization process to an unloading-depressurization process. This transition lasted for approximately 8–10 million years. As the reservoir rocks in the slope zone experienced elastic energy release (sandstone rebound), a substantial suction force emerged, facilitating the secondary migration and accumulation of oil from the oil source rocks to the reservoir rocks in the slope zone. This uplift process initiated petroleum accumulation and persisted until the comprehensive contraction adjustment stage of the basin, which extends to the present day. As a result, the uplift and erosion stage of the basin played a pivotal role in the migration and accumulation of lacustrine source rocks, creating favorable conditions for significant oil expulsion from the source rocks.

9.1.3 Comprehensive Contraction Stage of the Basin—Process of Energy Adjustment

After the overall uplift and erosion phase during the late stage of the Dongying Formation, the basin underwent a period of deposition compensation with the Guantao Formation and Minghuazhen Formation being deposited. During this phase of energy adjustment, by the late stage of the Minghuazhen Formation, the lower and middle subunits of the Shahejie Formation, acting as the source rocks, had fully matured. This led to a significant increase in the production of oil in the source kitchen. Furthermore, a certain amount of elastic energy accumulated within the sandstone reservoirs. Subsequently, during another uplift phase in the late deposition of the Minghuazhen Formation, not only were new reservoirs formed, but the existing reservoirs were also enriched and adjusted, further enhancing the potential for oil accumulation.

9.2 Characteristics of the Paleogene Main Source Rocks of the Bohai Bay Basin

This study primarily focuses on the characterization of the Paleogene main source rock (the Shahejie Formation) of the Bohai Bay Basin, taking the Jiyang Sub-basin for an example.

9.2.1 Depositional Environment and Paleontological Characteristics

The Paleogene sedimentation in the Bohai Bay Basin is significantly influenced by paleostructures and paleogeography. Let's take the example of the eastern part of the Jiyang Sub-basin, where the Shahejie Formation exhibits tripartite characteristics. In the lower sub-member of the Shahejie Formation, there are red gypsum-bearing salt sandy mudstones containing small-sized algae fossils and drought sporopollen fossils like Gyrogona qianjiangica and ephedripites. Moving to the middle sub-member, it consists mainly of gray gypsum salt-bearing mudstones and contains fossils of ostracods like Cyprinotus igneus and deflandrea, as well as dinoflagellate fossils. As we reach the upper sub-member, we find thin carbonate layers and oil shales with typical local ostracod assemblages including Chinocythere and Austrocypris.

In contrast, the lower and middle sub-members of the 3rd member of the Shahejie Formation represent thick deep lacustrine mudstones, shales, and marls interbedded with turbidites, reflecting the deposition during a rapid subsidence phase of the basin. Towards the late stage of the 3rd member and early stage of the 2nd member, uplift occurred, and the sediment mainly consisted of sandstones. The 2nd member, in comparison to the 3rd member, has a smaller extent. Its lower sub-member is characterized by deltaic river-mouth sandbars and floodplain swamp environments, with fossils of lakeshore swamp biota like Huabeinia unispinata, Camarocypris elliptica, Cypris shenglicunensis, Cyprois palustris, Tulotomoides terrassa, Comasphaeridium, Alnipollenites, and Polypodiaceaesporites. The upper sub-member, known as the "Red Coarse Member," is dominated by red clastic rocks in a fluvial environment and contains arid biota such as Camarocypris elliptica, Cyprinotus xiaozhuangensis, Charies producta, Ephedripites, Rutaceoipollis, and others.

In the 1st member of the Shahejie Formation, there is a clear tripartite division as well. The lower sub-member consists of gray mudstone interbedded with dolomite and oil shale. The middle sub-member is characterized by bioclastic carbonate rocks, cladochite dolomite, and oil shale. The upper sub-member is composed of deep gray mudstone. The fossil assemblages in this formation also exhibit tripartite characteristics. The lower and middle sub-members, especially the middle one, contain abundant small-sized ostracods with smooth shells adapted to strong hydrodynamic conditions, like Phacocypris huiminensis, as well as gastropods adapted to brackish environments. In the mudstone of the upper sub-member, larger-sized ostracods adapted to deeper water environments, such as Xiyingia and Guangbeinia, are present.

In the Dongying Formation, which is predominantly characterized by gray-green and variegated mudstones and gray-white sandstones, the lower part of the 3rd member contains weakly hydrodynamic fauna, including the flat-shelled ostracod Chinocythere unicuspidata and the simple shell-decoration ostracod Dongyingia laticostala, both adapted to weak hydrodynamic conditions. The middle part of the 2nd sub-member features massive robustly ornamented ostracods, such as Dongyingia inflexicostala and D. florinodosa, adapted to strong hydrodynamic conditions. Demise of the lacustrine basin occurred during the deposition of upper 1st sub-member, and there is a practical extinction of local aquatic species in the Paleogene of the Bohai Bay Basin. Palynological analysis indicates a cooling trend in the paleoclimate from the deposition of the Shahejie Formation to the Dongying Formation, with a transition from a predominance of Quercoidites to Ulmipollenites in the pollen assemblages.

9.2.2 Characteristics of Strata and Sedimentary Facies of the Shahejie Formation

The Shahejie Formation in the Bohai Bay Basin is widely distributed and primarily consists of dark-colored mudstone, siltstone, interbedded sandstone, shale, coal, oil shale, evaporite, argillaceous dolomite, dolomite, limestone, and basalt, with a thickness ranging from 2000 to 6500 m. It unconformably contacts the underlying Kongdian Formation. From bottom to top, it can be divided into four members: 4th member, 3rd member, 2nd member, and 1st member, with varying lithology and thickness from the central part of the basin to the edges (Fig. 9.2).

The lower sub-member of the 4th member originated during the early stage of basin subsidence when the subsidence rate was notably reduced, leading to a relatively flat terrain. The climate at that time was arid, and the primary formations were alluvial fans. As the basin expanded, it witnessed the sequential development of fan-deltas, nearshore subaqueous fans, and deep-water turbidite fans. The prevailing lithology comprises purplish-red and greyish-green mudstone, interspersed with purplish-red sandstone, siltstone, pebbled sandstone, gypsum-bearing mudstone, and occasional thin carbonate layers. Some localized salt accumulations can also be found, although the development of source rocks was limited. Moving to the upper sub-member of the 4th member, it represents the early stage of basin subsidence characterized by the presence of fan-deltas, nearshore subaqueous fans, deltas, and lakes. The sedimentary succession consists of a stable sequence of alternating "mudstone and sandstone" units. In well-log responses, mudstone exhibits low resistivity and relatively flat spontaneous potential curves, while sandstone shows high resistivity responses with finger-, bell-, and funnel-shaped patterns and high spontaneous potential responses. Within the upper sub-member of the 4th member, there exists a gray mudstone interbedded with siltstone or fine sandstone, forming an effective source rock interval. This mudstone is rich in organic matter, providing a substantial material basis for hydrocarbon generation. Moreover, the interbedded siltstone and fine sandstone create favorable spaces for both the generation and primary migration of petroleum.

The deposition period of the 3rd member marks the most intense phase of tectonic activity in the evolution of the Bohai Bay rifting basin. During this time, the basin experienced deeper subsidence and a significant increase in water depth, resulting in the development of subaqueous fans, fan-deltas, and slump turbidite fans. This period coincided with a major expansion of the lake basin and represented the main stage

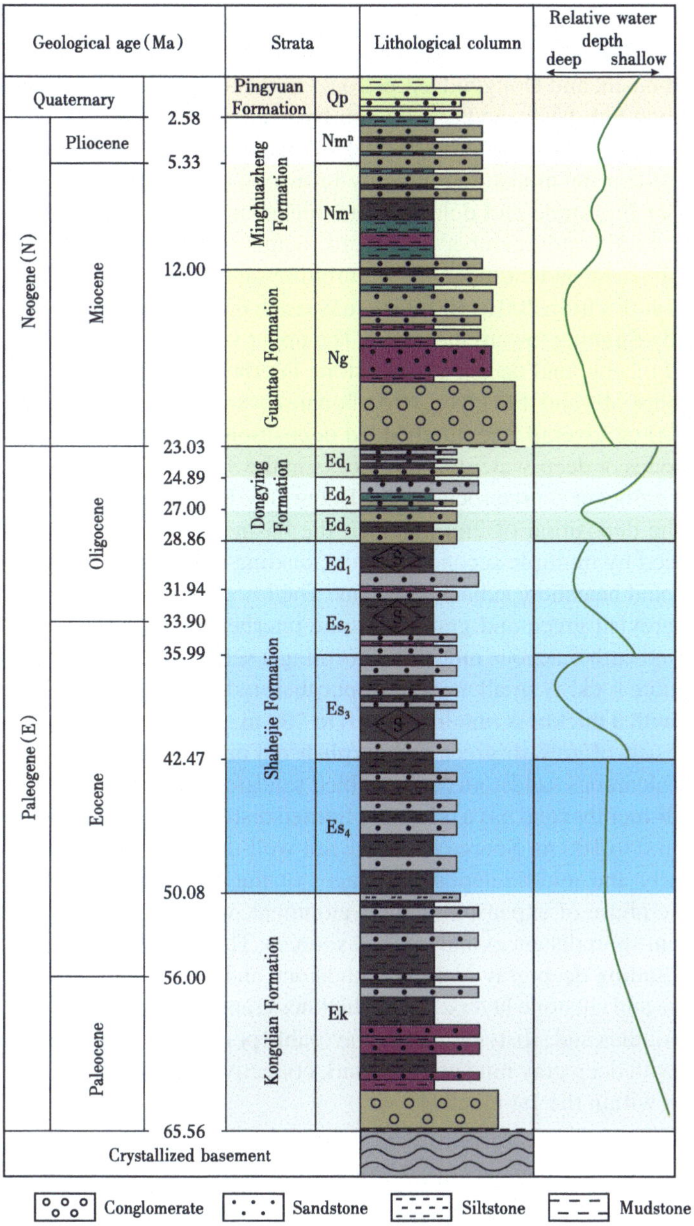

Fig. 9.2 Cenozoic strata diagram of Dongying sag. Modified from Tao (2020)

of source rock development. The 3rd member is distributed across various major sub-basins, with the most extensive development observed in the Jiyang Sub-basin, Jizhong Sub-basin, and Dongpu Depression. On the other hand, the Huanghua Sub-basin, Bozhong Sub-basin, and Liaohe Sub-basin have relatively thinner intervals of this member. The lower sub-member of the 3rd member is characterized by deep gray mudstone, oil-bearing mudstone, and gray-brown oil shale interbedded with varying thicknesses of limestone and dolomite. Its thickness generally ranges from 100 to 300 m. The middle sub-member of the 3rd member mainly consists of deep gray mudstone, oil shale, and multiple sets of turbidite sandstone or thin carbonate rocks. Its thickness varies from 200 to 500 m but may reach up to 700 m in deeper areas of the sag, gradually thinning toward the edges. The upper sub-member of the 3rd member is composed of gray and deep gray mudstone interbedded with siltstone, dolomitic sandstone, oil shale, and thin layers of carbonate rocks. Its thickness typically ranges from 0 to 500 m. Overall, the interbedded deposition of lacustrine shale and delta-front sandstones or deep-water turbidite rocks in the 3rd member plays a crucial role as a set of significant source rocks in the Bohai Bay Basin.

During the deposition of 2nd member, the basin experienced convergence and was influenced by multiple secondary faults, leading to the development of alluvial fans, deltas, and nearshore subaqueous fans. The lower sub-member of 2nd member consists of greyish-green and gray mudstone interbedded with sandstone, pebbled sandstone, and carbonaceous mudstone, forming a self-sourced and self-reservoired effective source rock. A small amount of purplish-red mudstone can be found in the upper part, with a thickness ranging from 0 to 200 m. The upper sub-member of 2nd member consists of greyish-green and purplish-red mudstone interbedded with gray sandstone, calcareous sandstone, and pebbled sandstone. It unconformably contacts the lower sub-member and has a relatively limited distribution range, with a thickness ranging from 0 to 100 m. Source rocks are not well-developed in this member.

In the early and middle deposition stages of the 1st member, the basin underwent a stable phase of expansion and development, where sedimentation primarily occurred from short distances and internal sources. The basin received a sequence of deposits, including deep gray and gray mudstone and marl, interspersed with granular dolomite and siltstone layers, with a thickness ranging from 150 to 400 m. These granular dolomites and siltstones exhibit favorable porosity and permeability and are juxtaposed with deep gray mudstone or marl, collectively forming an effective set of source rocks within the 1st member.

9.2.3 Characteristics of Source Rocks of the Shahejie Formation

The Shahejie Formation, which serves as the primary Paleogene source rock in the Bohai Bay Basin, exhibits a wide distribution and is found in nearly all depressions of the basin. Among the areas with relatively high exploration activity in the

Bohai Bay Basin is the Jiyang Sub-basin, encompassing the Dongying, Huimin, Chezhen, and Zhanhua depressions. In the Jiyang Sub-basin, the source rocks of the Shahejie Formation, from bottom to top, are as follows: the upper sub-member of the 4th member, lower sub-member of the 3rd member, middle sub-member of the 3rd member, upper sub-member of the 3rd member, lower sub-member of the 2nd member, and 1st member. Notably, both the lower sub-member of the 3rd member and the middle sub-member of the 3rd member serve as regionally favorable source rock intervals. Additionally, the upper sub-member of the 4th member is considered a good source rock for the Dongying and Zhanhua depressions, while the lower part of the 1st member is a favorable source rock layer for the Zhanhua and Chezhen depressions.

The source rocks of the Shahejie Formation in the Bohai Bay Basin have high organic matter abundance and good types, with high oil generation potential. In the 4th member of the Jiyang Sub-basin, most of the TOC content is above 2%, and the type of organic matter is mainly the Type I or Type II_1. The spatial distribution pattern of TOC content in the dark mudstones of the 3rd member is similar to that of the 4th member, with higher TOC content than the 4th member in areas other than the Dongying Depression. Especially in most areas of the Jiyang Sub-basin, the TOC content is between 2.0 and 5.0%. The maceral compositions of the 3rd member source rocks are mainly composed of amorphous sapropelic and exinite, with a content generally over 80%. The content of vitrinite and inertinite derived from higher plants is generally low, typically less than 10%. The main organic matter type is Type II_1, followed by Type II_2. This is because during the deposition of the 3rd member, algae were abundant, and the source input mainly originated from aquatic lower organisms, while the input of terrestrial organic matter was relatively limited. In comparison to the overall high organic carbon content in the 3rd member of the Jiyang Sub-basin, the high TOC zone of the 1st member shifts northward to the Zhanhua Depression, with Type I organic matter being dominant.

Furthermore, according to the analysis of X-ray diffraction of the whole rock minerals, the source rocks of the Shahejie Formation contain a variety of minerals, mainly including clay minerals, carbonates, pyrite, and detrital minerals. The relative content of various minerals varies greatly among different source rocks. Taking the Jiyang Sub-basin as an example, the clay mineral content in the oil source rocks of the Shahejie Formation ranges from 9 to 56%, the carbonates content ranges from 5 to 85%, the pyrite content ranges from 0 to 28%, and the content of detrital minerals such as quartz ranges from 0 to 51%. It can be shown that clay minerals and carbonates are the two most important minerals. Numerous studies have shown that these two mineral types have complex interactions with organic matter during the burial and diagenesis of source rocks, and they play an important role in the oil generation and expulsion processes of source rocks.

References

Guoyu L, Minggang L (2002) Atlas of China petroliferous basins, 2nd edn. Petroleum Industry Press, Beijing

Tao (2020) Paleogene-Neogene sedimentary and provenance characteristics in the East of Bohai Bay Basin. Master dissertation, China University of Geosciences Beijing

Chapter 10
Favorable Conditions for Oil Generation in Nonmarine Source Rocks

Abstract The formation and development of source rocks, as well as the processes of petroleum generation and accumulation under subsurface geological conditions, are influenced by a complex interplay of geological factors. These include the tectonic evolution of the basin, depositional environment, water medium, source material for oil generation, available space for petroleum generation, and transformation conditions. This chapter adopts an organic geochemistry perspective to systematically and comprehensively summarize the favorable conditions for petroleum generation in lacustrine source rocks under subsurface geological settings. The findings provide valuable insights into evaluating the petroleum generation potential of source rocks and enhancing exploration strategies.

Subsurface geological conditions play a comprehensive role in the formation, development, oil generation and accumulation processes of source rocks. These conditions include the tectonic evolution of the basin, depositional environment, water medium, source parent material, oil generation space, and transformation conditions. In this chapter, we adopt an organic geochemical approach to summarize the favorable conditions for oil generation in lacustrine source rocks based on the subsurface geological conditions.

D. Guan et al., *Theory and Simulation Experimental Technology of Oil Generation From Nonmarine Source Rocks in China*, https://doi.org/10.1007/978-981-96-3145-2_10

10.1 Favorable Tectonic Conditions for Oil Generation in Nonmarine Source Rocks

10.1.1 Continuous Subsidence Stage of the Basin is the Key Period for the Formation and Development of Nonmarine Source Rocks

Analysis of the petroleum geological evolution history of the Junggar Basin, Ordos Basin, Sichuan Basin, Songliao Basin, and Bohai Bay Basin reveals that these petroliferous basins have undergone three main development stages: continuous subsidence stage, overall uplift stage, and comprehensive contraction stage (Table 10.1).

The continuous subsidence stage in a basin refers to the accumulation process of various deposits within the basin. For continental petroliferous basins, the main components of sedimentary materials during continuous subsidence are argillaceous- and sandy sediments. The entire process involves the continuous filling and accumulation of these two materials. Sand sediments exhibit elastic characteristics, exhibiting elastic rebound when unloaded, as long as the loading does not exceed the limit of elastic deformation. On the other hand, clay sediments possess plastic characteristics, undergoing plastic deformation upon loading and not returning to their original state upon unloading. During sediment loading, these deposits undergo energy conversion and accumulation through physical and chemical processes. The physical and chemical effects are characterized by continuous loading and increased pressure, increasing formation temperature with increasing burial depth. Under elevated temperature and pressure conditions, as well as catalytic conditions, the organic matter in source rocks gradually matures and undergoing thermal cracking to generate petroleum. In the development zone of clay sediments, under the effect of hydrocarbon generation and increased pressure, abnormally high pressure is formed, driving the generated oil to migrate and charge the ambient sandstone. By the end of the continuous subsidence stage in the basin, most of the main source rocks have entered the oil generation stage, leading to a large-scale oil generation process. However, due to the overall closure of the basin lacking unloading and depressurization conditions, the generated petroleum mainly accumulates in sandstone layers within the source rocks and has not yet undergone significant migration outside the source rocks.

Taking the Songliao Basin as an example, the Late Jurassic to Cretaceous Nenjiang Formation represents the sedimentary period of continuous subsidence of the basin. Among them, the Late Jurassic to the Quantou Formation represents the early rift subsidence period characterized by rift-filling deposition, while the Qingshankou Formation to the Nenjiang Formation represents the period of continuous subsidence characterized by rapid deposition. Although the subsidence characteristics of rift and depression are different, the analysis of the petroleum geological evolution characteristics of the Songliao Basin reveals that this development stage represents a

Table 10.1 Division of tectonic evolution stages of continental petroliferous basins

Evolution stages	Physical field properties	Characteristics of petroleum geology	Junggar Basin	Ordos Basin	Sichuan Basin	Songliao Basin	Bohai Bay Basin
Contraction and adjustment stage	Short-term subsidence and uplift, adjustment of balance	Enrichment and adjustment of existed reservoirs; formation of new reservoirs,	Quaternary Paleogene–Neogene Cretaceous	Quaternary Paleogene–Neogene Late Cretaceous	Quaternary Paleogene–Neogene Cretaceous Late Jurassic	Quaternary Paleogene–Neogene Sedimentary period of Mingshui and Sifang formation	Quaternary Sedimentary period of Minghuazhen and Guantao formation
Uplift and erosion stage	Unloading-depressurization	Process of accumulation	Late Jurassic-early Cretaceous	From the end of Middle Jurassic to before Late Jurassic	The end of Nenjiang period to before Sifangtai period ago	From the end of Dongying period to before Guantao period	
Continuous subsidence stage	Pressure loading	Process of oil generation	Jurassic Triassic Permian	Sedimentary period of Yan'an, Fuxian, Yanchang	Middle Jurassic Sedimentary period of Ziliujing	Sedimentary period of Nenjiang, Yaojia, Qingshankou	Sedimentary period of Dongying, Shahejie

significant material accumulation process with petroleum geological significance. It involves the accumulation of organic-rich clay sediments and the interbedding of sand sediments with dark-colored mudstones. These sedimentary materials also undergo energy conversion and accumulation through physical and chemical processes, manifested as the physical field characteristic of continuous loading and increased pressure. During the loading and increased pressure process, the dark-colored organic-rich mudstones of the Qingshankou Formation, have undergone compaction and thermal evolution processes, resulting in becoming mature source rocks and entering a peak oil generation stage. Therefore, by the late Nenjiang Formation sedimentation period, source rocks of the Qingshankou Formation in the Songliao Basin have undergone oil generation, completing the accumulation of organic biogenic energy and its transformation into chemical hydrocarbon energy. With great amount of generated oil, the primary pores within the mudstones are continuously filled with oil. This not only hinders further compaction of the mudstones but also gradually subjects the interior of the mudstones to abnormally high pressures under the overlying strata's compaction. At the same time, the main reservoir sandstones such as Saertu, Putaohua, and Gaotaizi oil layers, during the late Nenjiang Formation sedimentation period, experience increasing density due to the compaction of the overlying strata. The characteristics of a rigid object gradually become apparent within the sandstones, and the overall interior of the sandstones exhibits a gradual increase in elastic energy, provided that the pressure from the overlying strata has not reached the point of sand grain fracturing. Therefore, during the continuous subsidence stage in the Songliao Basin, not only the accumulation of major materials such as source rocks, reservoir rocks, and cap rocks, as well as overlying strata, are completed, but also through a series of physical and chemical processes, the source rocks of the Qingshankou Formation undergo oil generation and development of abnormally high-pressure conditions, leading to petroleum accumulation. Additionally, significant elastic energy is accumulated within the main reservoir sandstones such as Saertu, Putaohua, and Gaotaizi oil layers.

In summary, the continuous subsidence stage in continental petroliferous basins represents an important phase in the formation and development of source rocks, creating favorable conditions for the accumulation of materials and energy in the source rocks.

10.1.2 Uplifting and Erosion Stage of the Basin is the Key Period for the Oil Migration and Accumulation of Nonmarine Source Rocks

Once a basin enters the uplift and erosion stage, the physical field properties of the basin change from a process of loading and increased pressure to unloading and decreased pressure. The overlying strata of the source rocks undergo erosion, resulting in a reduction in thickness and a decrease in effective stress, leading to the release of energy within the strata. Due to the elastic properties of sandstone, as long as the loading does not exceed the limit of elastic deformation, a rebound phenomenon that restores the original state occurs after the external forces are unloaded. This phenomenon creates negative pressure within pores in the sandstone (Fig. 10.1). The negative pressure generated by sandstone rebound is positively correlated with the amount of rebound, which, in turn, is positively correlated with the erosion thickness. Typically, the maximum erosion thickness occurs on the basin's peripheral slopes due to tectonic uplift, while the erosion thickness gradually decreases toward the center of the basin. Consequently, the sandstone rebound is greater on the basin's peripheral slopes, whereas the actual rebound in the sandstone of the source rock development zone is smaller. This results in a greater pressure drop in the sandstone layers of the basin's peripheral slopes compared to the sandstone layers in the source rock development zone. Under the pressure difference, the petroleum generated in the source rock migrates within the connected sand bodies toward the direction of decreasing pressure. As the petroleum is continuously expelled from the pores, the source rock undergoes further compaction and hydrocarbon expulsion under the load of the overlying strata. This process leads to the redevelopment of abnormally high pressures within source rocks due to hydrocarbon generation and increased pressure. The pressure difference formed between the overpressure in the source rock development zone and the low pressure in the reservoir rock development zone becomes the main driving force for petroleum accumulation. Therefore, the uplift and erosion stage of the basin is a key period for the migration and accumulation of lacustrine source rocks. It provides the dynamic conditions for large-scale oil expulsion from the source rocks.

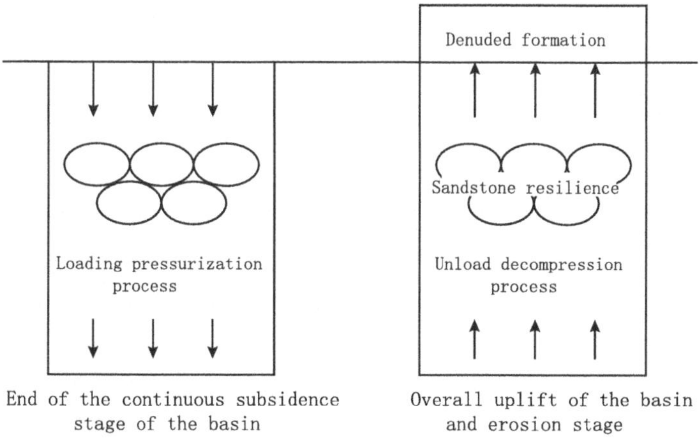

Fig. 10.1 The physical field characteristics of sandy rocks development area in the stage of erosion during the overall uplift of the basin

10.1.3 Differential Compaction Structure is Closely Related to the Process of Oil Generation, Expulsion, Migration and Accumulation

During the process of deposition, there are significant variations in the sand-to-mud ratio among different regions during the same deposition period. After compaction, regions with a higher sand-to-mud ratio experience smaller changes in original thickness, while regions with a lower sand-to-mud ratio experience larger changes in original thickness. This results in significant differences in thickness of the same sedimentary strata across different regions. The anticlinal structures observed in regions with a higher sand-to-mud ratio are referred to as differential compaction structures (Fig. 10.2). Differential compaction structures are a type of sedimentary structure that is primarily influenced by factors such as depositional environment and paleogeography, and they are directly controlled by the distribution and physical properties of the sediment. Large-scale differential compaction structures are most likely to form in areas with extensive ancient slopes extending into the center of a basin or in transitional zones between rivers and lakes, such as ancient uplifts (subsidence zones or buried hills). Along the long axis of the basin, it is favorable for the development of large rivers and large, well-constructed deltas, which often contribute the largest sand bodies that enter the lake. The large sand bodies exhibit significant differences in sand-to-mud ratio compared to the lacustrine mud sediments on both sides. Under the overall differential compaction, a series of compacted structures dominated by minor anticlines and nose-shaped uplifts are formed, while growth faults and their associated anticlines do not develop. The sand sediment that constitutes differential compaction structures is predominantly composed of quartz and feldspar particles, which have good sorting and rounding characteristics, and they are generally less cemented, with contact cementation being predominant. After diagenesis, these sandy sediments clearly become the best reservoir rocks in terms of storage capacity.

While tectonic stress does not directly form differential compaction structures, their formation and development closely relate to the overall uplift or subsidence movements of a basin. Large-scale differential compaction structures are more likely

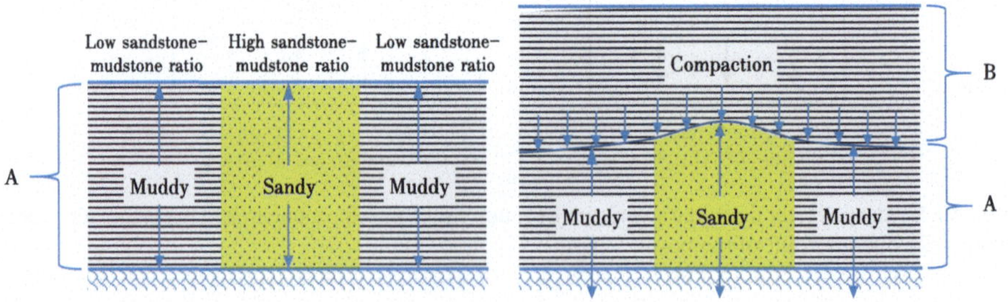

Fig. 10.2 Schematic diagram of differential compaction and "sedimentary structure"

to develop in sedimentary systems within large, stable, subsiding basins. When a basin experiences a slow uplift phase, the sediment load gradually decreases, and compaction weakens significantly. Mudstone layers, due to plastic deformation, cannot "recover" after unloading. On the other hand, sandstone exhibits elasticity and rebounds when the overlying load decreases, resulting in the enlargement of differential compaction structures both in magnitude and extent. Consequently, an effective pressure difference is established between the region where differential compaction structures develop and the region where source rocks are formed. This creates an ideal petroleum accumulation trap in the area with differential compaction structures. In essence, differential compaction structures form a comprehensive petroleum geological system that unifies petroleum generation, migration, and accumulation. They are highly productive in terms of oil reserves and often give rise to large oil fields.

The Daqing Placanticline proto-structure in the central depression of the Songliao Basin is the most complete differential compaction structure discovered in inland basins in China. During the early sedimentary period of the Qingshankou Formation to the Nenjiang Formation, the Daqing Placanticline area was located in the transition zone between a large river-delta facies and a lacustrine facies. Sandy sediments was well-developed, while the sag areas on both sides were characterized by lacustrine sedimentation with well-developed mud deposits. The regional differences in lithology between the Daqing Placanticline area and the surrounding sag areas were significant (Fig. 10.3). In the northern part of the Daqing Placanticline area, it was dominated by delta plain and delta-front facies, with well-developed sandy sedimentation. In the southern part, it was dominated by prodelta and delta front facies, with less developed sandy deposits compared to the northern part. The sand-mudstone sedimentation of the river-delta facies formed typical differential compaction structures under differential compaction, shaping the proto-structure of Daqing Placanticline (Fig. 10.4). These structures played an important controlling role in the petroleum generation and accumulation process of the oil-bearing formations in the central part of the Songliao Basin.

10.2 Favorable Sedimentary Conditions for Oil Generation in Source Rocks

10.2.1 The River-Delta Sedimentary System is a Favorable Target for Oil Accumulation of Lacustrine Source Rocks

Most of China's continental sedimentary basins are inland lake basins with fully enclosed, strongly reducing environments, either saline or brackish. Typically, a basin is composed of several sags and uplifts. During the continuous subsidence stage of the basin, source rocks develop in the sag. Due to significant differences in geomorphic conditions between steep-slope and gentle-slope areas, source rocks formed in these

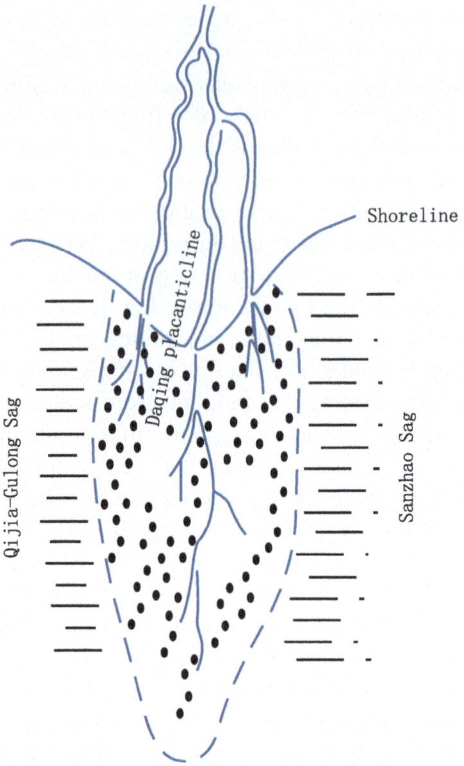

Fig. 10.3 Composite map of structure and sedimentary facies of Daqing Placanticline

two environments are noticeably different. In the steep-slope zone, the terrain has a steeper slope, often resulting in sudden, rapid accumulations of alluvial or floodplain deposits with varied particle sizes and poor sorting and rounding. These deposits create less favorable conditions for oil generation. In the gentle-slope zone, the terrain is relatively flat, and river-delta sedimentation is more common. River channels are classified into straight rivers and meandering rivers, while multiple river channels form braided rivers and network rivers. Since river sedimentation mostly occurs along the long axis of the basin, the clastic particles are better sorted and rounded after being transported over relatively long distances. The river-delta sedimentation system includes three sedimentary zones: (1) Delta plain zone, which is the part between the river and the lake shore, dominated by sand, silt, and mud deposits from distributary channels; (2) Delta-front zone, located near the river mouth in the beach and shallow lake area, primarily comprising sheet sand and shallow-lake mudstone deposits; (3) Prodelta mud zone, where sediments deposit into the semi-deep lake and gradually transition to deep lake conditions, characterized by dark mudstone with thin layers of siltstone. Due to the basin's tectonic movement and the influence of climates of various geological periods, the river-delta sedimentation system may advance or retreat, showing a combination of mudstone and sandstone layers with varying proportions in the vertical sedimentary profile. The region closer to the lake

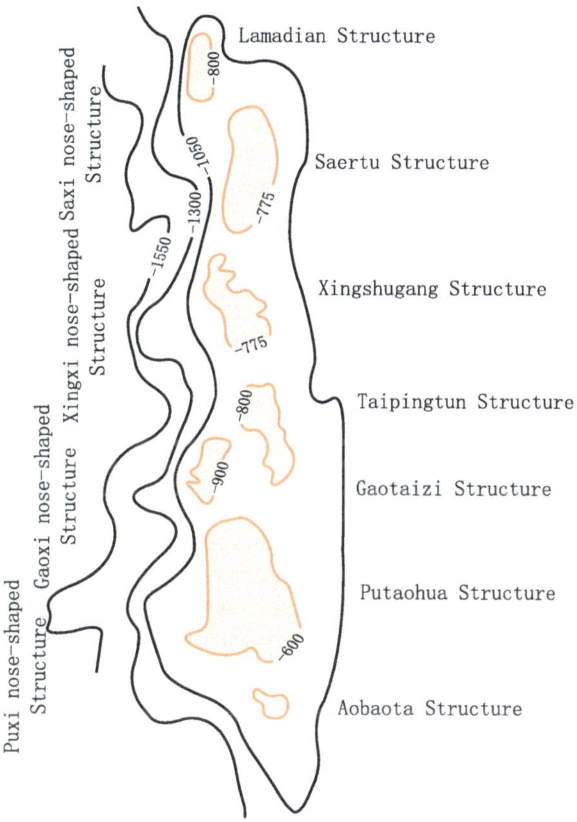

Fig. 10.4 Top surface structure diagram of the Yaojia Formation of Daqing Placanticline

center is mainly composed of pre-delta (or lacustrine) mudstone with interbedded sandstones, providing favorable conditions for the development of source rocks. The slope zone near the basin's edge is characterized by delta front sand bodies interbedded with thin mudstone layers, offering favorable conditions for reservoir development. If the delta-front sand bodies in the slope zone extend continuously toward the lake center and are connected to the source rock development area through an effective carrier system, the oil already generated and effectively expelled from the source rocks can migrate and accumulate in the favorable traps of the delta-front sand bodies under the driving force of secondary migration.

The 4th to 3rd sub-member of Paleogene Shahejie Formation in the Bohai Bay Basin, the Cretaceous Qingshankou Formation to the Nenjiang Formation in the Songliao Basin, the Triassic Yanchang Formation in the Ordos Basin, and the Jurassic Fengcheng Formation in the Junggar Basin all formed and developed in the river-delta depostion system. The lithological profiles of these source rocks show a combination of mudstone and sandstone layers in a ratio of approximately 7:3. The mudstone layers contain abundant organic matter, providing sufficient material basis for oil generation, while the sandstone layers provide ample storage space for the oil generated from source rocks and facilitate continuous pyrolysis and oil generation in the adjacent mudstone or shale layers. Therefore, the river-delta sedimentation system is a favorable area for the formation and development of lacustrine source rocks and also the main target for petroleum generation and accumulation in lacustrine source rocks.

10.2.2 Depositional Environment of Lacustrine Carbonate Rocks is also Favorable to the Formation and Development of Oil Source Rocks

In a pure lacustrine sedimentary environment without external systems of provenance, the predominant sediment consists of chemically and biogenically formed carbonate rocks (such as shell limestone, reef limestone, oolitic limestone, and dolomite) and mudstone. Carbonate rocks are often interbedded with mudstone, while some carbonate rocks occur as isolated features like mounds, lenses, or nodules without sedimentary structures within black mudstone or black shale. The combination of mudstone and carbonate rocks, characterized by favorable porosity and permeability conditions, constitutes an effective oil source rock with the potential for oil generation, expulsion, and accumulation. Oil generated within the mudstone can migrate short distances under pressure differentials and accumulate within reservoirs developed in carbonate rocks with dissolution pores and fractures. The Upper Jurassic Da'anzhai Member of the Ziliujing Formation in the Sichuan Basin serves as a typical example, where interbedded shell limestone with well-developed pores and fractures and dark-colored mudstone form an effective source rock and a self-generating and self-storing shale oil reservoir. Therefore, apart from fluvial-deltaic sedimentary systems with external sediment sources, this type of pure lacustrine sedimentary environment without external sediment sources also provides favorable conditions for the formation and development of source rocks.

10.3 Favorable Material Conditions for Oil Generation in Source Rocks

The main source rocks developed in continental sedimentary basins in China are formed in a brackish- to saline-water depositional environment (Table 10.2). Salinization of water not only facilitates the preservation and enrichment of organic matter in sediments, but also results in density stratification of the water column and promotes the upward migration of redox interfaces. These factors provide alkaline and weakly reducing conditions as well as greater accommodation space for the development of source rocks.

Table 10.2 The water environment of main oil source rocks in continental petroliferous basins

Basin	Source rock	Ratio of Sr/Ba	Palaeosalinity
Bohai Bay Basion	Paleogene-Neogene	0.62–11.31	Brackish—saline
Songliao Basin	Cretaceous	0.98–4.01	Brackish—saline
Sichuan Basin	Jurassic	0.37–4.9	Brackish—saline
Ordos Basin	Triassic	0.33–1.22	Brackish
Junggar Basin	Carboniferous-Permian	0.70–21.27	Brackish—saline

The development of planktonic algae in brackish- to saline-water environments is the primary source of organic matter in lacustrine source rocks. Among the planktonic algae, diatoms, dinoflagellates, and coccolithophores are the major contributors of organic material in source rocks. They are characterized by high lipid and protein contents. Lipids in particular contain fatty substances, waxes, and lipid-like components, which play an important role in petroleum generation.

In the Mesozoic–Cenozoic continental brackish- to saline-water lacustrine basins, the dominant types of planktonic algae are dinoflagellates and coccolithophores (Table 10.3 and Fig. 10.5). Dinoflagellates belong to the class Dinophyceae and appeared in the Late Jurassic. They are widely developed in brackish- to saline-water basins during the Cretaceous and Paleogene–Neogene periods. Coccolithophores, on the other hand, are the primary source of organic matter in Jurassic continental brackish-water lacustrine basins. They contain 30–70% hydrocarbons (based on plant dry weight) and are therefore known as "oil algae." They are significant components of algal coal, lignite, and oil shale.

Table 10.3 Phytoplankton in Meso-Cenozoic continental saline lake basin

Basin	Strata	Phytoplankton
Bohai Bay Basin	The Paleogene Shahejie Formation	Rich in brackish dinoflagellates assemblage, are the major contributors of organic material in main source rocks of the basin
Biyang Sag	The Paleogene Shahejie Formation	
Jianghan Basin	The Paleogene Qianjiang Formation	
Baise Basin	The Paleogene Nadu Formation	
Beibuwan Basin	The Paleogene Liushagang Formation	
Songliao Basin	The Cretaceous Qingshankou and Nenjiang Formation	
Tuha Basin	The Jurassic Sangonghe and Badaowan Formation	Exceptionally abundant Coccolithophores

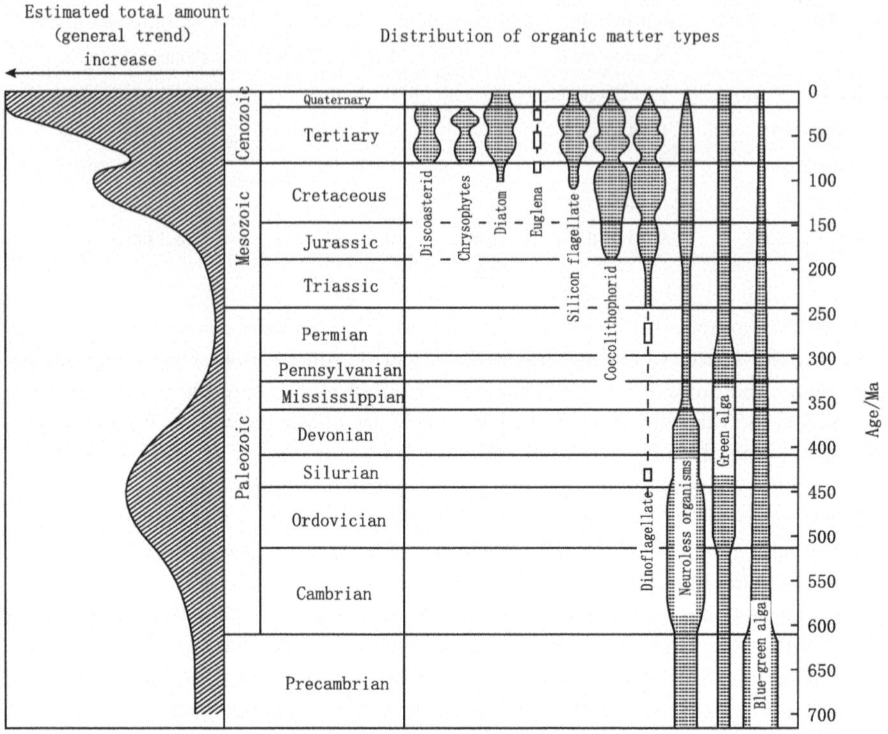

Fig. 10.5 Abundance changes of Phytoplankton in different geological history periods

The Shahejie Formation in the Bohai Bay Basin is rich in planktonic algae and particularly abundant in dinoflagellates, which are the primary suppliers of organic matter in the main source rocks of the basin. They have a decisive influence on the oil-generating capacity of the source rocks. During the deposition of the upper part of the 4th member to the 3rd member of the Shahejie Formation in the Dongpu Depression, the basin was divided into two parts by a nearly east-west uplift. The northern part represents a typical saline-water environment with an exceptionally abundant dinoflagellates assemblage. The main types of organic matter in the source rocks are sapropelic- and humic-sapropelic types, with Type I and Type II kerogen accounting for 84.7% and Type III kerogen only accounting for 15.3%. In contrast, the southern part represents a freshwater environment where humic substances dominate, with Type III kerogen accounting for as high as 56% and no Type I kerogen present. This is why over 90% of the discovered oil reserves in the Dongpu Depression are concentrated in the northern part, while the southern part accounts for less than 10%. In fact, in the Paleogene source rocks of eastern China, any source rock rich in dinoflagellates is a highly prolific main source rock. Examples include the Qianjiang Formation in the Jianghan Basin, the Hetaoyuan Formation in the Biyang Depression, and the Nadu Formation in the Baise Basin. Without exception, these main oil source rocks are rich in diatom fossils (Table 10.3).

Therefore, in a brackish- to saline-water environment, the development of planktonic algae serves as the primary source of organic matter in lacustrine source rocks, providing favorable material conditions for the oil generation.

10.4 Favorable Spatial Conditions for Oil Generation in Source Rocks

During the process of oil generation, the reaction space provided by mudstone for the thermal cracking and oil generation reactions of organic matter is limited. If the generated oil cannot be expelled from the mudstone in a timely manner, the oil generation reaction will be inhibited. However, this situation can be significantly altered if there are sandstone (or carbonate) interlayers with good porosity-permeability conditions between the mudstone layers. On one hand, the well-developed intergranular pores and fractures in the sandstone interlayers can provide sufficient storage space for the oil generated from source rocks, facilitating the primary migration. On the other hand, the favorable physical properties of the sandstone interlayers promote the conduction of heat energy, thereby sustaining the thermal cracking and oil generation reactions in the adjacent shale or mudstone layers.

According to theoretical calculations based on the primary migration model and geological conditions, a sand-to-mud ratio of 3:7 in the source rocks is most conducive to the primary migration. Moreover, the greater the porosity of the sandstone and the more balanced its distribution within the mudstone, the higher the efficiency of oil expulsion from source rocks.

The exploration practices in China's continental petroliferous basins have also confirmed that the main source rocks in these basins are not solely composed of dark-colored shales but rather consist of multiple layers of mudstone (or shale) and permeable layers. For example, the 3rd and 4th member of Paleogene Shahejie Formation in the Bohai Bay Basin, the 1st member of Lower Cretaceous Qingshankou Formation in the Songliao Basin, the Da'anzhai member of Jurassic Ziliujing Formation in the Sichuan Basin, the Triassic Yanchang Formation in the Ordos Basin, and the Permian Fengcheng Formation in the Junggar Basin are all examples of main source rocks composed of mudstone (or shale) and sandstone (or carbonate rock with similar physical properties). The ratio of sandstone to mudstone is mostly around 3:7.

Based on the concept of limited space oil generation, the Wuxi Geological Research Institute (Research Institute of Petroleum Exploration and Development, SINOPEC) developed a simulated experimental apparatus for reservoir pore-thermal-pressurized hydrocarbon generation and expulsion in 2009. Corresponding simulation experiments were conducted. The experimental results indicate that the process of oil generation from source rocks in a basin during continuous subsidence can be divided into three stages. Prior to maturity (% Ro) reaching 0.7%, source rocks are in a slow oil generation stage. Between 0.7 and 0.9% maturity (% Ro), source rocks enter a rapid oil generation stage, and by 0.9%, they have essentially reached the peak of oil generation. After exceeding 0.9% maturity (% Ro), the transformation of kerogen into hydrocarbons in source rocks is greatly inhibited by the internal fluid pressure. For source rocks with high organic richness and good organic matter types, abnormal pressure leads to a plateau in oil generation productivity. During the continuous subsidence stage of the basin, the fracturing of source rocks can result in a significant amount of trapped oil on the surface of the source rocks and in interconnected micro-fractures. Some expelled oil may enter thin sandstone layers interbedded with mudstone and potentially form shale oil reservoirs. Even if the source rocks in a basin (or sag) enter the peak oil generation stage, if there is no obvious tectonic uplift during the evolution of the basin (or sag) and no pressure difference exists between the oil generation zone and the main reservoirs, the generated oil is unlikely to migrate long distances to the reservoir for effective accumulation. Only when the pressure difference between the oil generation zone and the main reservoir reaches a critical pressure difference (approximately 4–5 MPa) during the uplift stage of the basin, can the oil generated by the source rocks effectively migrate long distances to the reservoir for effective accumulation.

Therefore, for an effective set of source rocks, it is not only important to have organic-rich mudstones but also crucial to have sandstone layers (or carbonate rocks) with good porosity-permeability conditions. The interlayering of sandstone and mudstone in a ratio of approximately 3:7 provides favorable spatial conditions for the oil generation and expulsion processes of source rocks.

10.5 Favorable Catalytic Conditions for Oil Generation in Source Rocks

During the tectonic evolution of a basin, volcanic eruptions of varying scales along major fault zones play a macroscopic role in controlling the temperature variations in different depressions within the basin. The differences in temperature fields lead to significant variations in the maturity thresholds of source rocks in the vicinity of deep-seated faults. Simultaneously, the abnormal thermal fluids triggered by deep-seated faults contain abundant minerals, metallic, and non-metallic elements. On one hand, these fluids can promote the proliferation of aquatic organisms, providing favorable conditions for the accumulation of organic matter in source rocks. On the other hand, they can actively catalyze the thermal cracking and oil generation processes of source rocks, facilitating the thermal cracking and oil generation of organic matter (Guan et al. 2004, 2008).

Both lacustrine source rocks and crude oils in China contain a considerable amount of metallic, rare metal, and rare radioactive metal elements. For example, in crude oils from petroliferous basins such as the Songliao, Bohai Bay, and Junggar, the content of metal elements such as Fe, Mn, Zn, and Cu is higher by one to two orders of magnitude compared to the average content in Chinese sedimentary layers. These metal and rare metal elements may originate from deep mantle sources and enter the source kitchen along nearby deep-seated faults, exerting a favorable catalytic effect on the thermal cracking and oil generation of organic matter. For instance, during the development period of the main source rocks of the Shahejie Formation in the Dongpu Depression, the basin was divided into two parts by a near-east-west uplift. The southern part was dominated by volcanic activity, while the northern part was characterized by deep-seated thermal brine upwelling. The brine upwelling in the northern lake area brought essential elements such as nitrogen, phosphorus, and potassium required for biological growth. The suitable climate, combined with abundant nutrients, promoted the rapid growth of halophilic microorganism such as algae. After their death, these microorganisms settled to the bottom and were protected from destruction by high salinity water, forming thick layers of source rocks. The interlayering of salt rocks with organic-rich source rocks actively catalyzed the thermal cracking and oil generation of organic matter, accelerating the thermal evolution of the 1st and 3rd members of Shahejie Formation and advancing the oil generation process of source rocks into an earlier "oil window."

References

Defan G et al (2004) Theoretical thinking of basin-forming hydrocarbon accumulation: from basin to oil and gas reservoir. Petroleum Industry Press, Beijing

Defan G, Xuhui X, Zhiming L et al (2008) Theoretical thinking of basin-forming hydrocarbon accumulation and hydrocarbon generation model in limited space. Oil Gas Geol 29(6):709–715

Chapter 11
Lacustrine Source Rock Petroleum Generation Simulation Instrument

Abstract To investigate the processes of petroleum generation in source rocks under actual geothermal and pressure conditions, as well as to examine the mechanisms of low-temperature catalytic petroleum generation, a novel simulation experimental apparatus was developed by the research team. This apparatus differs significantly from traditional in-situ thermal and pressure hydrocarbon generation simulation instruments. It includes three independent petroleum generation simulation units and a centralized product collection system. Each simulation unit operates independently, while data is uniformly collected and recorded. This chapter provides a comprehensive overview of the apparatus, detailing its main components, functions, technical specifications, primary applications, experimental procedures, and operational considerations. These insights demonstrate the apparatus's capability to advance research on petroleum generation mechanisms under realistic subsurface conditions.

Following the establishment of the organic origin theory for petroleum, research on source rocks has focused on the correlation between the chemical composition of petroleum and the origin of organic matter. This involves studying the depositional environment, formation and accumulation of organic matter, analysis and identification of organic matter in source rocks, conversion of organic matter to petroleum, organic geochemical indexes, types of organic matter, and evaluation of oil and gas potential in source rocks. The primary research methods involve analytical chemistry and organic chemistry laboratory experiments, which are used to provide qualitative evaluations of source rocks.

11.1 Overview

B. P. Tissot and D. H. Welte gave lectures in China in 1978, promoting the kerogen pyrolysis oil generation theory and introducing Rock-Eval pyrolysis instruments for evaluating oil shale. The same evaluation parameters used in oil shale sample

D. Guan et al., *Theory and Simulation Experimental Technology of Oil Generation From Nonmarine Source Rocks in China*, https://doi.org/10.1007/978-981-96-3145-2_11

pyrolysis evaluations were adopted for evaluating the oil and gas potential in source rocks (as shown in the table below). Since the 1980s, petroleum geologists have primarily used rock evaluation pyrolysis instruments to conduct oil generation simulation experiments, which use high-temperature pyrolysis to simulate the process of organic matter transforming into oil under high temperatures (Table 11.1).

Laboratory simulation experiments are conducted to replicate the geological conditions involved in the physical and chemical processes that generate petroleum from organic matter. In 2009, the Wuxi Research Institute of Petroleum Geology, Sinopec Petroleum Exploration and Production Research Institute developed a Formation Pore Pressure Oil Generation Simulation Experimental Instrument. This instrument mainly simulates oil generation processes and characteristics under temperature and pressure conditions when source rocks are filled with oil-generating depressions or troughs. Detailed information on this topic can be found in the book "Theory and Practice of Hydrocarbon Generation within Space-Limited Source Rocks" (Petroleum Industry Press 2014). This oil generation simulation instrument is noteworthy as the first significant simulation instrument in China and abroad, signifying the beginning of research into oil generation simulation experiments. Over nearly ten years of simulation experiments, medium-temperature pyrolysis of organic matter at 360 °C has been achieved. The entire simulation temperature ranges from 250 to 360 °C, yielding the S1 oil yield. Additionally, the samples were re-analyzed for high-temperature pyrolysis, in accordance with the requirements of the organic geochemistry lab specification, to obtain the S2 value and calculate the total oil potential of S1 + S2.

Since 2018, the authors and our research team have conducted comprehensive simulation oil generation experiments. These experiments involve analyzing the whole rock composition of source rocks, preparing samples based on the analysis data, and conducting simulation experiments under the real temperature and pressure conditions of subsurface geology. The primary objective is to explore the low-temperature catalytic oil generation process of source rocks. To achieve this, a new source rock oil generation simulation instrument suitable for simulation experimentation purposes and requirements was developed, building upon the original Formation Pore Pressure Oil Generation Simulation Experimental Instrument. The new instrument uses three sets of oil and gas generation simulation devices and a product collection system. Each set of devices can be controlled separately, with unified data collection and recording. Multiple samples can be tested simultaneously, enabling single-factor oil generation experiments and addressing the challenges of large workload and long duration in previous experiments.

In terms of simulation experimental methods, we have conducted basic research in various areas, including the analysis of the whole rock and whole component of source rock samples, the catalytic effects of metal and non-metal elements, low-temperature catalytic oil generation simulation experiments, and the acquisition of major parameters for S1 oil generation resources. Although many research efforts have not yet yielded systematic and complete results, they have laid the foundation

Table 11.1 Basic data for Chinese oil shale pyrolysis (after Zhao Longye 1991)

Sampling location	TOC/%	*T*max/°C	S1/(mg/g)	S2/(mg/g)	S3/(mg/g)	S1 + S2/(mg/g)	Petroleum generation index	Type index	HI mg/g (TOC)	OI mg/g (TOC)	CP/%	CP/TOC/%
Fushun (poor mine)	8.8	433	1.7	51.92	1.98	53.7	0.03	26.22	589	22	4.46	50.65
Huadian (the 8th layer)	21.1	432	1.98	76.72	6.72	70.7	0.03	11.42	363	11.42	6.53	30.96
Wangqing	12.85	431	6.8	79.2	5.6	86.98	0.08	14.16	616	14.16	7.14	55.6
Xiaoxia	30.45	429	14.46	185.12	4.16	159.98	0.07	44.58	481	44.58	16.57	43.03
Wucheng	10.33	419	10.12	75.68	3.28	85.72	0.12	23.05	731	23.05	7.11	68.07
Tongchuan	15.48	433	1.56	39.92	8.12	41.48	0.04	4.92	259	4.92	3.44	22.36
Urumqi	13.1	440	1.1	71.92	1.66	73.02	0.02	43.33	549	43.33	6.07	46.3
Yilan	4.29	437	0.66	17.13	1.3	17.04	0.04	9.54	400	9.54	1.48	34.52
Fengning	23.9	433	2.85	130.6	3.05	133.45	0.02	42.82	548	42.82	11.88	46.54
Maoming	4	434	0.25	2.7	0.7	2.95	0.03	3.86	96	3.86	8.24	5.1
Huating	39	426	4.35	152	3.4	156.35	0.03	44.71	389	44.71	12.9	33.27
Haishiwan	9.48	434	0.55	13.48	1.7	13.95	0.04	7.83	142	7.83	1.16	12.32
Tongchuan	11.9	427	0.3	27.95	2.7	28.75	0.03	10.35	234	10.35	2.39	20.05
Maoming (gold mine)	20.15		1.35	158.9	2.11	160.25	0.01	75.31		75.31		66
Maoming (Yangjiao)	22.29		2.14	135.24	5.49	137.38	0.02	24.63		24.63		51
Fushun (rich mine)	18.92		1.98	204.71	0.71	286.69	0.01	288.32		288.32		91

(continued)

Table 11.1 (continued)

Sampling location	TOC/%	*T*max/°C	S1/(mg/g)	S2/(mg/g)	S3/(mg/g)	S1 + S2/(mg/g)	Petroleum generation index	Type index	HI mg/g (TOC)	OI mg/g (TOC)	CP/%	CP/TOC/%
Fushun (poor mine)	9.28		0.82	87.67	5.39	83.49	0.01	224.8		224.8		
Fushun (1)	8.83	446	1.61	46.32	0.58	47.93	0.03	79.67	524.5	79.67		
Fushun (2)	16.52	445	1.7	103.72	0.57	105.48	0.02	181.96	627.8	181.93		
Fushun (3)	12.53	442	2.15	74.94	1.16	77.89	0.05	78.9	579.6	78.9		
Maoming (1)	14.96	431	1.53	64.39	1.39	65.42	0.02	46.31	521	46.31		
Maoming (2)	7.14	419	3.09	75.19	4.59	76.2	0.04	16.37	584.6	16.37		

for further in-depth work in this field. This book will briefly introduce some preliminary results and insights from low-temperature catalytic oil generation simulation experiments on source rocks.

11.2 The Main Components of Oil Generation Simulation Experiment Instrument

The instrument comprises several main components, including three sets of oil and gas generation simulation devices, a fluid injection system, a product collection and separation system, an automatic control and data acquisition system, and peripheral auxiliary equipment (Figs. 11.1, 11.2 and 11.3).

11.2.1 Oil Generation Simulation Experiment Instrument

The main equipment comprises three parts, namely an oil generation reaction kettle, an electric furnace, and a hydraulic control system.

The oil generation reaction kettle is a crucial component of the entire instrument (Figs. 11.4 and 11.5) and is utilized for placing the source rock samples during simulation experiments. This kettle is equipped with various components such as

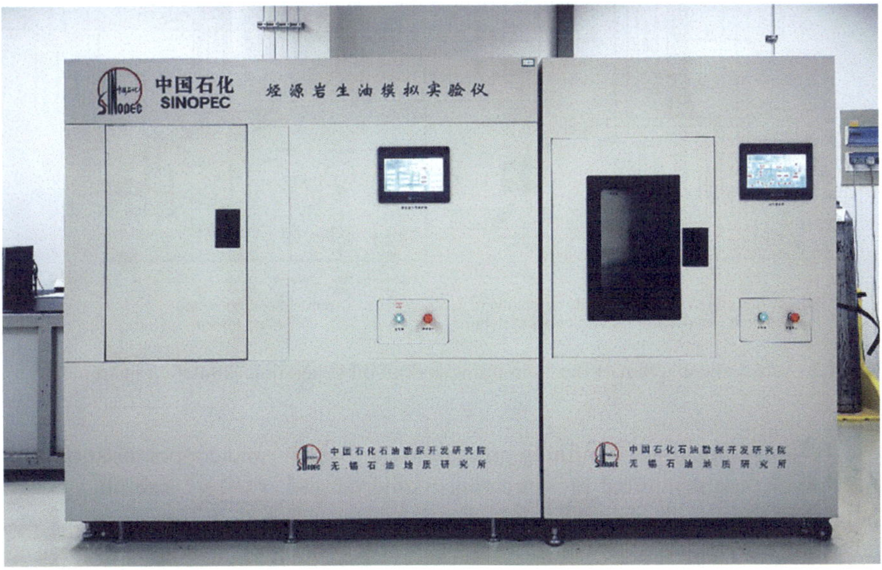

Fig. 11.1 Oil generation simulation experiment instrument

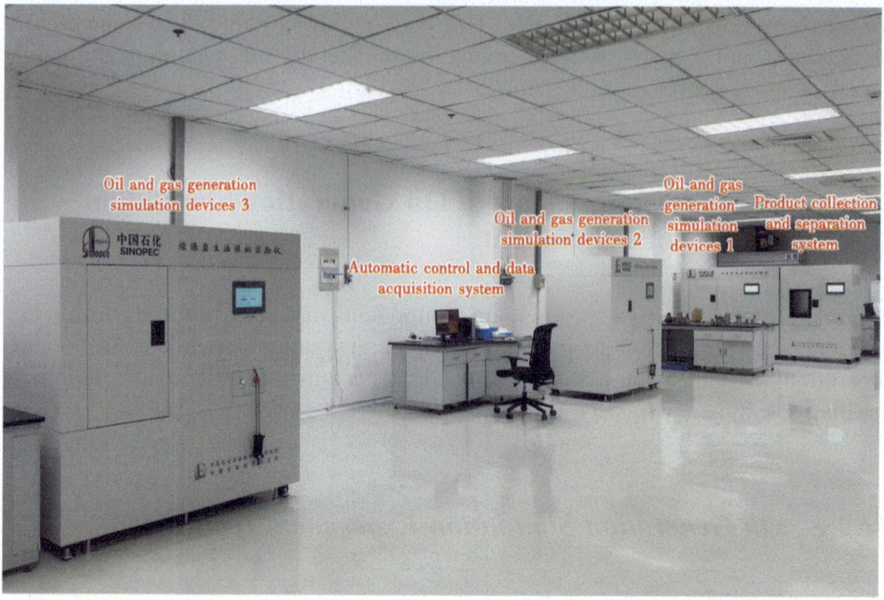

Fig. 11.2 Overall physical diagram of the oil generation simulation instrument

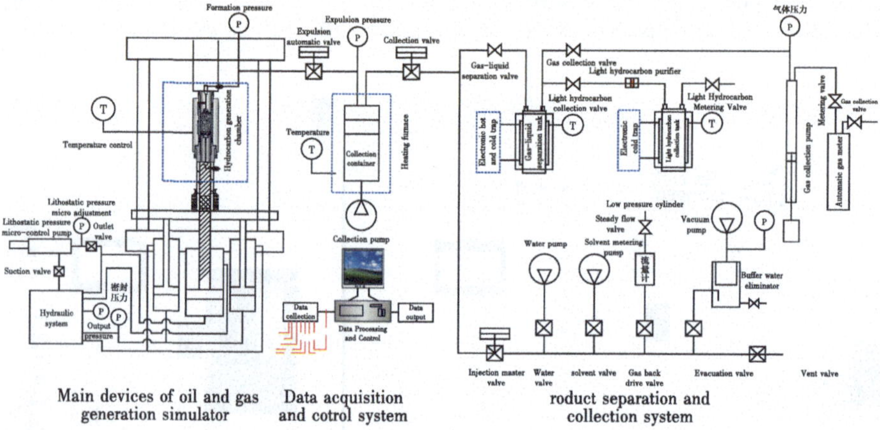

Fig. 11.3 Schematic diagram of the main framework of oil generation simulation instrument

base, bottom bolts, lower confining pressure rods, graphite and copper gaskets, fluid injection channels, internal pressure rings, sintered gasket sheets, sample sleeves, kettle body, fluid guide channels, upper confining pressure rods, top pressure sleeves, and small copper pressure rings. To seal the rock samples, components such as base, bottom bolts, graphite and copper gaskets 1, internal pressure ring 1, sample sleeve, internal pressure ring 2, copper gaskets 2, graphite gaskets 2, and top pressure sleeves are used; upper confining pressure rod, sintered gasket sheet, and lower confining pressure rod are utilized to apply confining pressure to the rock samples. The fluid guide channel expels out oil, gas, and water from the oil generation reaction kettle, and the fluid injection channel injects various fluids into the rock samples in the oil generation reaction kettle. The kettle body material is a high-temperature, high-strength alloy steel, with greater yield and tensile strength values than ordinary stainless steel, ensuring the safety of the simulation experiment process.

Main features: There is a fluid pipeline quick connector on the top of the kettle body for convenient loading and unloading of samples. The sample chamber has an inner diameter of 35 mm and can hold no less than 150 g of samples, meeting the requirements for rock and mineral analysis tests after the experiment (as illustrated in Fig. 11.6).

The high-temperature electric furnace (as depicted in Fig. 11.7) uses a forced hot air circulation working mode and is primarily made up of a stainless-steel lining, replaceable iron-chromium metal resistance wire, special high-density ceramic fibre insulation layer, interlayer heating circulation system, forced air high-temperature motor, high-temperature hot air circulation channel, and other components.

The main features of this equipment are as follows:

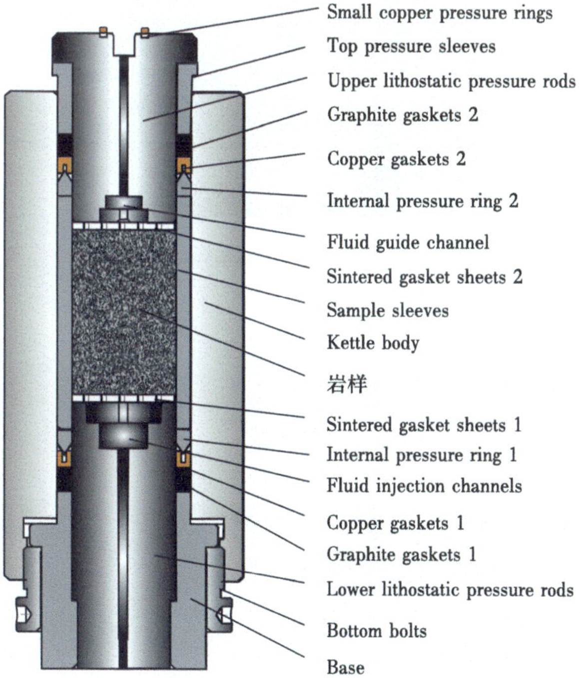

Fig. 11.4 Structural diagram of the oil generation reaction vessel

Fig. 11.5 Physical diagram of the oil generation reaction vessel

Fig. 11.6 Components of the oil generation reaction vessel

Fig. 11.7 Physical diagram of a chamber-type electric heating furnace

① The oil generation reaction kettle can maintain a constant temperature at the top and bottom, with a fast heating rate and adjustable heating speed.

② Utilizing the Programmable Logic Controller (PLC), it can accomplish 16-stage long-time continuous temperature control and constant temperature (Fig. 11.8), with automatic heating and steady temperature, process control (PID) parameter self-tuning, manual/automatic non-interfering switching, and over-temperature alarm functions.

③ It has high temperature control accuracy, with a temperature uniformity of ±3 °C, accuracy of ±2 °C, and an upper temperature limit of 350 °C.

④ The hydraulic control system mainly comprises a hydraulic system, cylinders, and micro-flow output control pumps. It adopts a coaxial three-cylinder technology (as depicted in Fig. 11.9), which automatically controls the pressure rise and fall, providing a maximum pressure of 150 MPa, equivalent to the overburden confining pressure experienced by rocks at a depth of 5000 m.

(1) The hydraulic system is responsible for controlling the hydraulic transmission of the oil generation reaction kettle's hydraulic drive device and managing the pressurization and depressurization of the three cylinders. The primary cylinder applies confining pressure to the rock samples in the oil generation reaction kettle, while the other two cylinders are utilized to seal the reaction kettle.

(2) The micro-flow output control pump is used for more precise pressure control inside the oil generation reaction kettle.

(3) The hydraulic device is concentrated at the bottom of the reaction kettle, making it convenient to load and unload samples, reducing the instrument's height, and enhancing its appearance.

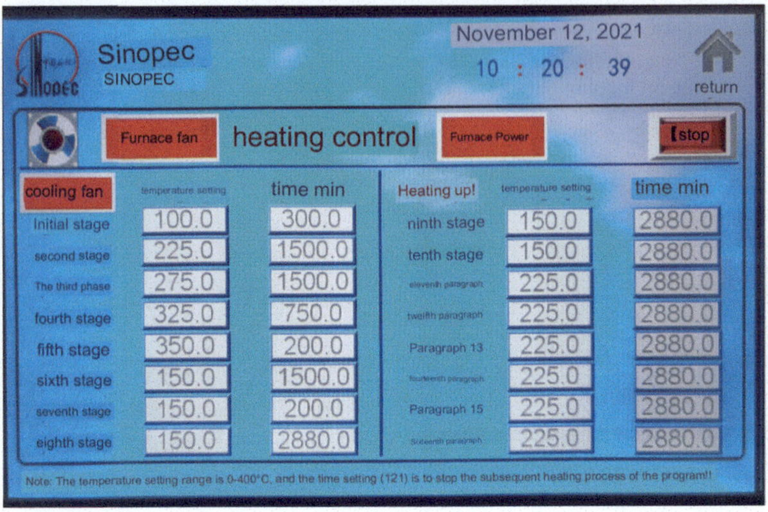

Fig. 11.8 Interface diagram of the high-temperature box-type electric furnace heating control

The equipment's main features are as follows:

① It can achieve continuous multi-stage automatic program pressure control, simulating the process of stratum uplift and subsidence.

② It ensures that the sample in the reaction kettle is subjected to overburden confining pressure that is stable and direct for an extended period.

③ The top plate device comprises a top plate and cylinders, which push the top plate upward to seal the oil generation reaction kettle. The top plate device adopted can increase the heat dissipation area, reduce the cylinder temperature, effectively slow down the pressure fluctuations and aging of the cylinder's internal seals caused by overheating and expansion of oil in the cylinder during long-term heating in the simulation experiment, and protect the cylinder's service life.

④ The entire hydraulic transmission system has automatic, manual, and remote control functions (Fig. 11.10). The Programmable Logic Controller (PLC) controls the hydraulic transmission system, and through parameter settings, it can automatically control the lifting of the cylinders (Fig. 11.11). The device is easy to operate. Since all three cylinders are placed at the bottom of the oil generation simulation device, the environmental requirements are much lower, making it suitable for laboratories with lower heights.

11.2.2 Fluid Injection System

The system is comprised of various components, including high-pressure constant metering pumps, water storage tanks, solvent tanks, gas flow meters, high-pressure piston containers, organic solvent cleaning devices, vacuum devices, and various

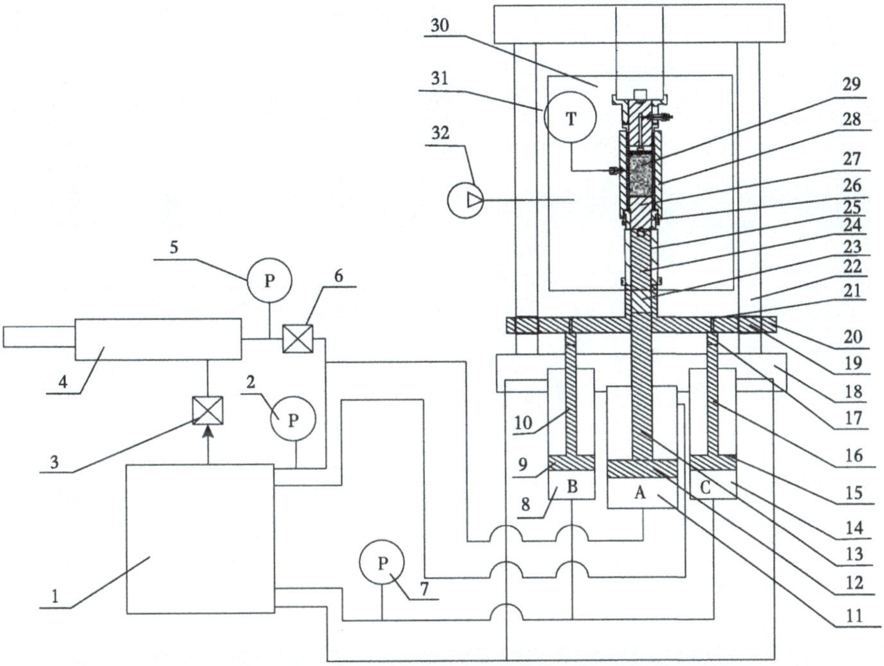

Fig. 11.9 Hydraulic control system flowcharts. 1-Hydraulic system; 2-output pressure controller A; 3-liquid suction valve; 4-pressure micro-control pump; 5-micro-control pressure controller; 6-liquid expulsion valve; 7-output pressure controller B; 8-cylinder B; 9-piston B; 10-cylinder top pillar B; 11-cylinder A; 12-piston A; 13-cylinder top pillar A; 14-cylinder C; 15-piston C; 16-cylinder top pillar C; 17-insulation block; 18-gantry base plate; 19-sliding sleeve; 20-top plate; 21-bolt; 22-fixed bracket; 23-insulation ceramic block; 24-top rod; 25-external pressure sleeve; 26-sealing pressure ring; 27-central pillar; 28-oil generation reaction vessel; 29-rock sample; 30-heating furnace; 31-temperature controller; 32-air cooling pump

pneumatic valves, etc. (Fig. 11.12). It is primarily used for equipment leakage testing, vacuum pumping, injecting high-pressure fluids into the oil generation reaction kettle, sustaining the fluid pressure in the reaction kettle, and online cleaning of created oil and gas products.

The equipment's main features are:

① The injection of different media fluids is regulated by high-pressure constant metering pumps and pneumatic valves, with a maximum injection pressure of 60 MPa for high-pressure fluids.

② The fluid injection system control has various functions, including real-time display of the system temperature and pressure, overpressure alarm, recording, and setting parameters, providing complete visualization control of the operation (Fig. 11.13).

③ During the simulation experiment, high-pressure fluids that contain metals and rare metal elements through artificial proportions can be injected to examine

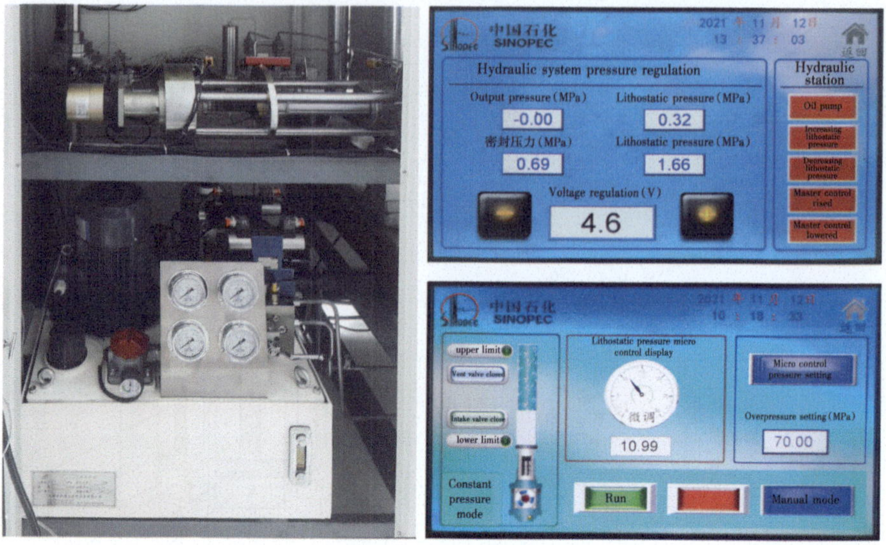

Fig. 11.10 Physical diagram of the hydraulic control system

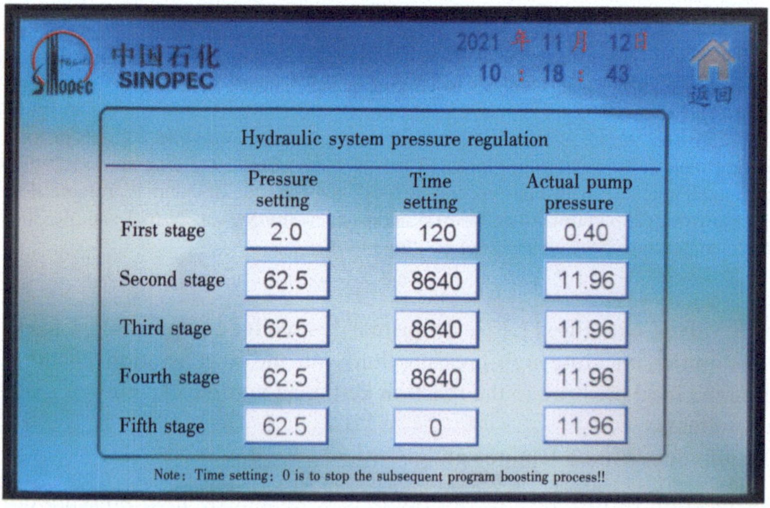

Fig. 11.11 Static rock micro-controlled pressure setting

the effects of single factors such as temperature (T), pressure (P), metal elements, formation water mineralization, and source rock composition on the oil generation process of source rocks.

Fig. 11.12 Physical diagram of the fluid injection system

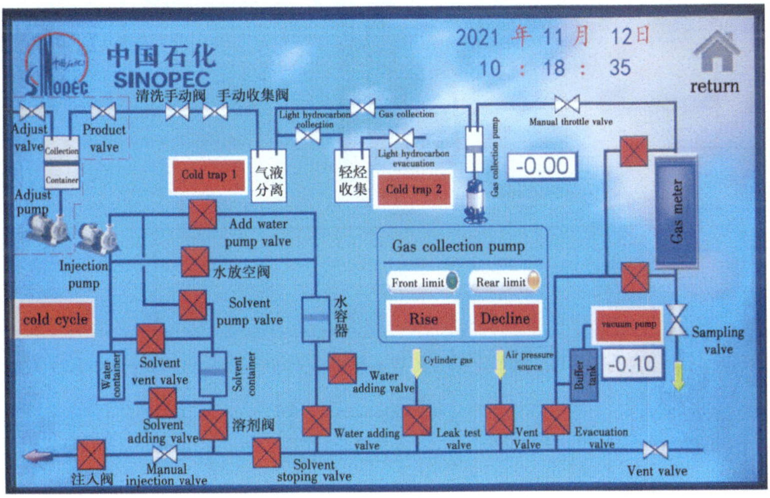

Fig. 11.13 Flowchart of the fluid injection system

11.2.3 Product Collection and Separation System

This mainly includes high-pressure pneumatic valves, high-pressure collection containers, balanced high-pressure pumps, gas-liquid separation devices, cold traps, hot and cold traps, automatic gas collection devices, and connecting pipelines, etc. (Figs. 11.13 and 11.14).

(1) The equipment utilizes a high-pressure pneumatic valve to regulate the expulsion of oil and gas from the pyrolysis reaction kettle, with a maximum controllable pressure of 150 MPa. The effective working pressure can be adjusted between atmospheric pressure and 120 MPa, with both manual and automatic functions. The high-pressure pneumatic valve can expel the generated oil and gas during

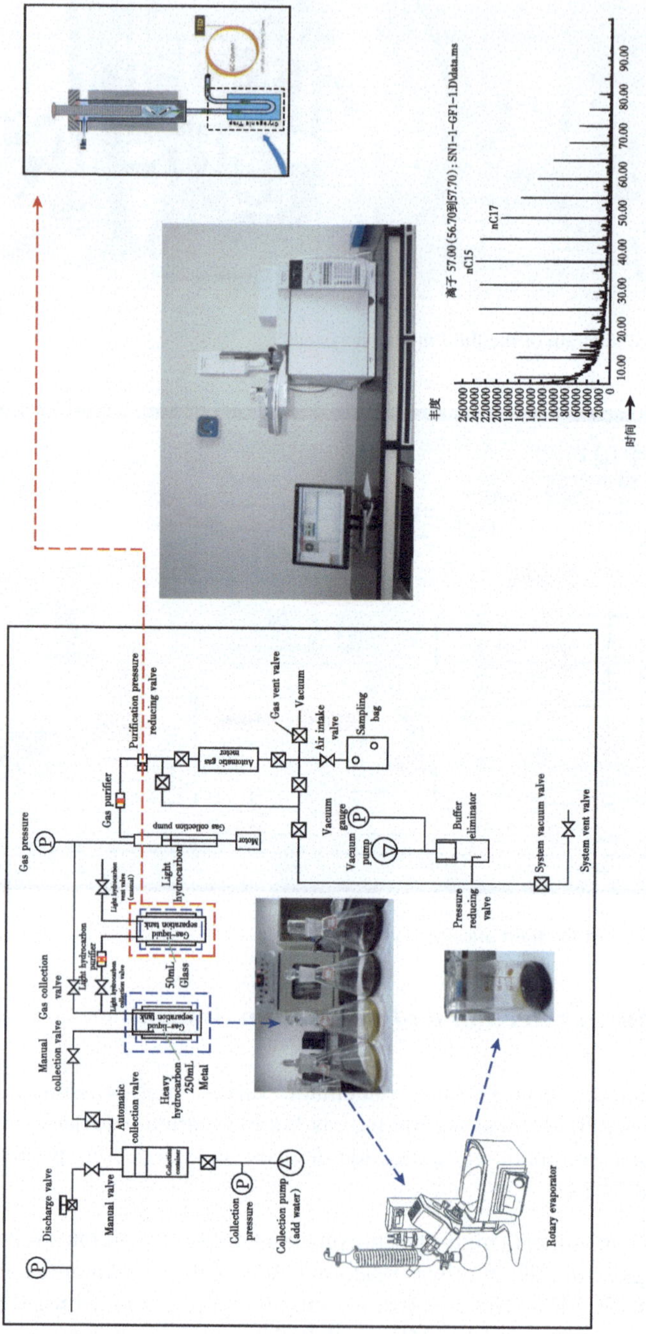

Fig. 11.14 Flowchart of the product separation and collection system

the process of oil and gas generation based on the set fluid pressure, enter the high-pressure collection container, and normalize the pressure inside the pyrolysis reaction kettle through the balanced high-pressure pump. Within a specific fluid pressure range, it can conduct oil and gas simulation experiments. Additionally, the valve serves as a safety valve. Its control signal is connected to an alarm, and when the pressure of the reaction system exceeds the maximum pressure point that the apparatus can withstand, it will emit an alarm signal and open the valve to release fluid and decrease pressure.

(2) The gas–liquid separation apparatus is utilized for the segregation of gas, heavy hydrocarbons, and light hydrocarbons. This device encompasses both heavy hydrocarbon separators and light hydrocarbon separators. The heavy hydrocarbon separator obtains and segregates fluids from the collection container, controls the temperature of the interior space through hot and cold traps, and collects heavy oil and liquid hydrocarbons within the separator. The temperature is elevated to segregate light hydrocarbons and gas. Meanwhile, the light hydrocarbon separator segregates light hydrocarbons within the separator via a cold trap and separates gas.

(3) The gas automatic collection unit makes use of a gas storage valve and gas collection solenoid valve to achieve automated, quantitative gas collection (Fig. 11.15). In tandem with the expulsion pneumatic valve, the product collection and separation system is capable of automatically collecting expelled fluids. The gas and liquid product metering device necessitates high precision to guarantee the accurate measurement of gaseous and liquid products, particularly light hydrocarbons.

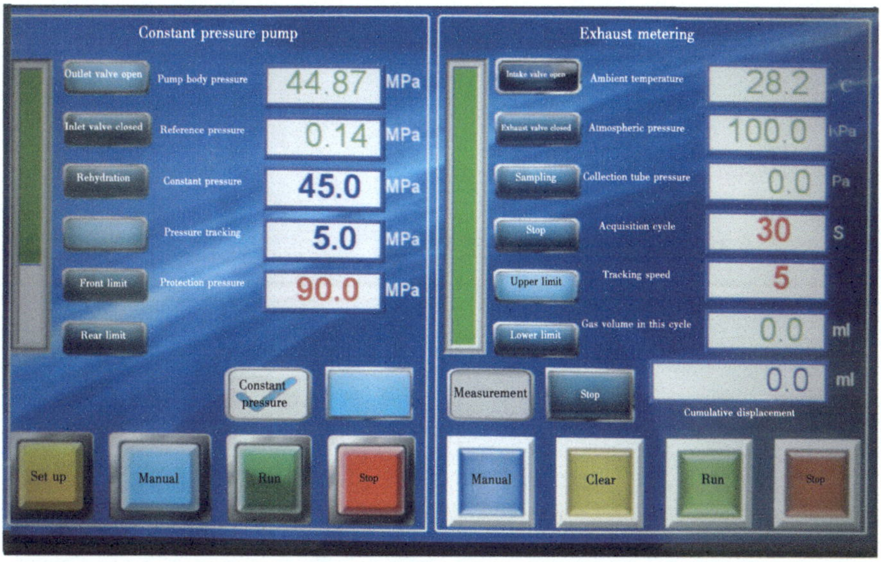

Fig. 11.15 Interface diagram for constant pressure control and gas automatic collection control

11.2.4 Automatic Control and Data Acquisition System

This system consists primarily of a PLC controller, touch screen, digital input module, temperature transmitter, high-temperature pressure sensor, pneumatic valve, solenoid valve, alarm devices, electrical components, and computer (as depicted in Fig. 11.16).

(1) The automatic control system incorporates servo technology, pneumatic technology, and programmable electrical control technology. Proportional servo control replaces conventional pneumatic switch control, enhancing control precision. With the touch screen's robust control functions and the PLC controller, the user-friendly human–machine interface of the touch screen is leveraged, making the system control more flexible, increasing the automation level, and improving system reliability. The number of switches on the operation panel is reduced, simplifying complex electrical wiring and enhancing user-friendliness. Operators can directly command the system's operation through the touch screen's buttons, streamlining operational complexity, and the running curve provides a more intuitive grasp of the system's operating status. The system features real-time display of controlled system parameters, curve display, control, alarm, recording, and parameter setting capabilities, ensuring control system safety and operational process visualization.

(2) The modular structure of the Programmable Logic Controller (PLC) is based on the instrument's functional requirements, dividing system functions into several relatively independent modules, which are in turn subdivided into smaller parts. In order to prevent errors in system program execution due to misoperation,

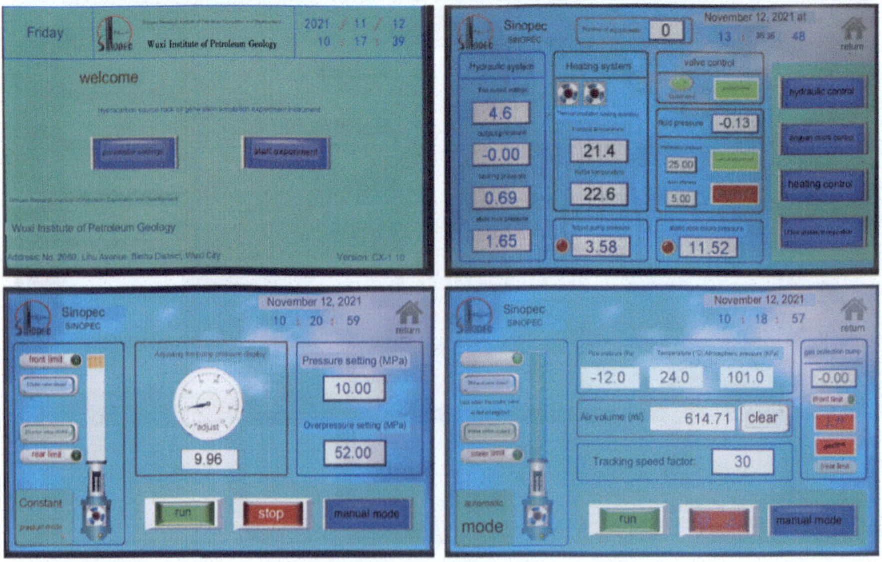

Fig. 11.16 PLC controller and touch screen control display interface diagram

execution condition judgments are incorporated into the program to achieve interlocking and prevent interference between subprograms.

The PLC acquires input signals from multiple sources, such as component detection results, cylinder and button switch signals, and analog signals representing temperature and pressure values. The analog signals enable specific temperature and pressure values to be monitored, while the button switch signals indicate various actions taken by the operator, and magnetic switch signals reflect the exact position of the piston rod. The system output information includes alarm buzzer signals, operation indicator lights, and valve signals, where the latter provide movement for components like pneumatic valves, solenoid valves, and cylinders, with the current operating status displayed via the operation indicator lights. In the event of operational errors, that is, when temperatures and pressures fail to meet system requirements, the alarm buzzer will sound an alert, ensuring the safe functioning of the system.

(3) The touch screen integrates the Windows CE operating system and configuration software, which enables selection of appropriate communication drivers, definition of configuration variables, establishment of logical connections, creation of process screens, and specification of system properties. This facilitates the automation and visualization of complex mechanical control, displaying screens via graphics and allowing human–machine interaction. Through RJ-45, USB, RS-422, and RS-232 interfaces, the touch screen exchanges data with the PLC and computer.
(4) Temperature control is achieved using temperature transmitters, which are critical for ensuring the accuracy of heating sample temperature and precision of temperature control in oil generation simulation experiments. These parameters are closely related to the temperature measurement accuracy of the temperature transmitter and the temperature control capability of the temperature controller. The temperature-sensitive element, which is the core component of the temperature transmitter, has an accuracy level of $\pm(0.15 + 0.2\%)$, meeting the national A-level standard with an error range of ±2 °C.
(5) Confining pressure and formation fluid pressure control is achieved using specially designed high-temperature pressure sensors. These sensors accommodate pressure in the entire system, thereby closely simulating confining pressure, formation fluid pressure and experimental conditions that replicate actual geological conditions. The pressure control range is 0–150 MPa, with an accuracy within ±0.2 MPa.
(6) The primary components of the data acquisition system are the touch screen, computer, printer and auxiliary equipment. The touch screen facilitates dynamic exchange and communication between the configuration software and the computer, while the computer controls and collects various types of data such as confining pressure, fluid pressure, simulation time, simulation temperature, and sample number (Fig. 11.17).

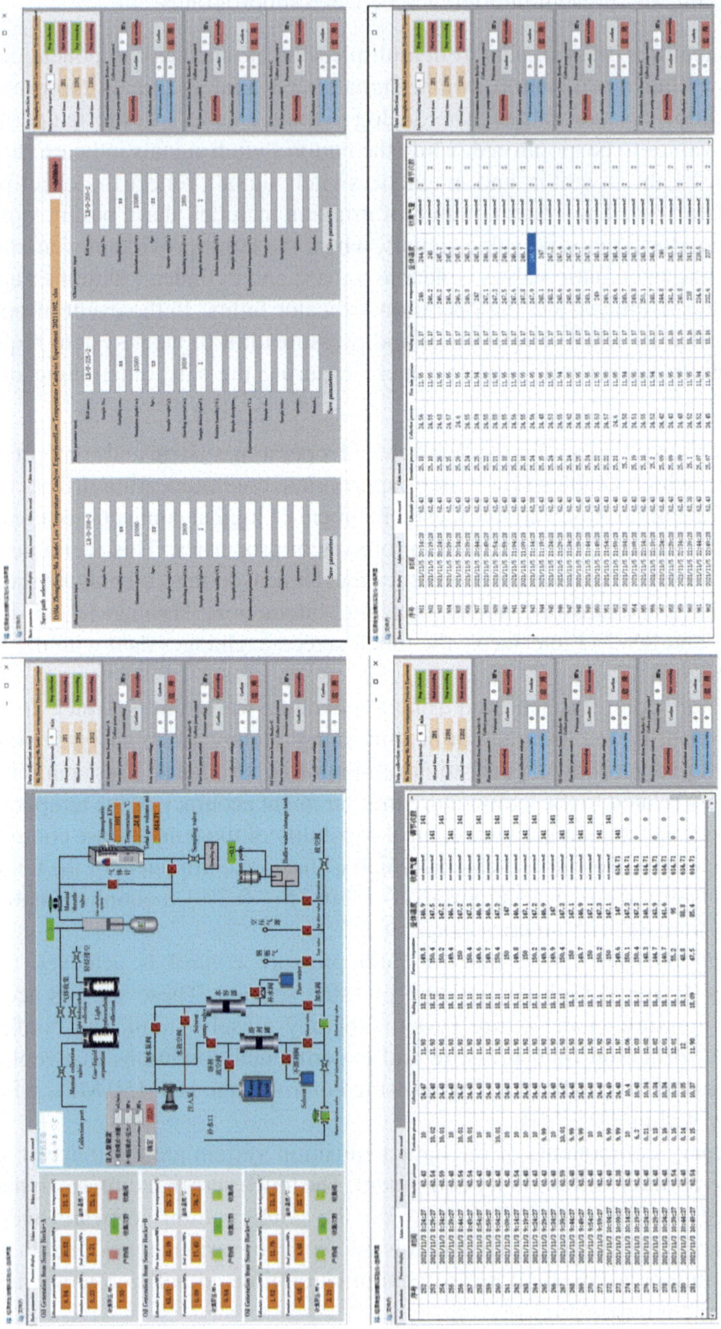

Fig. 11.17 Interface diagram of the data acquisition system

11.2.5 Peripheral Auxiliary Equipment

Peripheral auxiliary equipment mainly includes gantry frames, high-precision vacuum pumps, high-pressure buffer containers, various types of high-pressure automatic control valves, and various high-pressure connection pipelines.

11.3 Source Rock Pyrolysis Simulation Instrument Function and Technical Indicators

11.3.1 Main Functions

(1) Effectiveness: The equipment enables the simulation of source rock pyrolysis experiments under identical geological conditions, including temperature, pressure, sample composition, and rare metal element composition.
(2) Automation: The experimental process parameters, including pressure, temperature, and reaction rate, can be automatically controlled and adjusted, with real-time data recording. Two key components include the automatic control of all valves and automatic control of the oil generation and fluid product collection process.
(3) Safety: The experimental process is performed under high-temperature and high-pressure conditions to ensure process safety and stability.
(4) High-temperature, high-pressure, and corrosion-resistant technology for oil production reactors and pipelines: The oil generation reactions require high-temperature and high-pressure environments, which generate highly corrosive oil and gas products. Thus, the reaction vessel and its connecting pipelines must have high-temperature, high-pressure, and anti-corrosion properties to ensure that the reaction kettle maintains its seal and does not undergo any undesirable chemical reactions unrelated to oil generation.
(5) Uniform temperature control technology for the oil generation reactors: In a simulation experiment system, temperature is critical to the entire process. It is required to accurately detect the actual reaction temperature of the reaction kettle, ensuring that the reaction kettle temperature error is within ±2 °C.
(6) High-pressure sealing technology: Special design and materials are used to ensure the stability of the high-pressure system during the experimental process. The sealing ability of various seals, valves, and connection systems under high-temperature and high-pressure conditions is key to ensuring the system's stability.

(7) Hydraulic three-cylinder control technology: The main oil cylinder applies lithostatic al pressure on the rock samples in the oil generation reactor, while the other two cylinders ensure the reactor's sealing and automatically control the pressure rise and fall through a programmed process.
(8) PLC programming control technology: The entire control system can display real-time values, curves, control, alarms, recording, and parameter settings, ensuring the safety of the control system and the visualization of the operation process.
(9) Touch screen configuration control technology: The configuration variables are defined by selecting the appropriate communication driver program and implementing logical connections. This automates and visualizes complex mechanical control, displaying graphics on the screen and enabling human–machine interaction.
(10) Automatic control technology for experimental temperature and pressure: An automatic heating program controls the temperature, ensuring that it rises and falls according to the program without overshooting. Pressure control occurs by using a pressure pump to expel liquid automatically.

11.3.2 Major Technical Parameters

(1) Simulated temperature: The upper limit of the temperature is 350 °C, and it is typically used in the range of 60–250 °C. It has multi-stage long-term continuous heating, constant temperature, and cooling capabilities. The temperature control accuracy is ±2 °C, and the actual temperature shift should not exceed ±3 °C.
(2) Pressure: The fluid pressure can reach up to 45–120 MPa, while lithostatic al pressurecan go up to 100–150 MPa (simulating a maximum burial depth of 4500–5000 m). The device enables continuous multi-stage automatic pressure program control, simulating the process of strata uplift and subsidence.
(3) Sample quantity: The sample quantity is sufficient for analytical requirements, such as soluble organic matter extraction, scanning electron microscopy (sample weight 20 g), whole rock X-ray diffraction (sample weight 20 g), clay mineral X-ray diffraction (>50 g), whole rock reflectance, and metallic elements. The total quantity should not be less than 150 g, with a sample chamber inner diameter of 35 mm.
(4) Oil and gas generation simulation device: The device comprises three sets of reaction kettle bodies and gantry mechanisms, three sets of hydraulic workstations, and hydraulic transmission systems capable of automatic, manual, and remote control functions. Multiple sample oil generation simulation experiments can be carried out simultaneously.
(5) Sample chamber: The inner wall of the sample chamber is thickened, with an inner diameter of 35 mm and a loading capacity of not less than 150 g.

(6) Built-in small sample chamber: The built-in small sample chamber employs face sealing, with an outer diameter of 35 mm and a hollow inner diameter of 15–20 mm.
(7) Injection system: The device features one injection system, which controls fluid injection separately for the three sets of oil and gas generation simulation devices.
(8) Data acquisition and control system: The device allows for control and data recording of oil and gas generation simulation, automatic gas flow rate calculation, electronic condensation control (with a minimum temperature of −20 °C), quantitative weighing of liquid products, and chromatographic quantification of gas products.

11.4 Comparison of Source Rock Pyrolysis Simulation Instument and Formation Pore Pressure Thermal Simulation Instrument

For a long time, the Wuxi Petroleum Geological Research Institute has been actively studying source rock pyrolysis simulation experiments. Since 2005, their research has been guided by the basin-forming, hydrocarbon-forming, and reservoir-forming theory, using petroleum geochemistry disciplinary characteristics and research methods. They have independently developed the DK-I, DK-II, and DK-III types of Formation Pore Pressure Oil Generation Simulation Experimental Instrument. These instruments were designed to simulate the process and characteristics of organic matter pyrolysis and oil generation under the temperature and pressure conditions present when source rock fills the oil-generating depression (or sag). Compared to previous Formation Pore Pressure Oil Generation Simulation Experimental Instrument, the newly developed source rock pyrolysis simulation instrument has been comprehensively upgraded in terms of hydrocarbon generation simulation device, simulation method, and pressure application method. As a result, past issues such as large experimental workload and long-time consumption have been resolved. A detailed comparison of instrument performance and functions can be found in Table 11.2.

11.5 Main Applications of the Instrument

The hydrocarbon source rock pyrolysis simulation instrument is capable of reproducing simulation experiments to study the oil generation processes and characteristics in various oil and gas-bearing basins. It can simulate the oil generation process

Table 11.2 Performance and functionality comparison of the oil generation simulation instrument and formation pore pressure oil generation simulation experimental instrument (after Defan et al. 2008, 2011; Defan 2018; Zhiming et al. 2011; Zhongliang et al. 2012)

Experimental parameters	DK-I formation pore pressure oil generation simulation experimental instrument	DK-II formation pore pressure oil generation simulation experimental instrument	DK-III formation pore pressure oil generation simulation experimental instrument	Source rock oil generation simulation instrument
Experimental sample	Organic matter type, abundance, TOC, maturity level Ro	Organic matter type, abundance, TOC, maturity level Ro	Organic matter type, abundance, TOC, maturity level Ro	Organic matter type, abundance TOC, maturity level Ro, mineral composition, clay minerals, sedimentary water salinity, and mineralization degree
Oil and gas generation simulation device	Oil generation simulation for one sample	Oil generation simulation for one sample	Oil generation simulation for one sample	Simultaneous oil generation simulation for three samples (single-factor) experiment
Maximum simulation temperature/°C	550	550	550	350 (16-segment PLC long-term continuous temperature control and constant temperature)
Temperature range used/°C	250–550	250–550	250–550	60–250
Simulation mode	Constant temperature pyrolysis	Constant temperature pyrolysis	Constant temperature pyrolysis	Constant temperature pyrolysis, kinetic mode (variable heating rate)
Maximum static rock pressure/MPa	100	150	200	150
Maximum fluid pressure/MPa	60	90	120	60
Pressurization method	Static	Static	Static	Pressure applied dynamically by overlying static rock
Reaction space	Close to formation pore	Close to formation pore	Close to formation pore	Close to formation pore

(continued)

Table 11.2 (continued)

Experimental parameters	DK-I formation pore pressure oil generation simulation experimental instrument	DK-II formation pore pressure oil generation simulation experimental instrument	DK-III formation pore pressure oil generation simulation experimental instrument	Source rock oil generation simulation instrument
Hydraulic control	Control double-cylinder bidirectional	Control double-cylinder bidirectional	Control double-cylinder bidirectional	Three-cylinder in the same-direction
Formation fluid	Distilled water	Distilled water	Distilled water, formation water	Formation water containing different metallic elements
Oil and gas collection	Light hydrocarbon loss	Light hydrocarbon loss	Light hydrocarbon loss	Full component collection (gas, light hydrocarbon, heavy hydrocarbon)
Automation level	Manual, standalone	Semi-automatic, standalone	Button-controlled self-control	Standalone touch screen PLC self-control

of source rocks based on actual petroleum geological conditions (composition of reacting materials) and environmental factors (physical and chemical conditions of the reaction). The following experiments can be conducted using this instrument:

(1) Simulation experiments to examine the impact of clay mineral diagenesis transformation on organic matter oil generation;
(2) Simulation experiments to investigate the influence of transition metal group, radioactive metal elements, and other factors on organic matter oil generation; and
(3) Hydrocarbon generation kinetics simulation experiments.

11.6 Experimental Operation and Precautions

The source rock pyrolysis simulation instrument comprises of three oil and gas generation simulation devices (mainframes) and one product collection system. The product collection system is fitted with quick connectors, which allow it to be connected sequentially to each oil and gas generation simulation device (mainframe) during the experiment. It is important to take note of the following operation essentials during the experimental process.

11.6.1 Sample Loading

(1) Verify that all components of the pyrolysis reaction kettle are present: kettle body, base, bottom bolts, lower confining pressure rod, upper confining pressure rod, pressure-applying sleeve, two inner pressure rings, two copper gaskets, two sintered gaskets, two graphite gaskets, and a small copper pressure ring for the upper confining pressure rod.
(2) Prepare a backup sample by weighing an appropriate amount, preferably 80–150 g.
(3) Assemble the base, bottom bolts, and lower confining pressure rod sequentially. Follow this by adding graphite gasket 1, copper gasket 1, and inner pressure ring 1 onto the lower confining pressure rod in sequence and place the kettle body on top.
(4) Place sintered gasket 1 at one end of the sample sleeve, then insert the sample sleeve into the kettle body, and gradually add the sample in batches. Once all the samples are loaded, add sintered gasket 2 on top of the sleeve.
(5) Add inner pressure ring 2, copper gasket 2, and graphite gasket 2 on top of the sintered gasket 2. Then insert the upper confining pressure rod into the top of the sample sleeve and place the pressure-applying sleeve on top.
(6) Put the assembled reaction kettle in the heating furnace. Ensure that the side hole of the reaction kettle corresponds to the temperature sensor in the heating furnace.

11.6.2 Pres Ading

(1) To begin the experiment, switch on the hydraulic machine remote control, turn the red button clockwise and press the green "start" button. To switch off, press the red button again. It is important to ensure that the remote control used for each mainframe corresponds correctly and not mixed up.
(2) On the instrument mainframe screen, select "Start Experiment" —"Hydraulic Control"—"Voltage Adjustment" and adjust the voltage level to a suitable range of 6–7 V.
(3) Apply sealing pressure by pressing "Main Control Up" and controlling it to 20 MPa.
(4) Click on "Voltage Adjustment", lower the level to 3–5 V, and then press "Increasing Confining Pressure" for the actual confining pressure. During the initial stage of the experiment, the confining pressure can be lower than the experimentally set value, but it is necessary to observe the bottom of the reaction kettle and ensure that the pressure rod for applying confining pressure is not suspended.
(5) Select "Start Experiment"—"Confining Pressure Micro-control"—"Micro-control Pressure Setting" on the instrument mainframe screen. Choose the

confining pressure value in segments to maintain the set value and click "Run" after selection.

(6) Press the red button on the hydraulic machine remote control to turn off the pressure system once the procedure is over.

11.6.3 Leakage Test, Vacuum, and Water Injection

(1) On the instrument mainframe screen, select "Fluid Pressure Adjustment"—"Pressure Setting" and adjust it to 2 MPa. Start by pushing the adjustment pump piston to the top and then stopping its operation.

(2) For vacuuming, turn on "Product Valve"—"Manual Adjustment" on the mainframe screen and open "Injection Valve"—"Manual Injection Valve"—"Solvent Block Valve"—"Vacuum Valve" on the collection system screen. Click "Vacuum Pump" to initiate the process, which will usually take 3–5 min. Afterwards, close "Vacuum Pump" and "Vacuum Valve".

(3) To conduct a gas leakage testing, activate the "Leakage Testing Valve", open the high-pressure gas cylinder and the adjacent "High-pressure Gas Valve". Once the gas enters the high-pressure kettle system, shut off the "Leakage Testing Valve" and observe whether the "Fluid Pressure" displayed on the instrument mainframe screen drops. If it doesn't drop, it indicates a good seal. Manually open the "Vent Valve", release the nitrogen, then close the "Vent Valve" and repeat the vacuuming steps. Repeat this process 2–3 times, vacuuming the system for the last attempt.

(4) Next, perform a water leakage testing by turning on "Product Valve"—"Manual Adjustment" on the mainframe screen and opening "Injection Valve"—"Manual Injection Valve"—"Solvent Block Valve"—"Water Addition Valve"—"Water Pump Valve" on the collection system screen. Simultaneously, turn on the injection pump, set the "Constant Pressure Mode", and test for leakage at 30–50 MPa, not exceeding the previously set confining pressure value. Click "Run". When the "Fluid Pressure" reaches the injected value displayed on the mainframe screen, close the injection pump and the "Product Valve"—"Manual Adjustment" on the mainframe screen. Observe whether the "Fluid Pressure" drops for at least 30 min to complete the leakage testing.

(5) Once confirming that the system doesn't leak, turn on the "Product Valve"—"Manual Adjustment" on the mainframe screen and the "Injection Valve"—"Manual Injection Valve"—"Solvent Block Valve" and "Vent Valve" on the collection system screen. Adjust the "Fluid Pressure" to 2–3 MPa, and close the opened valves mentioned above.

11.6.4 Temperature Increment

(1) Set the experimental temperature and heating time by clicking "Heating Control" on the instrument mainframe screen.
(2) Before closing the furnace door of the heating furnace, ensure that the temperature sensor is inserted into the opening on the side of the high-pressure kettle.
(3) Start heating by selecting "Heating Control"—"Furnace Power"—"Heating" on the mainframe screen.

11.6.5 Pressure Setting

(1) On the instrument mainframe screen, input "Static Water Pressure" followed by "Back Pressure Difference" in the correct order.
(2) Click on "Fluid Pressure Adjustment"—"Pressure Setting" on the mainframe screen, and set a value slightly lower than the "Static Water Pressure", with a difference of less than 0.5 MPa.
(3) Apply sealing pressure by pressing "Main Control Up", typically set at around 20 MPa.

The instrument will then proceed with the heating and oil generation process until the experiment is complete.

11.6.6 Gas Collection

(1) Install and connect the gas–liquid separator tank and activate "Cold Circulation"—"Cold Trap 1" on the collection system screen to cool for 10 min.
(2) Insert the "gas sampling standby" into the sampling port, then move the "Gas Collection Pump" back to the "rear limit position" by clicking "Down".
(3) Start the "Vacuum Pump" and the upper and lower buttons on the left side of the "Gas Meter", including "Manual Throttle Valve", "Gas Collection Valve", "Manual Collection Valve", and "Cleaning Manual Valve" to vacuum the collection system for 3–5 min. Turn off the "Vacuum Pump" and switch off the two above-mentioned switches, along with the "Manual Throttle Valve".
(4) Reset the gas volume to "zero" by clicking on the "Gas Meter" and then click "Run".
(5) Open the "Product Valve" and gradually turn on the "Cleaning Manual Valve" to release the oil, gas, and water in the high-pressure kettle at a slow pace. Make sure that the pressure inside the gas collection pump is between 0.2 and 0.3 MPa. If the pressure is too high, immediately shut off the "Cleaning Manual Valve".

(6) Begin measuring the gas by gradually turning on the "Manual Throttle Valve". During the measurement, once the gas metering pump drops to the "lower limit position" and you hear the "intake valve" sound, stop turning the "Manual Throttle Valve" as the red arrow starts to move upward. The gas flows through the sampling valve into the "gas bag". Once it reaches the "upper limit position" and the "exhaust valve" sound is heard, and the red arrow begins moving downward, turn the "Manual Throttle Valve" back on to measure the gas.
(7) Repeat this process until you have completed the gas measurement. Finally, move the gas collection pump to the "front limit position".
(8) If only a small amount of gas remains, click "Manual Mode"—"Exhaust Valve"—"Up" to collect the gas.

11.6.7 Unloading Samples

(1) Use the remote control to activate the hydraulic machine. Turn the red button to the right and press the green start button.
(2) Access the instrument host screen and select "Start Experiment"—"Rock Static Micro-control"—"Micro-control Pressure Setting". Then disable the "Run" option.
(3) Go to the instrument host screen and click on "Hydraulic Control"—"Voltage Regulation". Set the voltage value to reach a sealing pressure greater than the pressure applied during the experiment.
(4) Push both the "Main Control Down" and "Lithostatic al pressureDown" buttons to stop pressurizing the high-pressure kettle and remove it.
(5) On the unloading table, take off each component sequentially and use aluminum foil to collect the samples.

11.6.8 Pipeline and Sample Chamber Cleaning

(1) After unloading, use chloroform to carefully clean the reaction kettle components.
(2) Remove the "gas–liquid separator" and thoroughly clean it with chloroform.
(3) To clean the pipeline, open the "Solvent Valve", "Solvent Pump Valve", "Manual Injection Valve", "Injection Valve", "Product Valve", and "Manual Adjustment". Close the "Cleaning Manual Valve", put the conical flask under the pipeline on the heating furnace, and start the "Injection Pump" (set pressure to 1–2 MPa). Keep cleaning until the outflowing fluid is colorless.
(4) In step (3), you can also close the "Manual Adjustment" and open the "Cleaning Manual Valve" to let the rinsing fluid flow out from the "gas–liquid separator".
(5) Collect the wash solutions from the above steps into a single conical flask.

11.6.9 Supplement Solvent

(1) Allow the water in the lower chamber of the "Solvent Tank" to flow out by opening the "Gas Drive Valve", "Solvent Block Valve", "Solvent Valve", and "Solvent Vent Valve". Wait until no more water flows out.
(2) Close all the valves mentioned above and open the "Supplement Solvent Valve" to release the gas.
(3) Close the "Supplement Solvent Valve" and open the "Solvent Valve", "Solvent Block Valve", "Vacuum Valve", and the "Vacuum Pump". Vacuum for 5–10 min.
(4) After shutting the valves used in step 3, insert the pipeline for the solvent supplement into the chloroform solvent. Open the "Supplement Solvent Valve" and add the solvent.

11.6.9.1 Supplement Formation Water

(1) Open the "Gas Drive Valve", "Water Addition Valve", and "Water Vent Valve" to allow the water in the lower chamber of the "Water Container" to drain until no more water flows out.
(2) Close all the valves mentioned above and open the "Supplement Water Valve" to release the gas.
(3) Close the "Supplement Water Valve" and open the "Water Addition Valve"; "Vacuum Valve", and the "Vacuum Pump". Vacuum for 5–10 min.
(4) After closing the valves used in step (3), insert the pipeline for water supplementation into the required formation water and open the "Supplement Water Valve" to add the water.

11.6.9.2 Data Recording

(1) Launch the recording interface and enter the designated save path.
(2) Select the interface with the basic parameters for the oil generation simulation host, and save the parameters by clicking on the "Save Parameters" button.
(3) Click on "Process Display," set the time interval for recording, choose the oil generation simulation host, and then click on "Start Recording" located in the corresponding interface.
(4) The running state is represented by a green button on the "Process Display" screen.
(5) Avoid turning off the power of any oil generation simulation host during the experiment as it will render the recording invalid.

References

Defan G (2018) Development and experiment of oil generation simulation experimental instrument for source rocks. Sino-Global Energy 23(5):29–36

Defan G, Xuhui X, Zhiming L et al (2008) Theoretical thinking of basin-forming hydrocarbon accumulation and hydrocarbon generation model in limited space. Oil Gas Geol 29(6):709–715

Defan G, Xuhui X, Zhiming L et al (2011) New progress in basic research on hydrocarbon generation and expulsion in limited space of source rocks. Pet Exp Geol 33(5):441–445

Defan G, Xuhui X, Zhiming L et al (2014) Theory and practice of hydrocarbon generation within space-limited source rocks. Beijing: Petroleum Industry Press, pp 1–182

Longye Z, Jiliang C, Tianshun W et al (1991) Grade dividing and composition of oil shale in China. Geoscience 5(4):423–429

Zhiming L, Lunju Z, Zhongliang M et al (2011) Simulation of hydrocarbon generation and discharge in limited space of source rocks and its significance. Pet Exp Geol 33(5):447–451

Zhongliang M, Lunju Z, Zhiming L (2012) Simulation experimental study on hydrocarbon generation and expulsion controlled by temperature and pressure in limited space of source rocks. Acta Sedimenta Sinica 30(5):955–963

Chapter 12
Lacustrine Source Rock Oil Generation Simulation Experiment Method

Abstract The development of petroleum generation simulation for lacustrine source rocks has evolved through three distinct stages: high-temperature simulations, limited-space simulations, and simulations under geological conditions. Increasing emphasis is being placed on techniques that more accurately replicate real geological environments. Notably, the catalytic effects of the "geological molecular sieve," a complex formed by organic matter, clay, and metal elements, have become a key area of focus. Building on an analysis of the "geological molecular sieve" characteristics in Chinese lacustrine source rocks, the research team systematically conducted medium- and low-temperature catalytic petroleum generation simulation experiments. The experimental products closely mirrored those formed under actual geological conditions. The observed changes in oil and gas yields highlighted the catalytic role of the "geological molecular sieve," which enhances petroleum and hydrocarbon evolution at low temperatures and accelerates gas formation processes at medium temperatures. This chapter provides a detailed account of the experimental findings, offering valuable insights into medium- and low-temperature catalytic petroleum generation processes and their implications for understanding hydrocarbon generation mechanisms.

12.1 Development History of Lacustrine Source Rock Oil Gene Ation Simulation Experiment Technology

The exploration of source rock simulation experiment technology is a crucial component of theoretical research on lacustrine source rock oil generation. In China, the development of lacustrine source rock oil generation simulation technology has been divided into three stages over the last five decades (Defan 2009, 2014, 2018; Defan et al. 2004, 2008, 2011; Zhiming et al. 2011; Zhongliang et al. 2012).

The first stage, spanning from the 1970s to 2005, focused on the development of kerogen pyrolysis oil generation experiment technology under high temperature conditions.

D. Guan et al., *Theory and Simulation Experimental Technology of Oil Generation From Nonmarine Source Rocks in China*, https://doi.org/10.1007/978-981-96-3145-2_12

The second stage, between 2005 and 2018, emphasized the development of oil generation simulation experiment technology within space-limited source rocks.

The third stage, which began in 2018 and continues to the present day, emphasizes the development of low-temperature catalytic oil generation simulation experiment technology under geological conditions.

12.1.1 High-Temperature Kerogen Pyrolysis Oil Generation Experiment Technology

The word "kerogen" derives from the Greek term "kerogen," which was initially coined to describe the organic matter present in Scottish oil shales. During the early stages of the European Industrial Revolution in the 17th century, European scientists managed to obtain petroleum from oil shale through distillation, calling it "artificial petroleum." As they continued to produce shale oil from oil shale processing, they discovered that the organic matter found in oil shale predominantly consisted of kerogen and bitumen. The former is insoluble in organic solvents and requires heating to 550 °C to generate "artificial petroleum," while the latter is soluble and can produce "artificial petroleum" at lower temperatures.

Following World War II, natural petroleum quickly replaced "artificial petroleum." With the establishment of petroleum geology as a discipline, research on oil generation became a primary area of interest within the field of oil and gas exploration. Leveraging their dominant position in kerogen research, European scientists rapidly recognized that there was no significant distinction between the kerogen found in source rocks and that in oil shale; all forms of kerogen could generate oil through thermal degradation. In 1978, Tissot and Welte book "Petroleum Occurrence and Distribution—New Approaches to Oil and Gas Exploration" officially proposing the kerogen thermal degradation oil generation theory. Chinese scholars quickly accepted this theory and actively incorporated it into scientific research, production, and teaching. Consequently, the development of lacustrine oil generation experiment technology in China has progressed to the stage of high-temperature kerogen pyrolysis oil generation experiments, guided by this theory.

12.1.1.1 Theoretical Basis and Research Methods

The basis of high-temperature kerogen pyrolysis oil generation experimental technology can be traced back to the kerogen thermal degradation oil generation theory. The central concept is to compare the oil generation mechanism of source rocks with the method of oil shale producing "artificial petroleum" under high-temperature conditions. The book "Petroleum Occurrence and Distribution—New Approaches to Oil and Gas Exploration" asserts that the organic matter found in oil shales primarily comprises an insoluble solid substance called kerogen. There is no natural petroleum in rocks, and extractable bitumen is rare. When heated to around 500 °C, oil shales will generate "artificial petroleum" through the pyrolysis process. The kerogen in oil shale does not differ significantly from that contained in source rocks, and, to some degree, the process of generating "artificial petroleum" from kerogen through the pyrolysis process is akin to the formation of petroleum by deeply buried source rocks due to high temperatures.

In 1978, Espitalié and his colleagues created the Rock–Eval pyrolysis analyzer (Fig. 12.1), which imitates the principle of oil shale retorts. The experimental process involves heating around 100 mg of the sample gradually up to 550 °C under a specific temperature program in an inert gas environment. During the heating process, the soluble organic matter first volatilizes at medium temperatures and is detected by a flame ionization detector as hydrocarbons (S1). As the pyrolysis temperature increases, kerogen thermally decomposes to generate hydrocarbons (S2) and oxygen-containing volatiles like carbon dioxide and water (S3). Another parameter is Tmax, which represents the peak temperature when the maximum hydrocarbon generation happens during pyrolysis (equal to the final temperature of oil shale pyrolysis). Ultimately, each pyrolyzed source rock sample produces S1, S2, S3, and Tmax records. Evaluating these four primary parameters helps analyse source rocks.

S1 signifies the proportion of the original oil and gas potential that has been effectively transformed into hydrocarbons, i.e., soluble organic matter. S2 indicates the oil and gas potential of kerogen, i.e., the potential that hasn't generated hydrocarbons yet. Therefore, S1 + S2 equals the ultimate oil and gas potential of source rocks expressed in "kg hydrocarbon/t rock." S3 represents oxygen-containing volatiles such as carbon dioxide and water in source rocks, while Tmax is the temperature where the maximum hydrocarbon generation occurs, mainly used for assessing the thermal maturity of source rocks. Moreover, S2/TOC is referred to as the hydrogen index, and S3/TOC is called the oxygen index, which can determine the type of kerogen. By analyzing the above parameters, the oil and gas potential of source rocks can be evaluated (Fig. 12.2).

12.1.1.2 Technical Advantages

Evaluating the oil generation potential of source rocks quantitatively has always been a crucial aspect of source rock investigation. The Rock–Eval pyrolysis analyzer, which utilizes high-temperature kerogen thermal degradation oil generation experiment technology, can rapidly measure the organic matter content, type, and maturity of source rocks. It can also calculate the oil yield of source rocks, providing geologists with a quantitative evaluation method to study the oil generation potential of source rocks. This advantage has led to the swift promotion and application of this experimental technology in scientific research, production, and teaching fields in China. For the following 30 years of petroleum exploration, this experimental technology served as the core technology for source rock oil generation experiments, with the principle of kerogen thermal degradation reaction under high-temperature conditions

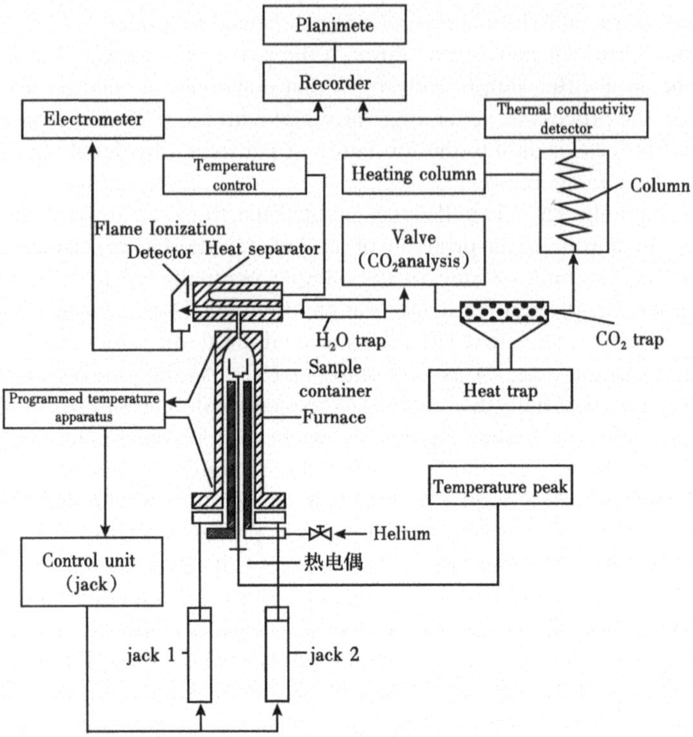

Fig. 12.1 Schematic diagram of the Rock–Eval pyrolysis instrument developed by Espitalié et al. (Tissot and Welte 1982)

directly serving as the oil generation mechanism of organic matter in source rocks under geological conditions.

12.1.1.3 Problems and Improvement Directions

Although the high-temperature kerogen thermal degradation oil generation experimental technology can quickly provide a quantitative evaluation of the oil generation potential of source rocks, the final oil yield obtained (the value of S1 + S2 when heated to 500–550 °C) cannot truly reflect the oil generation potential of source rocks under geological conditions. This is because the technology is essentially an organic chemical experimental method for petroleum generation from kerogen under the influence of high temperature. The experiment neglects the complex subsurface petroleum geological conditions and environments, and only applies high-temperature conditions that are impossible to exist under geological conditions, with the reaction process proceeding until all kerogens are thermally degraded to

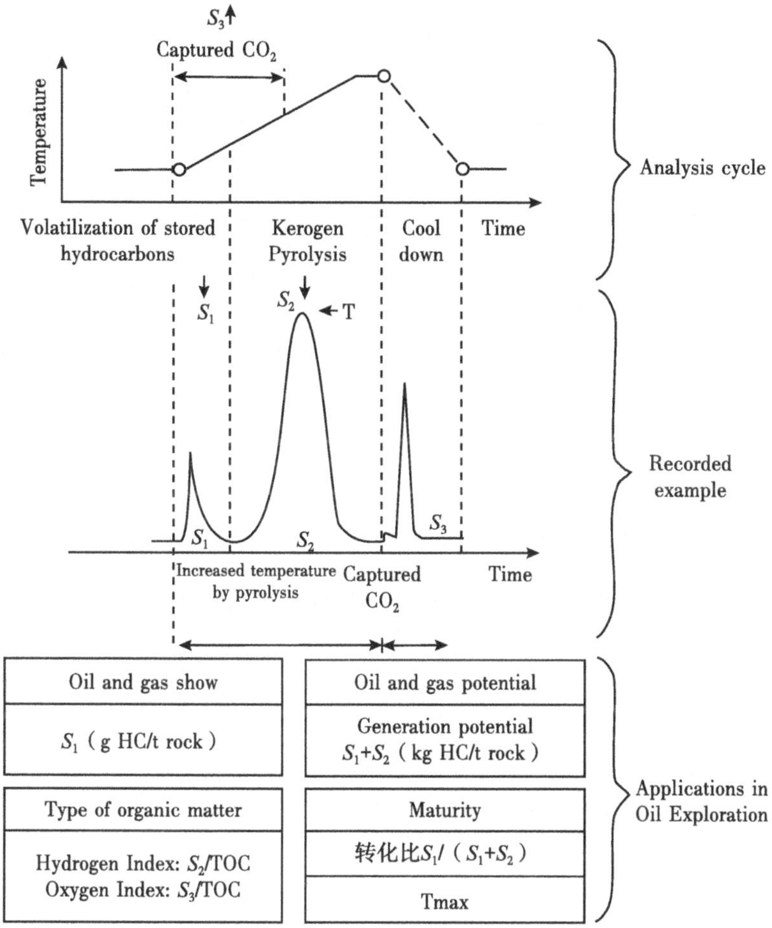

Fig. 12.2 Analysis cycle and recording obtained by the Espitalié pyrolysis method

generate petroleum. However, this reaction process is fundamentally impossible to occur under geological conditions due to the following reasons.

Firstly, all known source rocks contain residual organic carbon, indicating that only a portion of kerogen in source rocks can generate oil through thermal degradation under petroleum geological conditions and environments. Not all kerogens can thermally degrade to generate oil.

Secondly, the maturity (Ro) of most Mesozoic–Cenozoic lacustrine source rocks in China is currently between 0.8 and 1.0%, while the Middle and Upper Ordovician marine source rocks in the Tarim Basin have undergone about 500 million years, but the current Ro is mostly between 1.0 and 1.3%. These examples show that even after hundreds of millions of years of thermal evolution, the kerogen in source rocks cannot all thermally degrade to generate oil.

Lastly, the current burial depth and reconstructed paleo-burial depth temperatures of source rocks are usually no more than 200–250 °C. Under such low-temperature conditions, kerogen (S2) cannot generate oil through thermal degradation, which means that all kerogen in source rocks thermally degrading to generate oil under geological conditions does not exist.

Based on the analysis above, it is clear that the high-temperature kerogen thermal degradation oil generation experimental technology can quickly evaluate the oil yield of source rocks using the Rock–Eval pyrolysis analyzer and guided by the kerogen thermal degradation oil generation theory. However, this technology does not accurately simulate the oil generation process of source rocks under geological conditions as it neglects the complex subsurface petroleum geological conditions and environments. Instead, it is essentially a high-temperature organic chemistry experiment. Consequently, the key task is to analyze the petroleum geological conditions and environments of source rocks during the oil generation stage and develop experimental equipment that can simulate these conditions.

Through in-depth petroleum geological analysis and simulation experiments, we must study the oil generation process of organic matter in source rocks under subsurface petroleum geological conditions and environments. This will enable us to redevelop the lacustrine source rock oil generation simulation experimental technology in China, which accurately simulates the oil generation process of source rocks under geological conditions.

12.1.2 Limited Space Pore Pressure Oil Generation Simulation Experimental Technology

To develop source rock oil generation simulation experimental technology, we must adopt an organic geochemistry approach. This involves analyzing the petroleum geological conditions of source rock deposition, diagenesis evolution, and thermal evolution of chemical conditions. Sedimentary principles dictate that at the end of the continuous subsidence stage in a basin, clayey sediments and their organic matter would have filled the entire sedimentary space of the depression or sag.

Within this sedimentary space, the pore space developed during the diagenesis stage provides a reaction area for the thermal evolution of organic matter, as well as the space for generated petroleum. As the burial depth increases, the porosity of the source rock decreases due to the overlying rock layers' compaction. Meanwhile, the maturity of organic matter and the amount of generated petroleum also increase. When the oil volume fills all pore spaces, and oil saturation reaches equilibrium, porosity of the source rock ceases to decrease. Additionally, the thermal cracking of organic matter is inhibited, leaving unreacted organic matter in the source rock.

During the deposition and diagenesis evolution of source rocks, there is a limited pore space available for the thermal evolution of organic matter. To simulate the oil generation process under subsurface petroleum geological conditions accurately, research must first focus on studying the limited pore space's evolution characteristics and its relationship with organic matter's thermal evolution.

Since 2005, the authors and the research team have used organic geochemistry research methods and basin formation, hydrocarbon generation, and petroleum accumulation theories to examine actual data from Shengli, Zhongyuan, and Henan oilfields. This research led to the formulation of the Limited Space Oil Generation and Expulsion Theory, which guided the independent development of the DK-I, DK-II, and DK-III Formation Pore Pressure Oil Generation Simulation Experimental Instruments that suitably match the petroleum geological conditions and environmental features of source rocks subsurface.

It is important to note that the proposal of the Limited Space Oil Generation and Expulsion Theory and the development of the Formation Pore Pressure Oil Generation Simulation Experimental Instrument not only contributed significantly to the research results of source rock oil generation theory in China, but also initiated genuine source rock oil generation simulation experimental research in China.

12.1.2.1 Theoretical Basis and Research Methods

The process of source rock deposition and diagenesis, organic matter thermal evolution, oil generation, and primary migration occurs during the continuous subsidence stage of a basin, as per the theory of basin formation, hydrocarbon generation, and petroleum accumulation. From the perspective of source rock oil generation simulation experiments, the oil generation depression that forms during this stage is akin to an oil generation simulation instrument, with the source rock serving as the experimental sample. To replicate the oil generation process of source rocks in these depressions, the oil generation simulation instrument must be designed, and the simulation experiment carried out based on the petroleum geological characteristics of the oil generation depressions. Experimental samples should conform to the depositional and diagenetic evolution characteristics of source rocks in these depressions.

The physical field of oil generation depressions during the continuous subsidence stage of a basin is characterized by the continuous loading of sedimentary materials, increasing temperature and pressure. As the overlying rock layers compact the source rock, its porosity decreases, and organic matter gradually matures due to increasing temperatures. When the maturity (Ro) of the source rock reaches 0.5%, its porosity drops to 10–20%, and the entirety of the oil generation process from this point onwards occurs within this limited pore space.

When the generated oil and gas completely fill the pore space, they reach a relative equilibrium with oil–gas–water under the constraint of pore space and overlying rock pressure. This equilibrium inhibits the thermal degradation process of organic matter, resulting in unparticipated organic matter remaining in the source rock. If the oil generation depression during the continuous subsidence stage is viewed as a closed high-temperature, high-pressure chamber, then the source rock is akin to a distillation pyrolysis tower standing in the oil generation depression. As the organic matter in the source rock reaches the maturity threshold, it pyrolyzes to generate various hydrocarbons such as alkanes, cycloalkanes, and aromatic hydrocarbons, which constitute petroleum.

In 2009, the Wuxi Petroleum Geology Research Institute developed a formation pore pressure oil generation simulation experimental instrument based on the concept mentioned above. The instrument was designed to simulate the process and characteristics of organic matter thermal cracking and oil generation in source rocks under temperature, pressure, and limited space conditions at the end of a basin's continuous subsidence period.

The experimental instrument is made up of six components, including a high-temperature and high-pressure hydrocarbon generation system, a bidirectional hydraulic control system, a hydrocarbon expulsion system, an automatic control and data acquisition system, a product separation and collection system, and peripheral auxiliary equipment. The primary component is the high-temperature and high-pressure hydrocarbon generation system, which features a reaction kettle, electric high-temperature furnace, seals, connecting valves, and pipes. The reaction kettle serves as the core device where the samples undergo thermal and pressure simulation experiments. It is also the site where various products coexist (Fig. 12.3).

To ensure stable application of simulated overburden pressure directly to the samples for a long time, the bidirectional hydraulic control system can automatically control the rise and fall of pressure. The hydrocarbon expulsion system expels the generated oil and gas according to the set fluid pressure during the hydrocarbon generation process. This system realizes controllable generation and expulsion of hydrocarbons. The automatic control and data acquisition system comprises a pressure control system and fluid in/out alarm control. It makes sure that the temperature and pressure of the thermal and pressure simulation experiment work according to the automatic program while collecting data such as lithological pressure, fluid pressure, simulation temperature, and simulation time in real-time through a data acquisition card and computer. The product separation and collection system primarily

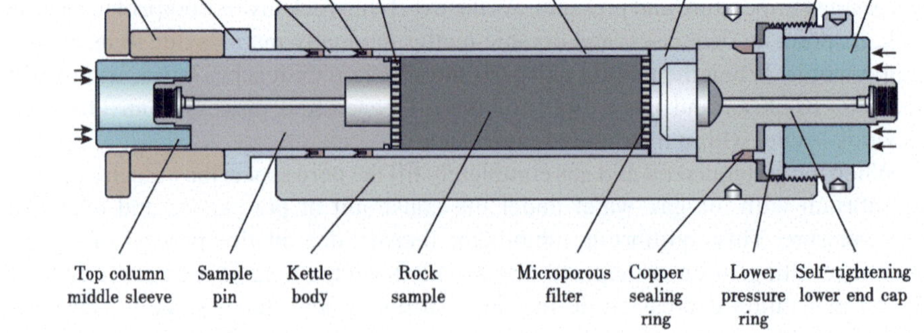

Fig. 12.3 Principle and structural diagram of the high-temperature and high-pressure reaction vessel

consist of automatic product collection and measurement devices, ensuring accurate collection and measurement of products. Peripheral auxiliary equipment mainly refers to supporting equipment and tools related to thermal and pressure simulation experiments. The specific experimental procedure involves the following steps:

(1) Sample selection: Choose source rock samples with different organic matter abundance, lithology, and low maturity.
(2) Preparation and loading of simulation samples: Based on the heterogeneity characteristics of source rocks, crush the samples to 40–60 mesh and mix them thoroughly. Divide the samples into several parts and compress a small portion from each temperature point into a cylindrical small core to ensure the uniformity and representativeness of the samples. The loading method should consider the lithological characteristics and the vertical lithological combination of the source rocks. This encompasses five types of lithological combinations: pure mudstone, lower mudstone-upper sandstone, upper mudstone-lower sandstone, upper sandstone-middle mudstone-lower sandstone, and interbedded mudstone-sandstone (Fig. 12.4).
(3) Setting of simulation experiment temperature and pressure conditions

    The selection of different simulation temperatures, corresponding burial depth, lithostatical pressure, and fluid pressure values for different evolution stages is mainly based on the actual burial history of the sampling area (the relationship between Ro and burial depth) and the Ro values at different temperatures in the simulation experiment.
(4) Heating and pressurizing simulation experiment steps

    ① Leakage test: Set up the sample chamber containing the core sample in the reaction kettle, apply pressure to seal it, and fill it with inert gas at 5–10 MPa for a leakage test. Repeat this process 3–5 times and finally evacuate the air.

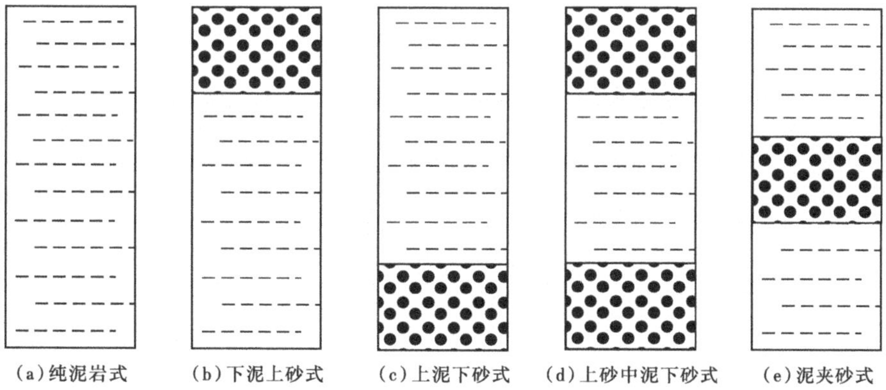

Fig. 12.4 Schematic diagram of the sampling method

② Water injection: Use a high-pressure pump to inject high-pressure water, either pure water, saline water, or formation water, at 60–80 MPa to fill the pore space of the compressed core sample completely.
③ Pressurization and temperature simulation: Apply lithostatic al pressure to the core sample using a hydraulic press to the predetermined value and compact the sample. Start the temperature controller and thermostatic furnace, heating it at a rate of 1 °C/min until reaching the set temperature. After reaching the desired temperature, maintain it for 48–96 h.
④ Quantitative sampling of gas and expelled oil (Fig. 2.2).

12.1.2.2 Technical Advantages

This technology simulates the oil generation process of source rocks within limited space formations during the continuous subsidence of a basin, based on petroleum geochemistry principles. By conducting these simulation experiments, data and oil generation curves can be obtained, which are consistent with the oil generation characteristics observed in modern oil-generating depressions. The experiments reveal that the primary stage of oil generation in source rocks under limited space geological conditions occurs within a Ro range of 0.6–0.9% and temperature range of 250–360 °C. When the simulation temperature reaches 360 °C (Ro = 0.9%), there is no further increase in oil generation amount or rate. Even when the simulation temperature increases from 360 to 385 °C and Ro increases from 0.9 to nearly 1.4%, the oil generation status remains constant. These results contrast with the "big belly" curve produced by traditional experimental methods (see Fig. 2.3), which suggest that the main stage of oil generation occurs between 200 and 550 °C. Compared to the traditional organic chemistry pyrolysis oil generation experimental methods, this technique reduces the simulation temperature by nearly 200 °C, representing the first attempt to simulate the oil generation process of source rocks according to subsurface geological conditions.

The independently developed "formation pore thermal pressure oil generation and expulsion simulation instrument" takes into full consideration the sedimentary diagenetic evolution characteristics of source rocks during the simulation experiment reaction process. This instrument not only preserves the original mineral composition structure and organic matter distribution state of source rocks, but also fills the oil generation space with high-pressure liquid water having the same mineralization degree as formation water. Moreover, it realizes the pyrolysis oil generation reaction of organic matter at medium temperature and short time under lithostatic al pressure and formation fluid pressure conditions that are similar to subsurface geological conditions. The following are the specific technical advantages of this instrument:

(1) It can conduct oil and gas generation and expulsion simulation experiments of source rocks under higher formation fluid pressure (up to 150 MPa), while applying overlying lithostatic al pressure and confining pressure (up to 200 MPa).

(2) By using original core samples or artificially compressed core samples, it retains the original porosity and structural composition of the samples as much as possible, ensuring that the oil generation reaction process occurs in the formation pore space and retains the original distribution state of organic matter.
(3) It can carry out both closed oil and gas generation and expulsion simulations under certain fluid pressure, as well as oil and gas generation and expulsion simulations under "pulse expulsion" conditions where the source-reservoir pressure difference is controllable.
(4) Water exists in the rock sample pores in the form of a high-pressure liquid, which represents a true water-added oil generation and expulsion simulation.
(5) It can simulate the oil generation capacity and expulsion efficiency of different types of source rocks under different temperatures, pressures, fluid media, and inorganic mineral conditions.
(6) It can selectively set and automatically adjust experimental conditions such as experiment time, experiment temperature, sampling amount, lithostatic al pressure applied to the core, formation fluid pressure and fluid properties, expulsion method, and expulsion pressure.

12.1.2.3 Problems and Improvement Directions

The technology for limited space formation pore thermal pressure oil generation and expulsion simulation experiments, combined with the petroleum geological characteristics of source rock evolution, has made initial progress in realizing the pyrolysis oil generation process of organic matter under medium temperature conditions (with a simulation temperature range of 200–360 °C) and has achieved innovative results and understanding. Nonetheless, there are still some issues that need to be further studied and improved in theoretical thinking, simulation conditions, and production applications. For instance, while this experimental technique has significantly reduced the simulation temperature by nearly 200 °C compared to traditional kerogen pyrolysis oil generation simulation experiments, the temperature is still relatively high compared to the oil generation temperature of source rocks in lacustrine oil-bearing basins under geological conditions (50–200 °C). This may affect the oil generation mechanism from the thermal evolution of organic matter in source rocks and obscure the role of other geological factors like pressure, clay minerals (montmorillonite, illite), formation water, and metal elements in the oil and gas generation process. To reveal the oil generation mechanism of source rocks under geological conditions, the simulation temperature must reflect the oil generation temperature during the geological history period, and an in-depth and comprehensive analysis of various factors affecting the oil generation of source rocks under geological conditions, including material basis, transformation conditions, catalytic mechanisms, and oil generation space, should be conducted.

Furthermore, although the limited space formation pore thermal pressure oil generation and expulsion simulation experiment technology has already obtained the S1 oil generation amount under medium temperature conditions (200–360 °C), the samples also need to be re-analyzed by high-temperature pyrolysis to obtain the S2 value to meet the requirements of organic geochemistry laboratory specifications. The ultimate goal is to use basin simulation techniques based on the kerogen pyrolysis oil generation theory to calculate the total oil generation potential of S1 + S2 and use it as the geological resource volume of oil-generating depressions. However, the S2 value is not geologically meaningful in practice.

As evident from the past fifty years of development, the lacustrine source rock oil generation simulation experiment technology has achieved numerous innovative results in theoretical research and experimental equipment development. Nevertheless, in terms of thinking and experimental methods, traditional approaches and methods are still habitually employed to address problems. Therefore, the oil generation simulation experiment technology and methods require constant innovation, improvement, and practical development.

12.1.3 Low-Temperature Catalytic Oil Generation Simulation Experiment Technology Under Geological Conditions

Since 2018, the development of lacustrine source rock oil generation simulation experiment technology has reached its third stage, namely the development stage of low-temperature catalytic oil generation simulation experiment technology under geological conditions. During this period, while addressing the current issues in the oil generation simulation experiment technology of source rocks, the authors and research team, through intensive research, discovered that crude oil in various oil and gas-bearing basins generally contains metals, non-metals, rare elements, and radioactive elements, all of which, along with clay minerals, are highly efficient catalysts. Moreover, Hunter's statistical results on global sedimentary basins indicate that the main reason why source rocks can generate oil by pyrolysis at temperatures ranging from 60 to 150 °C subsurface is that this temperature range constitutes the main stage of the catalytic cracking process. Meanwhile, the production practices of the refining and chemical industry have proven that catalysts play an essential role in accelerating crude oil processing.

Inspired by these phenomena, the authors and the research team conducted technical breakthroughs and theoretical innovations, comprehensively analyzing the oil-generating materials, oil-generating spaces, and catalytic transformation conditions of source rocks under geological conditions. They discussed the primary characteristics of the low-temperature catalytic cracking oil generation process of source rocks and the technical methods of simulation experiments. As a result, the DK-IV type source rock oil generation simulation experiment instrument was developed.

12.1.3.1 Theoretical Basis and Research Methods

The subsurface oil generation process of source rocks is essentially the organic geochemical reaction that transforms organic matter in source rocks into petroleum. Any chemical reaction requires three conditions: substances participating in the reaction, conditions applied to the reaction, and a container that accommodates both reaction-participating substances and reaction-generated substances. Therefore, when studying the source rock oil generation simulation experiment method, one should adopt the thinking mode of organic geochemistry and approach it from the three aspects of reaction substances, reaction conditions, and reaction space required for oil generation. A comprehensive analysis of the various factors influencing the oil generation process of source rocks is necessary to uncover the low-temperature catalytic oil generation characteristics of source rocks under geological conditions.

Analysis of Source Rock Composition

In the oil generation process of source rocks found in oil-generating depressions, all components of the source rock are involved (Seewald et al. 2003; Zhiming et al. 2010; Rahman et al. 2017; Huimin et al. 2018). Therefore, the initial step in the experimental process is to perform a whole-rock analysis. Such an analysis provides an overall grasp of the mineral composition of the source rock, including clay minerals, detrital components, metal and non-metal elements, rare radioactive elements, etc. This analysis also identifies the types, abundance, Ro values, soluble and non-soluble organic matter content of the source rock, as well as the porosity, permeability, and diagenetic evolution characteristics of the source rock.

Analysis of Physicochemical Conditions of Organic Matter Pyrolysis Oil Generation Process

Physicochemical conditions mainly refer to the temperature and pressure regime that occur when the burial depth of the source rock reaches above 60 °C. These conditions are associated with the catalytic cracking effects of clay minerals, metal and non-metal elements, rare radioactive elements, etc., on the organic matter pyrolysis process. Due to strong differences in lacustrine basins in China, thermal simulation experiments should be carried out based on the composition of source rocks and the temperature and pressure conditions of the geological evolution stage of the source rocks. This is according to a whole-rock analysis of source rocks from different geological periods. Such analyses help better understand the impact of various components in the source rock on the organic matter pyrolysis oil generation process. Additionally, it helps analysts to analyze the promoting or inhibiting effect of each substance in the oil generation process of the source rock. It also identifies the key role of which substance or combination of substances. Based on these results, source rocks can be screened from sedimentary rocks of different geological periods, and source rock evaluation and oil generation potential prediction can be conducted.

Providing Space for Organic Geochemical Reactions and Accommodating Reaction Products

Based on sedimentary principles, when the deposition stage of a source rock ends, the source rock body has already filled the entire space of the sedimentary depression (oil-generating depression). Upon reaching the oil generation stage, the space for organic geochemical reactions and accommodating reaction products (i.e., petroleum and natural gas) within the source rock is limited to the pore space at a certain burial depth. For source rocks composed of alternating sand and mudstone layers, this pore space includes the pore space in mudstones as well as the pore (or dissolution, fracture) space in the intercalated sandstones (or carbonates). During simulation experiments, it is important to use the sandstone–mudstone ratio and interlayer combination mode that reflect the material and spatial conditions of the source rocks. This is based on the sedimentary and diagenetic characteristics of the source rock.

In 2019, the Wuxi Petroleum Geological Research Institute of Sinopec independently developed the DK-IV type source rock oil generation simulation experiment instrument. The instrument mainly simulates the low-temperature catalytic pyrolysis oil generation process and characteristics of source rocks under geological conditions. Further details on the experimental instrument are provided in this chapter and will not be repeated here.

12.1.3.2 Technical Advantages

The low-temperature catalytic oil generation simulation experiment technology, developed based on organic geochemistry principles, aims to explore the low-temperature catalytic pyrolysis oil generation mechanism of source rocks under geological conditions. This is achieved through whole-rock analysis and low-temperature catalytic simulation experiments. Compared to previous source rock oil generation simulation technologies, this technology offers advantages such as the ability to simulate experimental conditions as close as possible to the subsurface geological conditions of the source rock oil generation stage. This can objectively reflect the oil generation potential of source rocks and has significant implications for source rock quality evaluation and resource potential assessment. The advantages of this technology are specifically reflected in the following aspects.

First, the experimental method is based on analyzing the whole-rock composition of source rocks and preparing samples according to the actual material composition of source rocks. This allows for the full reproduction of the sedimentary and diagenetic characteristics of source rocks.

Secondly, this method reduces the temperature of the oil generation simulation experiment from the previous range of 200–360 °C to a low-temperature range of

50–200 °C. This enables the simulation of the source rock oil generation process at temperatures similar to the true formation temperature of the oil generation stage.

Thirdly, the study of the enrichment characteristics of various special trace elements in source rocks has confirmed that some transition metal and radioactive metal elements can promote the oil generation process.

Since this experimental method is designed to simulate oil generation according to subsurface geological conditions, the obtained oil generation capacity (S1) of source rocks can objectively reflect their oil generation potential. This provides important technical support for evaluating petroleum resources and planning exploration activities.

12.1.3.3 Problems and Improvement Directions

The low-temperature catalytic oil generation simulation experiment technology, conducted under geological conditions, emphasizes the need to carry out oil generation simulation experiments that closely reflect subsurface geological conditions. This includes comprehensively considering various factors that can affect the oil generation process of source rocks. Currently, significant progress has been made in the development of oil generation simulation instruments and the study of low-temperature catalytic oil generation mechanisms. However, there are still some challenges to overcome. For example, in collected samples, most of the original organic matter may have evaporated, the catalytic effect on oil generation may not be evident in a short period of time, and determining the amount of catalytic substances used in the experiments can be difficult.

12.2 Catalytic Mechanism of Geological Molecular Sieve Oil Generation in Source Rocks

12.2.1 Concept of Geological Molecular Sieve

Molecular sieve is a hydrated aluminosilicate material with a selective adsorption function. As a solid acid catalyst, it has strong acid catalysis and has become the most important crude oil cracking catalyst (Kennedy et al. 2014; Li et al. 2015; Zhu et al. 2018). Clay minerals are a kind of layered aluminosilicate containing water. As the main component of source rock, they play an important role in the process of transforming organic matter into petroleum and can be called molecular sieve in geological body (Fig. 12.5).

The catalysts for oil refining are mainly divided into catalytic cracking catalysts and hydrogenation catalysts. The catalytic cracking catalyst is basically composed of three elements, namely, “aluminum silicate” catalyst. The catalytic activity is closely related to the internal surface area. For example, the internal surface of

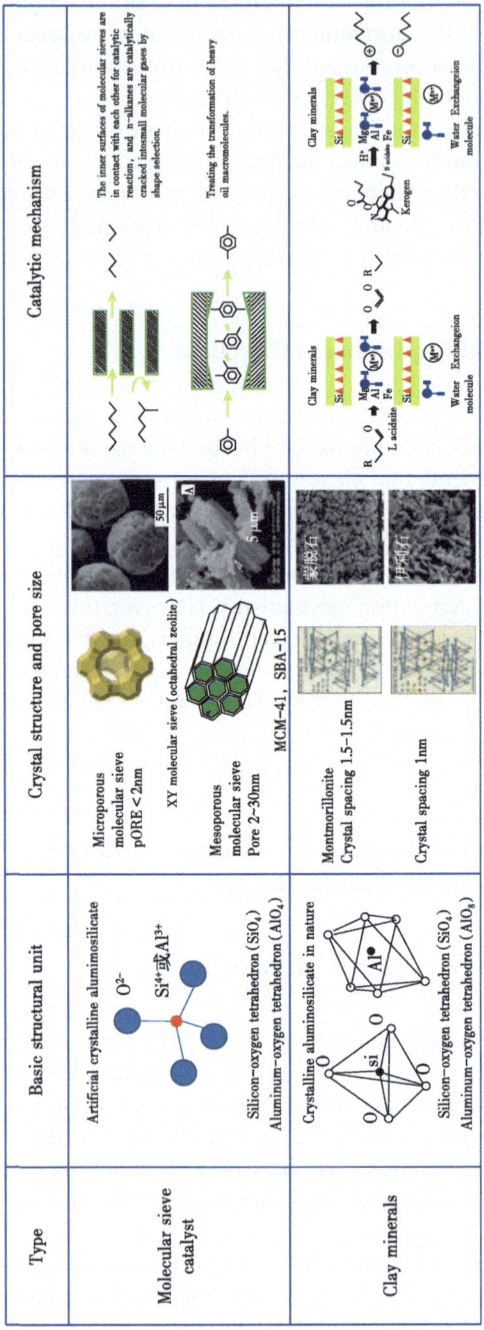

Fig. 12.5 Molecular sieve catalyst and structure of clay minerals

synthetic aluminum silicate catalyst is more than 10 times higher than that of natural clay catalyst, and it has higher cracking activity. The specific surface area of fully synthetic aluminum silicate catalyst can be as high as 600 m^2/g, and the average pore size is about 5 nm, which is helpful for distillate oil molecules to crack into smaller molecules in the pores of the catalyst. The catalytic cracking catalyst has an ideal pore distribution, so heavy oil molecules have accessible active centers. This pore distribution should include large, medium, and micropores in appropriate proportions to realize the orderly cracking process of heavy oil macromolecules. Macropores mainly come from the pores between bonded particles, micropores are provided by molecular sieve components, and mesopores come from secondary pores of molecular sieve, macroporous molecular sieve, and carrier. This is very similar to the "organic matter-clay mineral aggregate" in different occurrence states in lacustrine source rocks. The interlayer combination of clay is similar to mesoporous molecular sieve, and the intergranular of a single clay mineral is similar to microporous molecular sieve. Meanwhile, hydrogenation catalysts used in petroleum refining and petrochemical industry are basically supported catalysts. According to the different ways of introducing active components, hydrogenation catalyst generation methods are divided into impregnation method, coprecipitation method, and kneading method. The commonly used impregnation method is to load the active component on the carrier by fully contacting the solution (impregnation solution) containing the active component with the prepared carrier. This is similar to that in the process of sedimentary diagenesis and biological reproduction of source rocks, biological organisms can enrich a large number of metal elements through organ absorption, surface adsorption, and complex formation, and together with clay minerals form various combinations of "organic matter–clay minerals-special trace elements" loaded with different metal and nonmetal elements. The only difference is that the catalyst in the petroleum refining field has a specific ideal structure, while the catalyst in source rocks is not strictly defined because of different sedimentary environments and lithology. Therefore, the concept of "geological molecular sieve" is put forward.

12.2.2 Mechanism of Catalytic Hydrocarbon Generation by Geological Molecular Sieve

12.2.2.1 Catalytic Mechanism of Transition Metals

Among the metallic elements involved in catalysis in geological processes, the transition metal elements have the highest catalytic efficiency, such as chromium, manganese, iron, cobalt, nickel, copper, zinc, palladium, silver, platinum, gold, and mercury. The reason why transition metal elements have high catalytic efficiency is that the nuclear charge, electron affinity, and electron shell structure of the atoms in mineral composition directly affect the catalytic effect of minerals. Generally speaking, in gas–solid heterogeneous catalytic reactions, gas particles must react on the surface of a solid catalyst, and at least one of the gas particles involved in the reaction must be chemically adsorbed on the surface of the solid catalyst. For this reason, gas molecules are activated, and the bond of gas molecules becomes longer. Quantum chemical calculations show that chemical adsorption is equivalent to raising gas molecules to their first excited state, which leads to increased reactivity. The catalytic performance of metal catalysts is related to their ability to adsorb and activate reactant gas molecules. In the electronic structure of transition metal atoms, there are one or more unpaired electrons in the outermost electron shell, which can form coordination bonds with adsorbed gas molecules, and then strong chemical adsorption occurs. In the electronic structure of non-transition metal atoms (such as Al, K, Na, etc.), the outermost electron layer only has S or P electrons, so the chemical adsorption energy is weak, resulting in low activity of Al_2O_3 and NaCl and poor catalytic effect.

Under diagenetic reduction conditions, transition metals on kerogen-rich mineral substrates can become active catalysts for the genesis of light hydrocarbons and natural gas (Mango 1992; Mango et al. 2009). This catalysis is achieved through the cyclization of three different carbon numbers (Fig. 12.6) and the cleavage of carbon–carbon and carbon–metal bonds [as shown in Eqs. (12.1) and (12.2), in which R represents an alkyl or naphthenic group connected with metal M]. From the perspective of organic geochemistry, the mineral matrix type, organic matter type, organic matter content, or the ratio of mineral matrix (especially catalyst) to organic matter are the basic material conditions that determine oil and gas generation. On the other hand, catalysts are considered to play an important role in several important transformations of hydrocarbon formation, such as depolymerization, cracking, polymerization, isomerization, and hydrogen disproportionation. In the usual Ziegler–Natta (Z–N for short) catalytic process, the molar concentration ratio of olefin to catalyst differs by tens of thousands of orders of magnitude, allowing olefin polymerization to be catalyzed sequentially by hydrogenation through Eqs. (12.1) and (12.2).

However, in nature, the olefins generated by kerogen degradation are present in extremely small quantities. The molar concentration ratio of olefins to catalysts is approximately 1:1, and the catalytic center is typically fixed on the olefin matrix. As a result, some interchangeable intermediates can be established through the reversible steps in Eq. (12.1). The four interchanged intermediates in Fig. 12.7 are considered to be part of a total system. Comparing the irreversible changes of hydrogenation in Eq. (12.2), methane can be formed through a catalytic cycle of hydrogen and n-alkanes. The δ bond between the metal and carbon can move along the carbon chain through hydride addition and elimination steps. Thus, the alkyl group connected with the δ bond between metal and carbon at the β position can have one or more carbons, which mainly depends on the position of the δ bond on the carbon chain. In Fig. 12.6, when methyl is replaced by an alkyl group with a higher carbon number, the catalytic cycle indicates the formation path of natural gas and higher alkanes. Assuming that the stability of the metal–methyl bond is higher than that of the alkyl–metal bond, the steady-state concentration of the methyl phase (CH_3–M) is dominant. Therefore, the above reaction process can be used as a kinetic way to enrich methane in natural gas (Fig. 12.7).

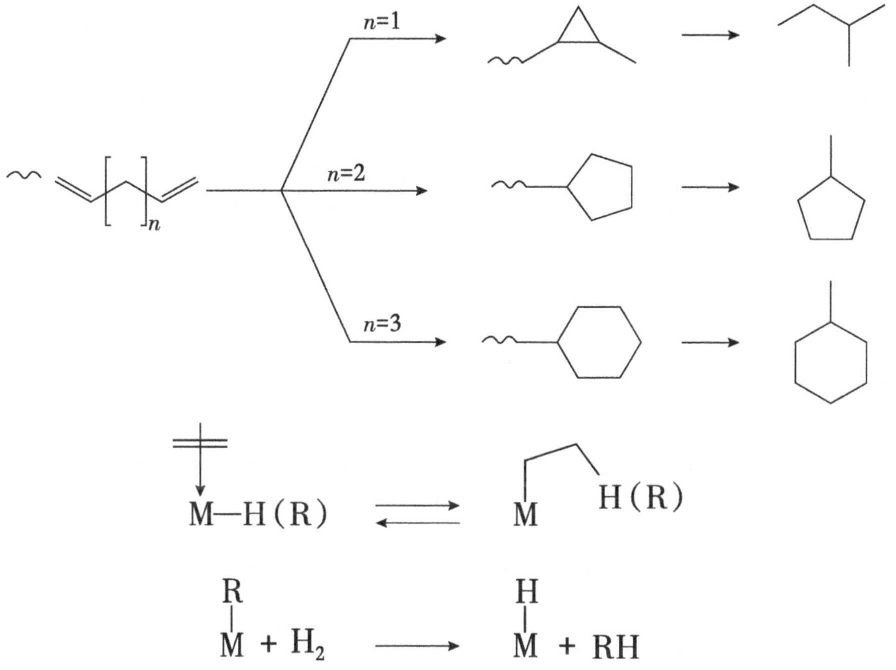

Fig. 12.6 Schematic diagram of three kinds of catalytic cyclization

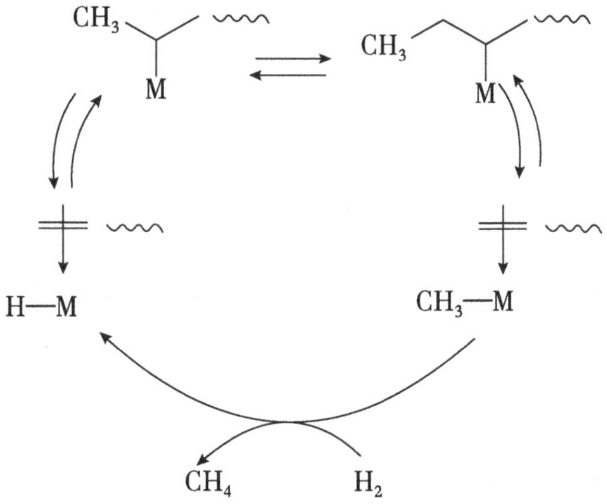

Fig. 12.7 Catalytic process of hydrogenation of long-chain n-alkanes to methane

There are also viewpoints that suggest transition metal elements can play a significant catalytic role. This may be due to the fact that the outermost electron shell of transition metal elements is easily susceptible to losing electrons, providing a carrier for electron transfer in organic reactions and an excellent source of electrons. Combined with the instability of the formed complex, charged groups generated in the reaction process can be released to play a catalytic role (Wang et al. 2006).

In light of the variations in mineral composition and the occurrence of specific trace elements in source rocks in lacustrine oil-bearing basins, four groups of dynamic experiments were designed. These groups include carbonate minerals (LX-TSY), clay minerals (LX-NT), clay minerals + transition metal elements (LX-TM), and clay minerals + radioactive elements (LX-U). The effects of different clay minerals, transition metal elements, and radioactive elements on source rocks were analyzed and discussed. The metal element contents loaded in the LX-TM series experiments are similar to those of crude oil in the Songliao Basin, while the radioactive metal element contents of U and Th loaded in the LX-U series experiments are similar to those of source rocks in the Ordos Basin.

Among the non-hydrocarbon products in different series of experiments, the LX-TM series of experiments loaded with transition metal elements have a higher CO_2 yield than other series of experiments (Fig. 12.8). The decarboxylation reaction or heteroatom removal reaction of organic matter closely related to L-acid can produce a large amount of CO_2. This suggests that a "geological molecular sieve" combined with transition metal elements can catalyze hydrocarbon generation by promoting heteroatom removal reaction or increasing the activity of L-acid in clay minerals. The yield of H_2 in the LX-NT series experiments without metal elements is significantly higher than that in other series of experiments, indicating that the experiments with metal elements have a certain consumption of H_2. This suggests that the transition metal elements in the "geological molecular sieve" are directly involved in catalytic hydrogenation for hydrocarbon generation.

12.2.2.2 Catalytic Mechanism of Radioactive Element Uranium

The atomic number of uranium is 92, and its relative atomic mass is 238, which includes three isotopes: ^{238}U, ^{235}U, and ^{234}U. Uranium's chemical properties are highly active, and it can react with almost all elements except for noble gases. Uranium has a strong reducing ability, and both metallic uranium and low-valent uranium are significantly reduced. Due to its reactivity, uranium exists in nature in the form of a strong complex. Uranium is generally considered to exist in nature as tetravalent and hexavalent uranium, with hexavalent uranium being the primary component and the uraniumphthalein-anion combination as the basic structural unit. Uranium has a unique and excellent coordination performance, which enables it to form coordination compounds with many ligands, giving it good complex catalysis and redox catalysis characteristics (Mao et al. 2014).

The formation of petroleum is a process in which organic matter combines with hydrogen to remove impurities such as oxygen and nitrogen (Taylor et al. 2003).

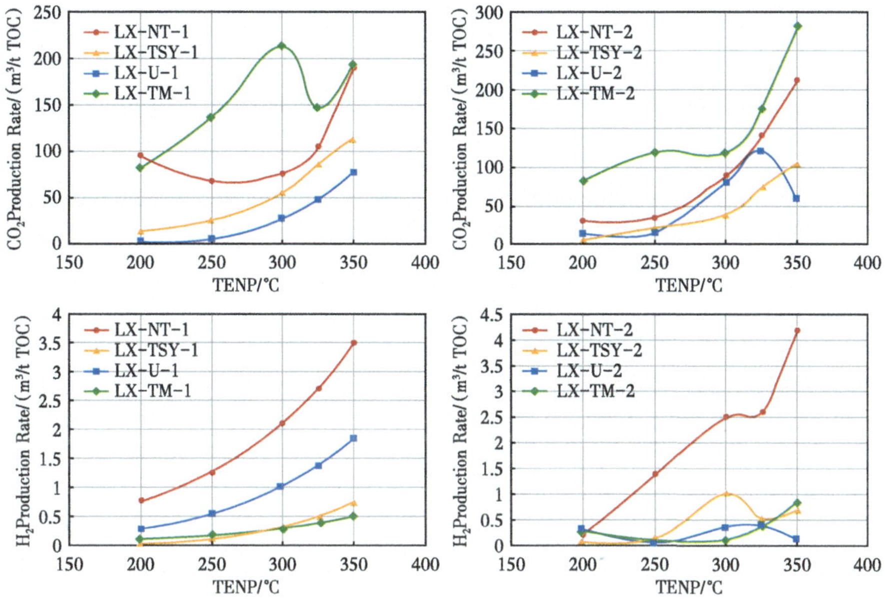

Fig. 12.8 Changes of CO_2 and H_2 generation rate in different series of experiments (H_2 generation rate is the data after fitting)

Therefore, the hydrogen content is considered the key to the hydrocarbon generation potential of organic matter. The conventional view is that part of the organic matter condenses to provide hydrogen for hydrocarbon generation (Hunt 1979). However, it is essential to note that exogenous hydrogen other than source rocks plays a vital role in hydrocarbon generation of organic matter. In the presence of an exogenous hydrogen source, the traditional hydrocarbon generation mode will change significantly. As long as carbon exists, oxidation products (organic acids and CO_2) and methane can be continuously formed.

In fact, the existence of U can promote the decomposition of water into H_2 through radiation and contact reaction and provide a hydrogen source for oil formation. On the one hand, elements such as U, Th, and K can release α, β, and γ rays, and the particles in these rays can decompose water into hydrogen and oxygen. These ions are decaying, and ions with energy values ranging from 4.15 to 4.50 MeV are continuously released (Wang 1986), which can ionize water molecules into free radicals such as e^-, H^+, H, and OH, forming H in microseconds:

$$H_2O + (\alpha, \gamma) \rightarrow H \cdot + OH\cdot \tag{12.3}$$

$$2OH\cdot \rightarrow HOOH \tag{12.4}$$

$$2HOOH \rightarrow 2H_2O + O_2 \uparrow \tag{12.5}$$

$$2H\cdot \rightarrow H_2 \uparrow \tag{12.6}$$

Among them, the concentration of H_2 is closely related to the concentration of U, Th and K, with an obvious increase in H_2 content as the concentration of U, Th and K is increased.

In addition, when metallic uranium comes into contact with water in a vacuum, water can react with uranium to generate UO_2 and release hydrogen (Feng et al. 1982). The reaction mechanism is as follows:

$$2H_2O + U \rightarrow UO_2 + 2H_2 \uparrow \tag{12.7}$$

When hydrogen is saturated, it reacts with uranium to form UH_3. Under the condition of high temperature heating, UH_3 will decompose and generate uranium and hydrogen again:

$$3H_2 + 2U \xrightarrow{80\,^{\circ}C} UH_3 \tag{12.8}$$

$$2UH_3 \xrightarrow{700\,^{\circ}C} 2U + 3H_2 \uparrow \tag{12.9}$$

Under the influence of U, water can be decomposed into H^+ and OH^- which are highly reactive and can easily combine with carbon to form hydrocarbons (Liu et al. 2013). Meanwhile, the presence of exogenous hydrogen can compensate for the hydrogen needed for the thermal degradation of organic matter, thereby greatly increasing the methane and total hydrocarbon yields (Mao et al. 2014).

In the one-month low-temperature catalytic and kinetic experiments, the samples loaded with radioactive U element and its combination (DYS-150-10U, LX-U) exhibited high H_2 yield. This further demonstrates that the presence of U can promote the decomposition of water into H_2 through radiation and contact reaction and provide a hydrogen source for oil formation.

12.2.2.3 Mechanism of Hydrocarbon Generation Catalyzed by Geological Molecular Sieve

The most important chemical reaction in the process of organic matter oil generation and oil refining is the catalytic reaction of acid and alkali catalysts. The acid–base catalytic theory of Brønsted and Lewis is usually adopted in the catalytic reaction of hydrocarbon generation from organic matter. In the study of surface acidity, it is generally accepted that the proton on the surface is the Brønsted acid site and the proton on the surface is the Lewis acid site, and the two can be converted to each other. The "geological molecular sieve" directly affects the transformation of Lewis acid site and Brønsted acid site centers, the reaction process of diagenesis and hydrocarbon generation, and the properties of oil and gas products due to the differences in the types, contents, and occurrence states of metal elements.

The interlayer combination of organic matter and clay is a necessary condition for catalytic hydrocarbon generation of clay minerals. Through many scanning electron microscope observations, the occurrence relationship between organic matter and clay can be divided into three categories: (1) the dispersed distribution of organic matter, clay minerals, and rigid particles; (2) organic matter and clay are distributed in layers; (3) organic matter occurs between clay minerals (Fig. 12.9).

Not all clay minerals with different modes of occurrence can catalyze the hydrocarbon generation and transformation of organic matter. Only when organic matter exists between clay minerals is the catalytic effect of clay minerals on organic matter obvious (see Fig. 12.10). In the immature–mature stage, organic matter that exists in between clay layers exhibits hydrocarbon generation phenomenon, forming a smooth and round oil film on the surface of clay minerals, while the hydrocarbon generation and transformation degree of layered organic matter is low.

Due to the formation of numerous organic acids during diagenesis and catagenesis stages, metal ions are enriched by the existence of organic acids, which exist in the form of complexes in source rocks. The combination of organic matter, clay, and metal elements creates a similar "geological molecular sieve". Corresponding to the three geological stages of transformation from smectite to illite–montmorillonite/montmorillonite–illite disordered interbedded interval, illite/montmorillonite ordered interbedded interval, and illite interval (Fig. 12.11)—the catalytic hydrocarbon generation mechanism of the geological molecular sieve can be divided into three stages:

(1) Non-covalent bond destruction: Sedimentary organic matter undergoes hydrothermal decomposition (destruction of non-covalent bonds, hydrogen bonds, van der Waals forces, etc.), which requires only low energy.
(2) Heteroatom removal reaction: The "soluble intermediate (asphaltene + resin)" removes heteroatoms (C–S/C–O/C–N fracture) to form oil (mainly hydrocarbons forming macromolecules), while generating inorganic substances such as CO_2 and H_2S.

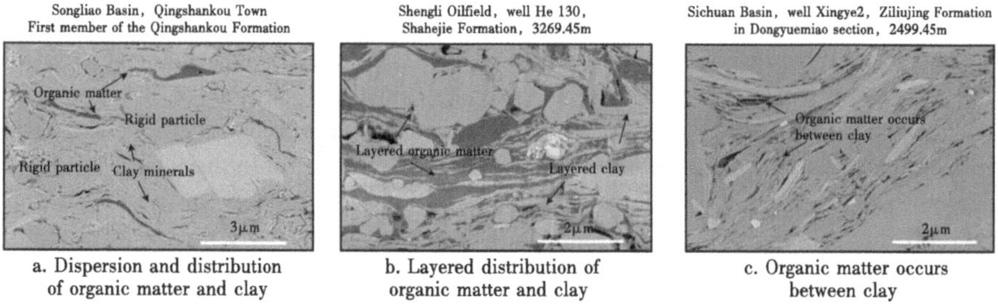

Fig. 12.9 Occurrence mode of organic matter and clay

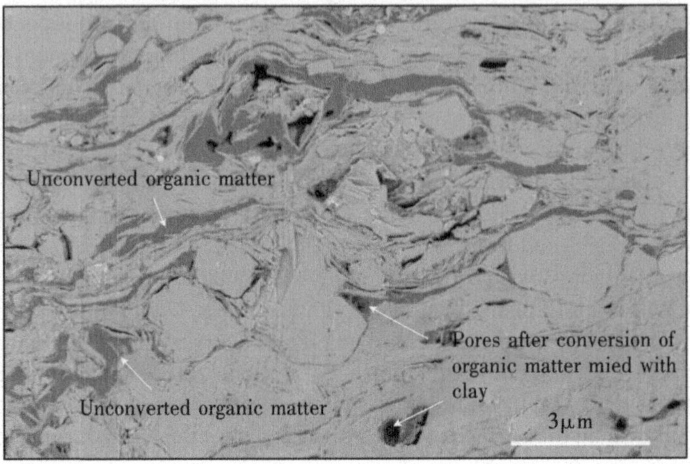

Fig. 12.10 Differential hydrocarbon generation of organic matter–clay complex Shengli Oilfield, well He 130, Shahejie Formation, 3269.45 m

(3) Condensation (synthesis) reaction of kerogen: Hydrocarbons (mainly C_1–C_{15} short-chain alkanes) are formed through a condensation (synthesis) reaction based on the bond energy of branched hydrocarbons.

12.2.3 Composition and Enrichment Characteristics of Special Trace Elements in Lacustrine Source Rocks

A great deal of statistical analysis shows that crude oil produced in different basins enriches in various metal and nonmetal elements. This phenomenon also exists in the source rocks of these crude oils (Fig. 12.12), and the content of these elements in crude oil is often higher than that in the source rocks. This indicates that migration and enrichment have occurred during the process of oil generation.

During the formation and evolution of sedimentary basins, volcanic activity and magmatic intrusion are the most common external material intervention phenomena, especially in rift basins (Table 12.1). Among these, various volcanic materials, particularly volcanic ash, have a significant influence on the sedimentary environment, water properties, and biological reproduction, and therefore play an essential role in promoting the development of source rocks and oil generation in the basin. Volcanic spewing substances such as volcanic ash contain many elements required for biological reproduction, and when they enter lake or ocean basins, they promote the rapid growth of aquatic organisms. In areas of volcanic eruption, volcanic ash falling into the water has a "fertilization" or "nutritional" effect, causing creatures in the water to grow rapidly in a short time. The addition of volcanic ash also increases the salinity of the water body, leading to the stratification of the water body and the formation of

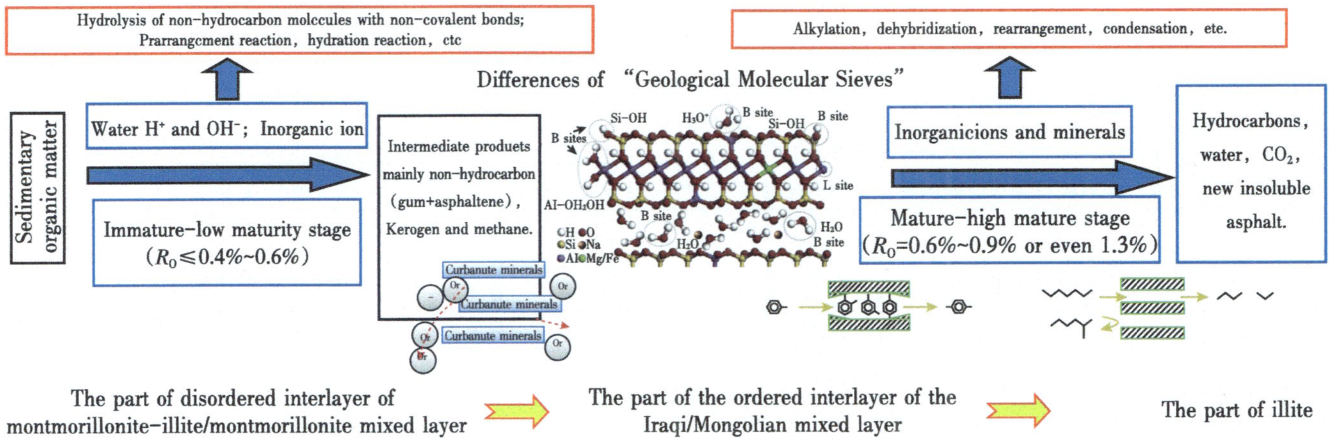

Fig. 12.11 Schematic diagram of oil generation catalyzed by "geological molecular sieve" in source rock

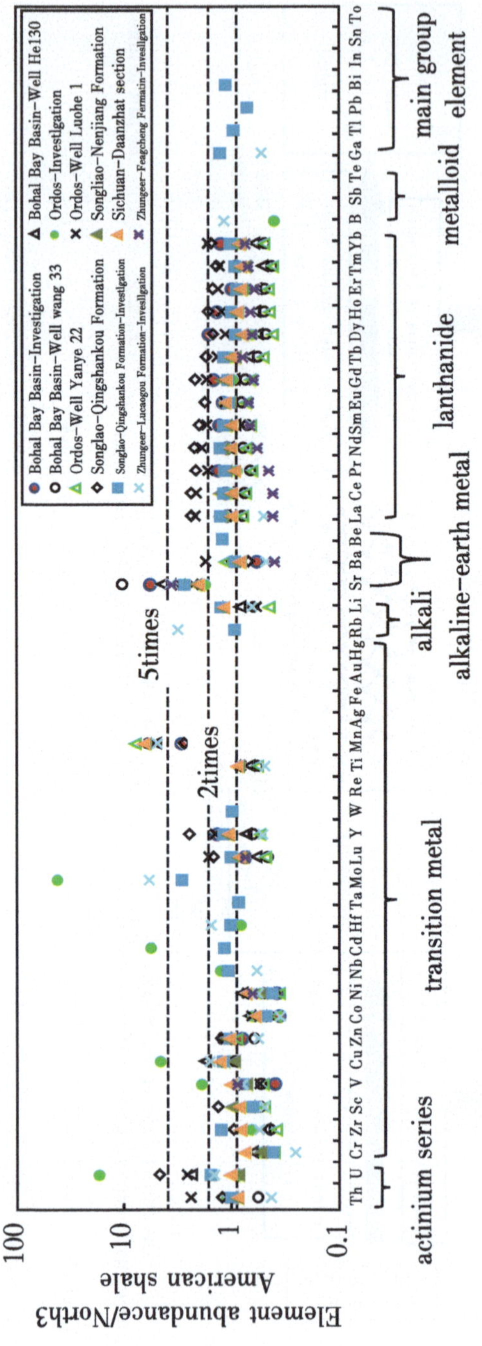

Fig. 12.12 Enrichment characteristics of special trace elements in main lacustrine source rocks

a reducing environment at the bottom. At the same time, it creates a favorable burial environment and provides the conditions necessary for the formation of high-quality source rocks. The study of lacustrine and marine oil-bearing basins reveals that high-quality source rocks are closely related to the temporal and spatial distribution of volcanic ash or tuff, with the TOC content increasing accordingly.

In the process of biological growth, organisms will enrich many metal elements through organ absorption, surface adsorption, and complex formation. During the geological evolution from primitive organisms to kerogen, although the organic body releases metal elements due to oxidative decomposition, many organic acids, such as amino acids, fulvic acids, and humic acids produced by the decomposition of organisms, have a strong complexing or chelating effect on the metal elements, which can lead to the enrichment of all kinds of metal elements (Fig. 12.13). For example, the abundance of organic matter in the source rocks of the Qingshankou Formation and Nenjiang Formation in the Songliao Basin increases with the enrichment of metal elements, and the abundance of metal elements increases with the increase of clay mineral content (Figs. 12.14 and 12.15).

The special trace elements contained in the main source rocks of major lacustrine basins were statistically analyzed (Fig. 12.16), which can be divided into transition metal elements, radioactive actinide metal elements, lanthanide metals, alkali metals and alkaline earth metals, and some main group elements. There are some differences in each basin, but in general, the contents are mainly Zn, V, Mn, and Mo of transition families, U and Th of radioactive actinides, and Sr, Ba, and Li of alkali metals and alkaline earth metals, which can reach 2–5 times of the crustal background value.

Table 12.1 Volcanic activity and distribution of source rocks in main lacustrine basins

Basin	Age of volcanic activity	Contact mode	Volcanic rock type
Bohai Bay Basin	Sedimentary periods of the 4th and 3rd members of the Shahejie Formation	Locally uneven distribution, upwelling along the fault	Mainly erupted basic basalt with little tuff
Songliao Basin	The second member of the Qingshankou Formation in the Songliao Basin underwent development through four volcanic events during the Cretaceous period	Interlayers	Olive andesite, local basalt, provenance is neutral-acid extrusive rock
Ordos Basin	Multi-stage volcanic activity in Chang 7 member, with the strongest at the bottom	Mark layer at the bottom of Chang 7, interbedded tuff	Tuff contrast layer Inflow dacite–dacite area, indicating its source is mainly neutral-acid rocks
Junggar Basin	Sedimentary period of Fengcheng Formation	Interlayers	Carboniferous volcanic lava, basic basalt and neutral andesite

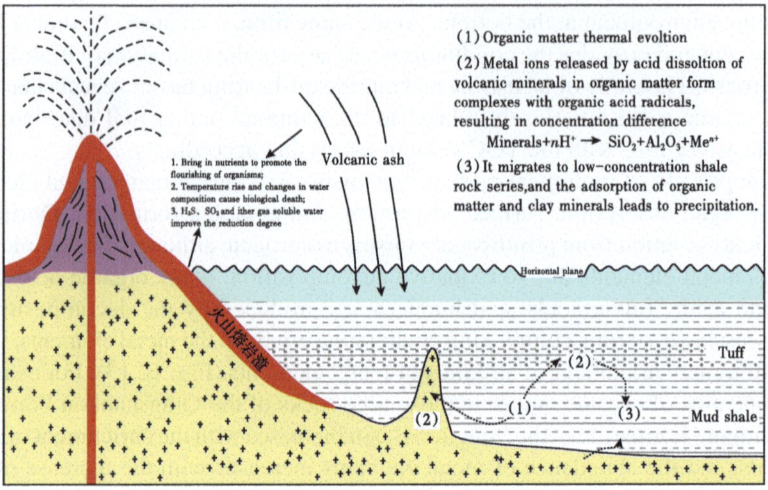

Fig. 12.13 Influence model of volcanic activity on hydrocarbon generation of source rocks

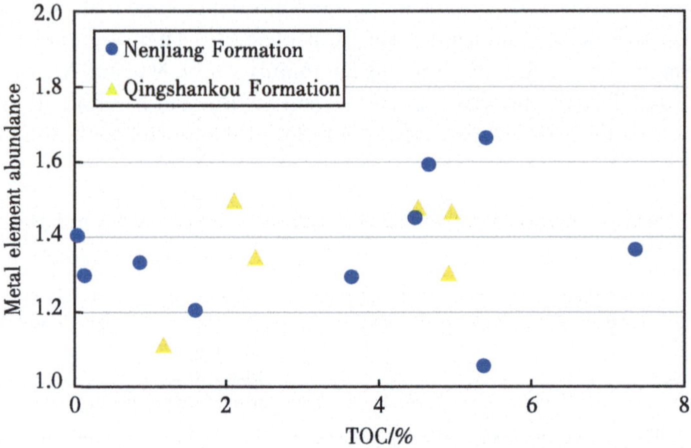

Fig. 12.14 Relationship between TOC and metal elements of main source rocks in the Songliao Basin

In summary, the abundance of organic matter in source rocks increases with the enrichment of metal elements and clay minerals, and the abundance of metal elements increases with the increase of clay mineral content and organic matter abundance. This shows that there is a synergistic evolution process among the three, and these rich special trace elements also exist in the "organic matter–clay mineral aggregate". A great deal of petrochemical experience shows that clay minerals have a high specific surface area and acid center and have a strong catalytic cracking reaction ability. Some transition metal elements are important catalysts for the deep processing of

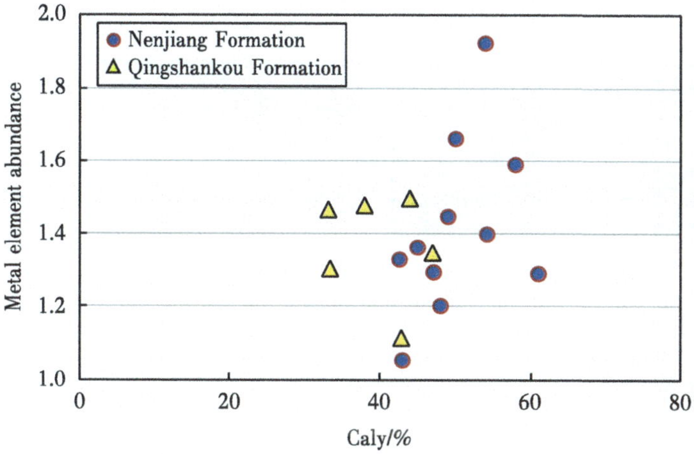

Fig. 12.15 Relationship between clay mineral content and metal elements of main source rocks in the Songliao Basin

crude oil, which are beneficial to the transformation of macromolecular hydrocarbons into small molecular hydrocarbons and mainly play the role of hydrogenation. The detailed analysis of lacustrine source rocks also shows that its organic matter mainly exists with clay minerals and these special trace elements, which is likely forming a natural "geological catalyst" and will inevitably have an important impact on the oil generation process of source rocks.

Hence, when conducting catalytic oil generation simulation experiments under geological conditions, it is crucial to comprehensively analyze the mineral composition of source rocks. This analysis should consider factors such as clay minerals, cuttings, metal and nonmetal elements, rare radioactive elements, and more. It should be based on the material composition of source rocks, as well as the temperature and pressure conditions during the geological evolution stage of source rocks. By doing so, we can gain insights into how various material components in source rocks influence the process of organic pyrolysis oil generation. This analysis helps to elucidate the catalytic oil generation mechanism of source rocks under geological conditions.

Bsains	Siurce rocks	Transition group metal			Actinide metal			Lanthanide metal			Alkali metals Alkaline earth metals			Main group element		
		1~2倍	2~5倍	>5倍	1~2倍	2~5倍	>5倍	1~2倍	2~5倍	>5倍	1~2倍	2~5倍	>5倍	1~2倍	2~5倍	>5倍
Bohai bay	Paleogene Shahejie Formation	Y、Ni、Cu Cr	Zn Mn、V Cd		Th	U		Ce、Pr、Sm、Eu、Gd、Yb Er	Dy		Li	Sr、Ba				
Songliao	Ceetaceius Qingshankou Formation, Nenjiang Formation	Ni、Ta、Cu、Cr、Nb、Ti、Zr、Sc、Hf、Lu、W、V、Y	Zn、Mn、Cd、	Mo	Th		U	Ho、Tb、Eu、Sm、La	Er、Ce、Pr、Nd、Gd、Dy、Yb、Tm		Sr	Li、Ba		Ga、Ti Pb	Bi	
Sichuan	Junassic Ziliujing Formation	Cu、Lu、Y、Ni	Zn、V、Mn		Th、U			Er、Ce、Tb、Eu、Gd、Tm、Dy、Yb、Er				Li、Ba				
Ordos	Permian Yanchang Formation	Zr、Ta、Ti、Cr、Y、Ni、Mn、Hf、Nb、Lu	Cu、Zn、V	Cd Mo		Th	U	Sm、Dy、Nd、Gd、La、Tm、Ce、Yb、Er			Sr	Ba				
Junggar	Permian Lucaogou Formation, Fengcheng Formation	Lu、Zn、Ni、Cu、V	Mn Hf		U			Tm、Tb、Er			Li、Sr、Ba	Rb				

Fig. 12.16 Distribution characteristics of trace elements in oil source rocks of main continental basins

12.3 Simulation Experimental Study on Moderate-Temperature Catalytic Oil Generation in Source Rocks

12.3.1 Conditions and Methods of Simulation Experiment for Moderate-Temperature Catalytic Oil Generation

Hydrocarbon generation kinetics falls under the realm of thermal analysis kinetics. Its fundamental principle involves conducting hydrocarbon generation simulation experiments with programmed temperature control. This entails using a specific heating rate to mimic subsurface conditions. Throughout the reaction process, changes in physical parameters like mass and temperature difference between samples and reference materials are recorded over time or temperature. Assuming that these monitored physical quantities are directly proportional to the progress of the reaction or reaction rate, it becomes possible to calculate the pre-exponential factor A, activation energy E, and the reaction mechanism function f(a) based on the collected data. The experiment is carried out by gradually heating the samples to a predetermined temperature at a controlled rate, thereby avoiding the potential influence of excessively high temperatures experienced through constant heating at elevated temperatures in the past.

Four groups of experiments, LX-TSY, LX-INT, LX-TM, and LX-U, used the same heating mode. The heating temperature ranged from 200 to 350 °C, and the heating rate was divided into two series: 1 and 2 °C/h, to obtain kinetic parameters of hydrocarbon generation in different simulation experiments. LX-TSY and LX-NT series of experiments are compared with each other (Tables 12.2 and 12.3). The former is mainly composed of carbonate minerals (45%), while the latter is mainly composed of clay minerals (50%), to clarify the influence of clay minerals on oil and gas generation. LX-NT, LX-U, and LX-TM experimentswere also compared. The LX-U series of experimental samples were loaded with radioactive elements represented by U and Th elements, while the LX-TM experimental samples were loaded with transition metal elements represented by Mn, Mo, and Zn, to clarify the mechanism of catalytic hydrocarbon generation by different metal elements.

Table 12.2 Kinetic experimental conditions of LX-TSY series

Serial number	TEMP/°C	Lithostatic pressure/MPa	Fluid pressure/MPa	Heating rate/°C/h	Heat-up time/day	Quartz/%	Feldspar/%	Carbonate/%	Clay mineral/%
1	200	62.5	25	1	8.33	25	10	45	20
2	250	62.5	25	1	10.42	25	10	45	20
3	300	62.5	25	1	12.5	25	10	45	20
4	325	62.5	25	1	13.54	25	10	45	20
5	350	62.5	25	1	14.58	25	10	45	20
6	200	62.5	25	2	4.17	25	10	45	20
7	250	62.5	25	2	5.21	25	10	45	20
8	300	62.5	25	2	6.25	25	10	45	20
9	325	62.5	25	2	6.77	25	10	45	20
10	350	62.5	25	2	7.29	25	10	45	20

Table 12.3 Kinetic experimental conditions of clay mineral containing catalyst series

Serial number	Temperature /°C	Lithostatic pressure/ MPa	Fluid pressure/ MPa	Heating rate/°C/h	Heating time/day	Quartz/ %	Feldspar/ %	Carbonate/ %	Clay mineral (LX-NT)/ %	Transition metal (LX-TM)/ (μg/g)	Radioactive metal (LX-U)/ (μg/g)
1	200	62.5	25	1	8.33	25	10	15	50	Mn, Zn (2105.26) Mo (31.58) Cd (94.74) Ni (968.42)	U (31.58) Th (10.53)
2	250	62.5	25	1	10.42	25	10	15	50		
3	300	62.5	25	1	12.5	25	10	15	50		
4	325	62.5	25	1	13.54	25	10	15	50		
5	350	62.5	25	1	14.58	25	10	15	50		
6	200	62.5	25	1	4.17	25	10	15	50		
7	250	62.5	25	1	5.21	25	10	15	50		
8	300	62.5	25	1	6.25	25	10	15	50		
9	325	62.5	25	1	6.77	25	10	15	50		
10	350	62.5	25	1	7.29	25	10	15	50		

12.3.2 *Analysis of Simulation Experiment Results of Medium Temperature Catalytic Oil Generation*

12.3.2.1 Variation Characteristics of Different Series of Experimental Products

Changes of LX-NT Series Simulation Experiment Products

The LX-NT series experiments used mudstone samples with mainly clay minerals (50%) and a relatively low content of carbonate minerals (15%), without the addition of metal element catalysts. The hydrocarbon gas yield in the experimental products varied in two stages, as shown in Table 12.4. In the first stage, with simulated heating temperatures between 200 and 325 °C, the experiment with a heating rate of 1 °C/h consistently yielded more hydrocarbon gas than the one with a heating rate of 2 °C/h. The hydrocarbon gas yield was relatively stable and weakly affected by temperature but increased rapidly after the simulated temperature exceeded 300 °C (Fig. 12.17). In the second stage, with simulated heating temperatures above 325 °C and a heating rate of 2 °C/h, the hydrocarbon gas yield was higher than in the experiment with a heating rate of 1 °C/h. Compared to the first stage, the hydrocarbon gas yield increased rapidly with temperature. The variations of CO_2 yield, methane, ethane, and non-hydrocarbon gases were consistent with those of hydrocarbon gases.

The experimental oil generation rate and gas generation rate show different changing characteristics, which generally increase with the increase of temperature. The experimental oil generation process is divided into two stages (Table 12.5). In the first stage, the simulated heating temperature is between 200 and 250 °C, and the heating rate of 2 °C/h is always higher than the heating rate of 1 °C/h experiment. The oil generation rate increases slowly and is weakly affected by temperature (Fig. 12.18). In the second stage, the simulated temperature was higher than 250 °C, and the heating rate was 1 °C/h, and the hydrocarbon gas yield was higher than that of the heating rate was 2 °C/h, which showed a reversal. The oil-generating yield increased obviously with the increase of temperature, and an inflection point appeared when the heating temperature was 300 °C. After that, the oil-generating yield increased rapidly and began to generate a lot of oil.

Changes of LX-TSY Series Simulation Experiment Products

The mineral composition of mudstone samples used in the LX-TSY series experiments is mainly composed of carbonate minerals (45%) and has a relatively low content of clay minerals (20%), and no metal elements were added as catalysts. The oil-generating characteristics of the experimental products are like those of gas generation, with oil and gas yields increasing gradually with the change of temperature. The oil–gas yield is always higher when the heating rate is 1 °C/h, and it increases significantly with the increase of temperature when the temperature is higher than

Table 12.4 Characteristics of gas generation changes in different series of experiments

Sample number	TEMP/ °C	Lithology	CO_2 generation rate/m^3/tTOC	H_2 generation rate/m^3/tTOC	Hydrocarbon gas generation rate/m^3/tTOC	CH_4/m^3/tTOC	C_2H_6/m^3/tTOC	C_3H_8/m^3/tTOC
LX-NT-200-1	200	Mudstone	94.37	0.22	1.78	1.34	0.33	0.11
LX-NT-250-1	250	Mudstone	68.22	23.1	1.41	0.94	0.31	0.16
LX-NT-300-1	300	Mudstone	76.36	0.25	1.72	1.35	0.25	0.12
LX-NT-325-1	325	Mudstone	105.75	2.83	3.64	2.96	0.4	0.27
LX-NT-350-1	350	Mudstone	190.59	5.29	8.49	5.85	1.11	0.7
LX-NT-200-2	200	Mudstone	31.24	0.2	0.8	0.5	0.2	0.1
LX-NT-250-2	250	Mudstone	35.3	1.38	0.91	0.68	0.12	0.11
LX-NT-300-2	300	Mudstone	89.64	2.51	1.06	0.79	0.13	0.13
LX-NT-325-2	325	Mudstone	141.6	2.63	4.03	3.1	0.47	0.19
LX-NT-350-2	350	Mudstone	210.85	4.19	9.54	6.81	1.36	0.73
LX-TSY-200-1	200	Mudstone	13.45	0.05	0.75	0.55	0.15	0.05
LX-TSY-250-1	250	Mudstone	25.54	0.16	0.32	0.21	0.05	0.05
LX-TSY-300-1	300	Mudstone	54.54	0.57	0.97	0.75	0.13	0.09
LX-TSY-325-1	325	Mudstone	88.5	0.61	2.79	1.91	0.37	0.24
LX-TSY-350-1	350	Mudstone	113.86	0.43	7.13	4.4	1.15	0.77
LX-TSY-200-2	200	Mudstone	8.6	0.09	0.1	0.05	0.05	0
LX-TSY-250-2	250	Mudstone	22.82	0.14	0.15	0.1	0.05	0
LX-TSY-300-2	300	Mudstone	38.92	1.03	0.76	0.63	0.09	0.04
LX-TSY-325-2	325	Mudstone	74.55	0.55	2.34	1.66	0.31	0.18
LX-TSY-350-2	350	Mudstone	104.86	0.72	4.72	2.95	0.8	0.46

(continued)

Table 12.4 (continued)

Sample number	TEMP/°C	Lithology	CO_2 generation rate/m^3/tTOC	H_2 generation rate/m^3/tTOC	Hydrocarbon gas generation rate/m^3/tTOC	CH_4/m^3/tTOC	C_2H_6/m^3/tTOC	C_3H_8/m^3/tTOC
LX-U-200-1	200	Mudstone	1.22	0.24	0	0	0	0
LX-U-250-1	250	Mudstone	5.36	0.89	1.34	1.12	0.11	0.11
LX-U-300-1	300	Mudstone	27.04	0.82	2.69	2.34	0.23	0.12
LX-U-325-1	325	Mudstone	49.27	1.08	8.75	7.22	0.72	0.45
LX-U-350-1	350	Mudstone	76.87	2.28	31.68	16.72	5.49	4.05
LX-U-200-2	200	Mudstone	14.1	0.31	0.93	0.85	0	0
LX-U-250-2	250	Mudstone	15.89	0.04	0.13	0.13	0	0
LX-U-300-2	300	Mudstone	79.65	0.36	2.65	2.25	0.14	0.09
LX-U-325-2	325	Mudstone	119.93	0.39	4.22	3.36	0.34	0.22
LX-U-350-2	350	Mudstone	57.93	0.11	15.23	11.35	2.52	1.36
LX-TM-200-1	200	Mudstone	81.95	0.16	0.36	0.24	0.04	0.04
LX-TM-250-1	250	Mudstone	136.18	0.12	0.32	0.32	0	0
LX-TM-300-1	300	Mudstone	214.11	0.23	3.1	2.79	0.12	0.08
LX-TM-325-1	325	Mudstone	147.25	0.27	3.75	3.38	0.18	0.09
LX-TM-350-1	350	Mudstone	194.24	0.93	8.98	7.02	0.79	0.47
LX-TM-200-2	200	Mudstone	82.09	0.29	0.43	0.19	0.1	0.1
LX-TM-250-2	250	Mudstone	121.01	0.11	0.22	0.22	0	0
LX-TM-300-2	300	Mudstone	116.1	0.11	1.3	0.92	0.16	0.16
LX-TM-325-2	325	Mudstone	174.45	0.4	5.55	5.05	0.2	0.1
LX-TM-350-2	350	Mudstone	281.78	0.82	11.05	8.29	1.13	0.66

Note LX-NT-200-1 refers to the experiment of LX-NT series at the temperature of 200 °C and the heating rate of 1 °C/h, the same below

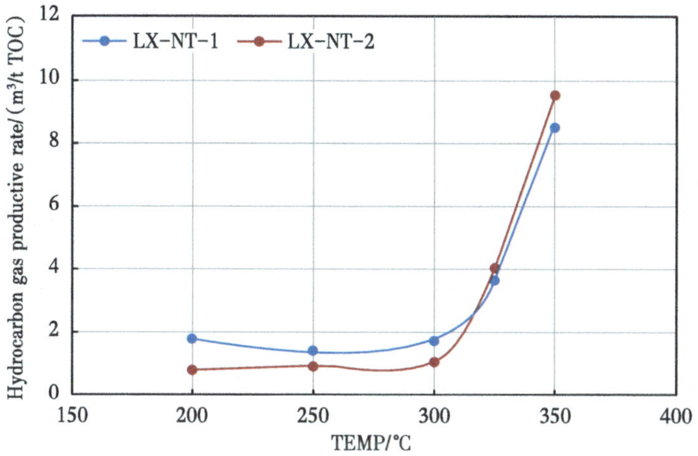

Fig. 12.17 Hydrocarbon gas yield change at different heating rates in LX-NT series experiments

300 °C (Tables 12.4, 12.5 and Fig. 12.19). Additionally, the variation in methane, ethane, and propane yields in hydrocarbon gas is consistent with the total hydrocarbon gas yield, and the CO_2 content in non-hydrocarbon gases increases significantly with the simulated temperature.

Changes of LX-U Series Simulation Experiment Products

The mineral composition of mudstone samples used in LX-U series experiments is primarily clay minerals (50%), with a relatively low content of carbonate minerals (15%). Additionally, U (31.58 μg/g) and Th (10.53 μg/g) were added, which is like the metal element content in source rocks in the Ordos Basin. When the simulated heating temperature is between 200 and 300 °C, the hydrocarbon gas yield in the experimental products remains relatively stable and is weakly affected by temperature (Fig. 12.20a). However, when the simulated heating temperature exceeds 300 °C, the hydrocarbon gas yield increases significantly with the increase in temperature. The variation in methane, ethane, propane, and non-hydrocarbon gas (H_2) yield is consistent with that of hydrocarbon gas (Table 12.4). Except for the simulation experiment at 200 °C, the hydrocarbon gas yield of the experiment with a 2 °C/h heating rate is higher than that of the experiment with a 1 °C/h heating rate.

The experimental oil generation rate and gas generation rate show different characteristics, which generally increase with an increase in temperature. The experimental oil generation process is divided into two stages (Table 12.5). In the first stage, the simulated heating temperature ranges from 200 to 325 °C, and the experimental oil generation rate with a heating rate of 2 °C/h is generally higher than that with a heating rate of 1 °C/h. The oil generation rate is slightly affected by temperature and changes slowly (Fig. 12.20b). In the second stage, the simulated heating temperature

Table 12.5 Characteristics of oil generation in different experiments

Sample number	Temperature/°C	Lithology	Hydrocarbon gas generation rate/m^3/tTOC	Expelled oil generation rate/kg/tTOC	Residual oil generation rate/kg/tTOC	Total oil generation rate/kg/tTOC	Total hydrocarbon generation rate/kg/tTOC
LX-NT-200-1	200	Mudstone	1.62	8.08	15.46	23.54	25.16
LX-NT-250-1	250	Mudstone	1.4	0.57	40.61	41.18	42.58
LX-NT-300-1	300	Mudstone	1.54	1.31	94.21	95.52	97.05
LX-NT-325-1	325	Mudstone	3.2	3.77	142.07	145.84	149.04
LX-NT-350-1	350	Mudstone	9.29	20.69	376.36	397.05	406.34
LX-NT-200-2	200	Mudstone	0.82	2.17	26.39	28.56	29.38
LX-NT-250-2	250	Mudstone	1.13	1.29	42.74	44.03	45.16
LX-NT-300-2	300	Mudstone	1	1.45	54.22	55.67	56.67
LX-NT-325-2	325	Mudstone	4.01	1.99	106.6	108.59	112.6
LX-NT-350-2	350	Mudstone	9.98	23.28	254.49	277.77	287.75
LX-TSY-200-1	200	Mudstone	0.7	4.36	27.98	32.34	33.04
LX-TSY-250-1	250	Mudstone	0.33	2.13	40.18	42.31	42.64
LX-TSY-300-1	300	Mudstone	0.89	2.85	85.54	88.39	89.28
LX-TSY-325-1	325	Mudstone	3.11	40.4	179.17	219.57	222.68
LX-TSY-350-1	350	Mudstone	8.39	11.02	258.28	369.3	377.69
LX-TSY-200-2	200	Mudstone	0.1	1.52	27.93	29.45	29.55
LX-TSY-250-2	250	Mudstone	0.13	5.4	22.66	28.06	28.19
LX-TSY-300-2	300	Mudstone	0.66	5.58	74.81	80.39	81.05
LX-TSY-325-2	325	Mudstone	2.49	9.37	133.2	142.57	145.06
LX-TSY-350-2	350	Mudstone	5.46	44.73	198.8	243.53	248.99

(continued)

Table 12.5 (continued)

Sample number	Temperature/ °C	Lithology	Hydrocarbon gas generation rate/m^3/tTOC	Expelled oil generation rate/kg/tTOC	Residual oil generation rate/kg/tTOC	Total oil generation rate/kg/tTOC	Total hydrocarbon generation rate/kg/tTOC
LX-U-200-1	200	Mudstone	0	1.14	41.29	42.43	42.43
LX-U-250-1	250	Mudstone	1.17	1.07	31.51	32.58	33.75
LX-U-300-1	300	Mudstone	2.22	2.05	58.56	60.61	62.83
LX-U-325-1	325	Mudstone	8.02	12.63	108.81	121.44	129.46
LX-U-350-1	350	Mudstone	41.69	74.51	228.35	302.86	344.55
LX-U-200-2	200	Mudstone	0.76	0.73	36.6	37.33	38.09
LX-U-250-2	250	Mudstone	0.09	0.79	42.86	43.65	43.74
LX-U-300-2	300	Mudstone	2.38	1.59	64.56	66.15	68.53
LX-U-325-2	325	Mudstone	4.04	2.18	108.05	110.23	114.27
LX-U-350-2	350	Mudstone	14.12	6.47	168.57	175.04	189.16
LX-TM-200-1	200	Mudstone	0.41	0.54	44.97	45.51	45.92
LX-TM-250-1	250	Mudstone	0.23	0.81	39.46	40.27	40.5
LX-TM-300-1	300	Mudstone	2.55	0.56	22.57	23.13	25.68
LX-TM-325-1	325	Mudstone	3.08	1.92	70.52	72.44	75.52
LX-TM-350-1	350	Mudstone	8.75	9.97	154.28	164.25	173
LX-TM-200-2	200	Mudstone	0.58	0.54	45.7	46.24	46.82
LX-TM-250-2	250	Mudstone	0.15	0.84	39.41	40.25	40.4
LX-TM-300-2	300	Mudstone	1.34	1.14	50.95	52.09	53.43
LX-TM-325-2	325	Mudstone	4.53	1.11	70.47	71.58	76.11
LX-TM-350-2	350	Mudstone	10.56	7.57	121.76	129.33	139.89

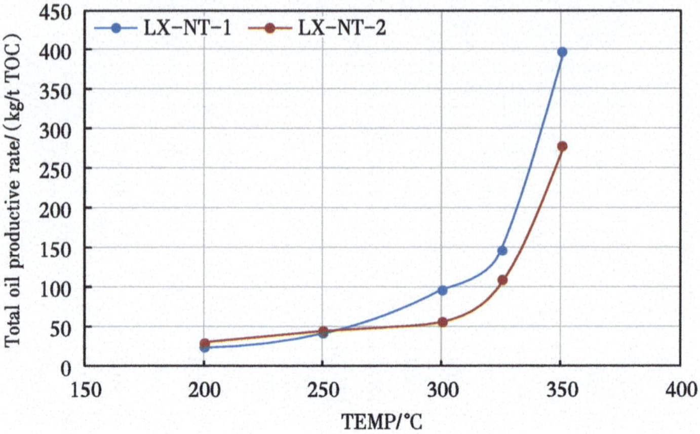

Fig. 12.18 LX-NT series experiments with different heating rates

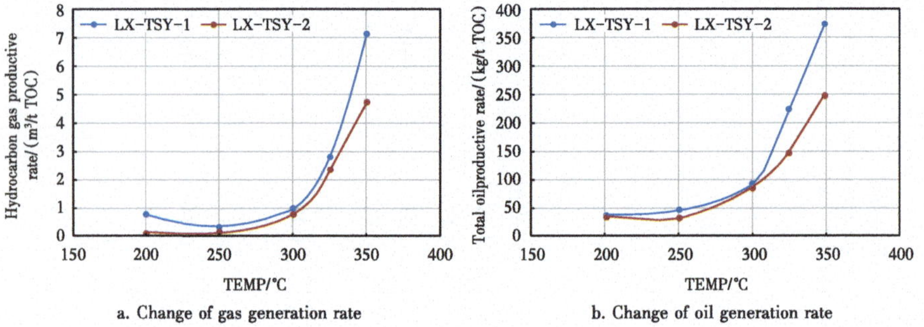

Fig. 12.19 Changes of oil and gas yields at different heating rates in LX-TSY series experiments

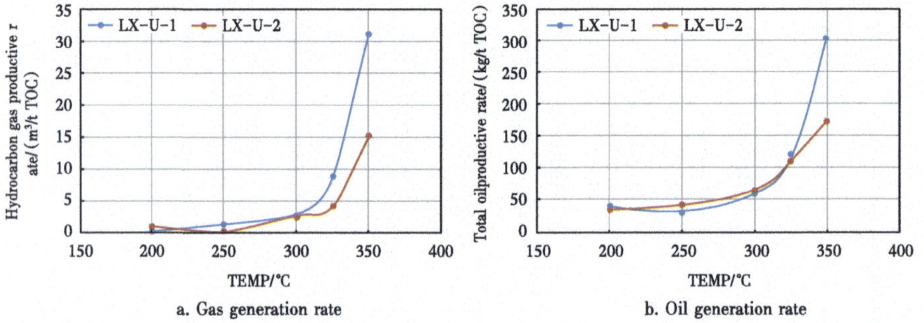

Fig. 12.20 Changes of oil and gas yields at different heating rates in LX-U series experiments

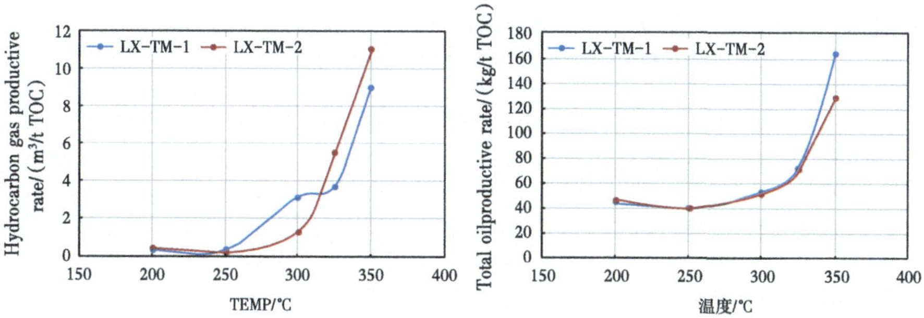

a. Gas generation rate

b. Oil generation rate

Fig. 12.21 Changes of oil and gas yields at different heating rates in LX-TM series experiments

exceeds 325 °C, and the heating rate is 1 °C/h, while the hydrocarbon gas yield is higher than that of the heating rate of 2 °C/h, which shows a reversal. The oil generation yield increases significantly with the increase of temperature.

Changes of LX-TM Series Simulation Experiment Products

The mudstone samples used in the LX-TM series experiments are mainly composed of clay minerals (50%) with a relatively low content of carbonate minerals (15%). Additionally, the samples contain Mo (31.58 μg/g), Zn (2105.26 μg/g), Mn (2105.26 μg/g), Cd (94.74 μg/g), and Ni (968.42 μg/g). The mineral composition is similar to that of crude oil in the Songliao Basin. The experimental yield of both oil and gas increases with the simulation temperature, with a slow growth rate when the temperature is lower than 300 °C, and a rapid increase as the temperature rises above 300 °C. Methane is the main hydrocarbon gas, with a content ranging from 50 to 100%. Different heating rates have a significant impact on the hydrocarbon gas yield. When the simulated temperature is above 325 °C, the experiment with a heating rate of 2 °C/h yields more hydrocarbon gas than the one with a heating rate of 1 °C/h (Fig. 12.21a). The experimental yield of oil is less affected by the heating rate in the simulation experiment when the temperature is below 325 °C, and the oil yield is similar at different heating rates. However, when the temperature is above 325 °C, the experimental oil yield is higher with a heating rate of 1 °C/h (Fig. 12.21b). In the experiment, the CO_2 yield of non-hydrocarbon gas is high, ranging from 81.95 to 293.09 m^3/tC, and the variation trend of CO yield is similar to that of the hydrocarbon gas.

12.3.2.2 Comparison of Different Series of Hydrocarbon Generation Kinetics and Characteristics of Dynamic Parameters Change

The variations in oil and gas yield observed in different experiment series show some similarities, with an inflection point at a temperature of 300 °C. At temperatures below 300 °C, the oil and gas yield changes gradually with temperature, while at temperatures above 300 °C, the yield increases rapidly with temperature (Fig. 12.22). Furthermore, the process of oil and gas generation is complementary to some extent, as the higher the oil generation rate at a given temperature, the lower the hydrocarbon gas yield.

In the experiment series with a heating rate of 1 °C/h, the LX-U series loaded with a catalyst containing radioactive metal elements exhibits a higher oil generation rate than other experiment series at a simulated temperature of 200 °C, and almost no hydrocarbon gas is generated (Fig. 12.22c). This suggests that radioactive metal elements are more conducive to organic oil generation at lower temperatures in the experiments. After the simulated experiment temperature exceeded 250 °C, the gas yield of LX-U series experiments began to exceed that of other experiment series. At 350 °C, the gas yield was three times higher than that of other series, while the oil yield was always lower than that of other series, indicating that radioactive metal elements are more conducive to the generation of hydrocarbon gases at higher temperatures (>300 °C). Regardless of whether the heating rate is the same or not, the

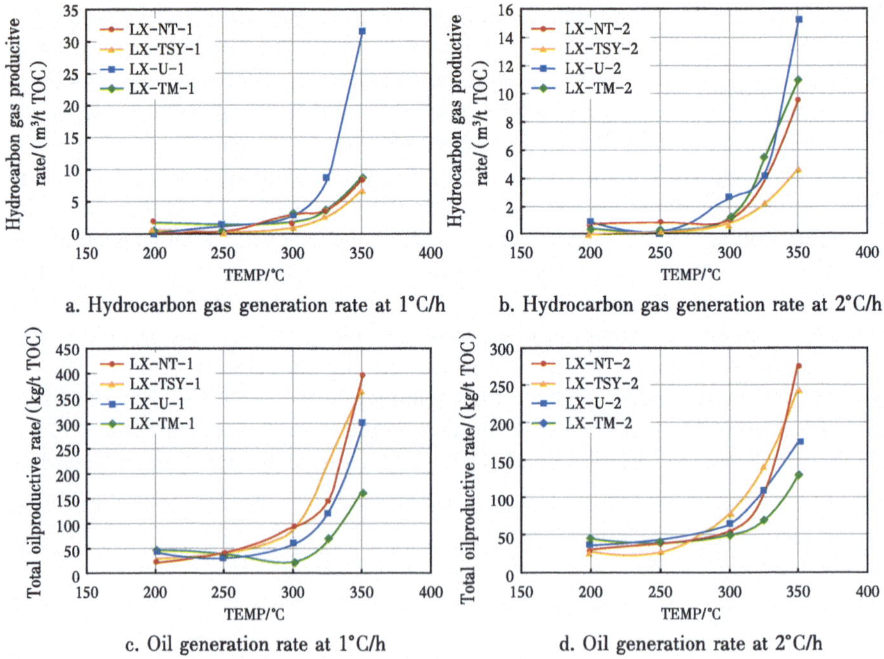

a. Hydrocarbon gas generation rate at 1°C/h

b. Hydrocarbon gas generation rate at 2°C/h

c. Oil generation rate at 1°C/h

d. Oil generation rate at 2°C/h

Fig. 12.22 Comparison of oil and gas generation rates at different heating rates hydrocarbon gas yield of the LX-TSY series with lower clay mineral content in the experimental samples is always lower than that of the LX-NT series with higher clay mineral content (Fig. 12.22a, b), indicating that an increase in clay mineral content promotes the gas generation process of source rocks.

A comparative analysis of the different series of experimental products at the same temperature shows that the characteristics of catalytic hydrocarbon generation can be divided into three stages within the simulated experimental temperature range of 200–350 °C: the oil generation stimulating stage, the transition stage, and the gas generation stimulating stage (see Fig. 12.23).

Oil generation stimulating stage (T < 250 °C): In this stage, the combination of metal elements mainly promotes oil generation and inhibits the generation of hydrocarbon gas. Among the metal element combinations, transition metal elements have the strongest effect on oil generation, with an increase in oil generation rate of about 93%. An increase in the content of carbonate minerals shows an inhibitory effect on hydrocarbon gas generation, with a yield decrease of about 88%. The oil generation rate of the radioactive element combination is higher than that of other series of experiments, and almost no hydrocarbon gas is generated, which shows that radioactive elements are more helpful for organic matter oil generation in experiments at lower temperatures. By analyzing the experimental data and extrapolating to lower temperatures through mathematical fitting of data points, the effect of stimulating oil generation and inhibiting gas generation becomes more evident (Fig. 12.24).

The characteristics of catalytic hydrocarbon generation can be divided into three stages within the simulated experimental temperature range of 200–350 °C: oil generation stimulating stage, transition stage, and gas generation stimulating stage, as shown in Fig. 12.23, which shows the comparative analysis of different series of experimental products at the same temperature.

In the oil generation stimulating stage (T < 250 °C), the combination of metal elements mainly promotes oil generation and inhibits the generation of hydrocarbon gas. The combination of transition metal elements has the strongest effect on oil generation, increasing the oil generation rate by about 93%. The increase of carbonate mineral content also inhibits hydrocarbon gas generation, with the yield decreasing by about 88%. The oil generation rate of experiments with radioactive elements is higher than that of other series of experiments, and almost no hydrocarbon gas is generated, indicating that radioactive elements are more helpful to organic matter oil generation in lower temperature experiments. After sorting out the experimental data and extrapolating to a lower temperature through mathematical fitting of data points, the effect of promoting oil generation and inhibiting gas generation becomes more apparent, as shown in Fig. 12.24.

In the transition stage (250 °C ≤ T < 300 °C), the addition of metal elements has no effect on oil generation, and the generation of hydrocarbon gas is still limited. With the increase of temperature, the inhibition of hydrocarbon gas gradually changes into promotion, and oil generation is gradually inhibited.

In the gas generation stimulating stage (T ≥ 300 °C), the combination of metal elements has an obvious effect on promoting gas generation, while the generation of oil is inhibited. At 300 °C, metal elements have the strongest stimulating effect

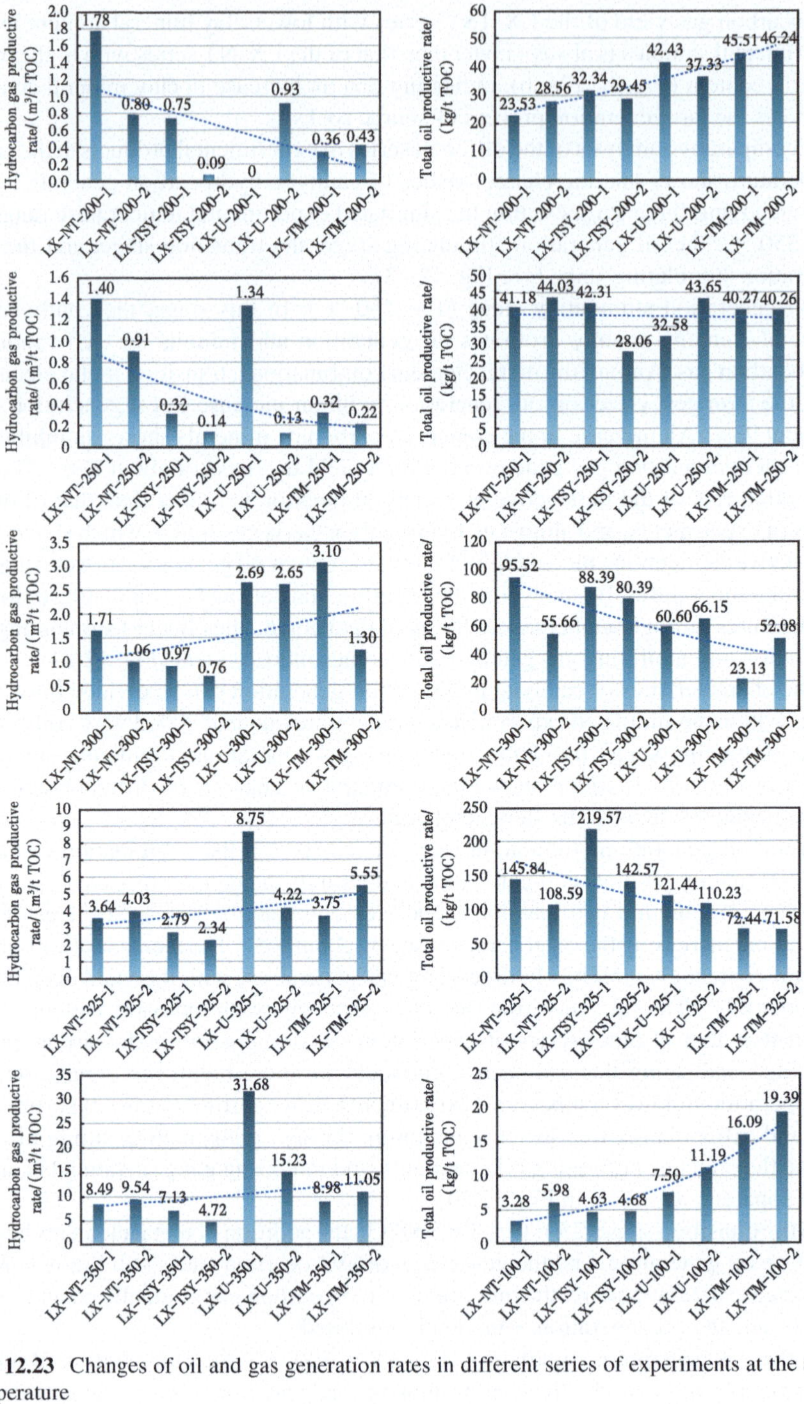

Fig. 12.23 Changes of oil and gas generation rates in different series of experiments at the same temperature

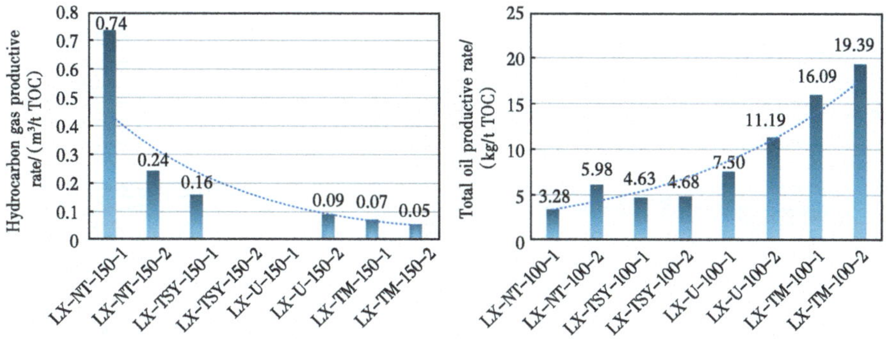

Fig. 12.24 Different series of experiment fitting to infer the changes of low temperature oil and gas generation rates

on hydrocarbon gas. With the increase of temperature, the stimulating effect on hydrocarbon gas becomes weaker, and the inhibition effect of metal elements on oil generation tends to be stronger. The combination of transition metal elements has the strongest inhibition effect on oil generation.

Based on the yield of LX-NT series of experimental products with no added metal elements and with clay minerals as the main mineral composition, the change rates of oil and gas products in different series of experiments are calculated, respectively. The change characteristics of oil and gas products with temperature are shown in Fig. 12.25. The characteristics of the change rate of oil and gas products are consistent with the characteristics of the three stages of catalytic hydrocarbon generation. With the increase of temperature, the change rate of hydrocarbon gas gradually increases, and the inhibitory effect of metal elements on hydrocarbon gas generation becomes weaker and weaker until it shows a promoting effect. The change rate of total oil gradually decreases, and lower temperatures promote oil generation more strongly. The hydrocarbon gas yield of LX-TSY series is always lower than that of LX-NT series, indicating that clay minerals in source rocks can promote hydrocarbon gas generation, while the change rate of total oil is always small, indicating that the influence of clay minerals on oil generation needs to be combined with related transition metal elements and radioactive elements.

12.3.3 Summary of Simulation Experiment of Moderate Temperature Catalytic Oil Generation

(1) Within the simulated experimental temperature range of 200–350 °C, catalytic hydrocarbon generation exhibits three distinct stages: the oil generation stimulating stage, the transition stage, and the gas generation stimulating stage.
(2) The experimental results within the simulated temperature range of 200–350 °C demonstrate that while certain trace elements can still have an effect at moderate

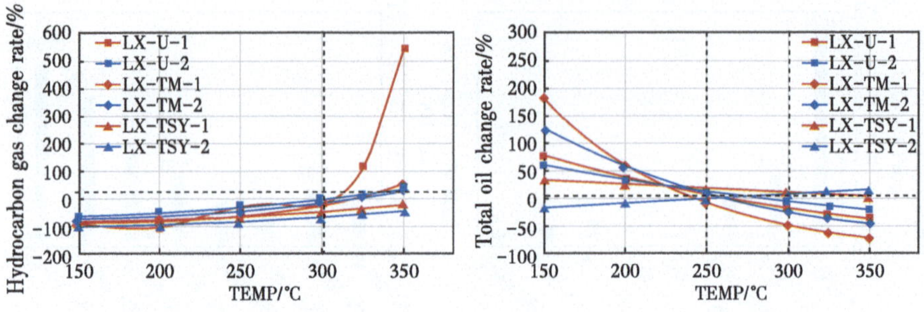

Fig. 12.25 Fitted changes of oil and gas generation rates of different series of experiments at low temperature (data after fitting)

temperatures, they are more likely to promote hydrocarbon gas generation and their mechanism of participation in organic matter hydrocarbon generation has changed. However, such reactions are not observed in actual geological situations. Therefore, when studying the oil generation of lacustrine source rocks, it is advisable to conduct experiments at relatively lower temperatures.

12.4 Simulation Experimental Study on Low Temperature Catalytic Oil Generation of Source Rocks

12.4.1 Conditions and Methods of Low Temperature Catalytic Oil Generation Simulation Experiment

12.4.1.1 Sample Selections

The mineral composition of the experimental samples was designed to reflect the basic mineral composition characteristics of organic-rich source rocks (TOC $\geq$ 2%) in the main lacustrine basins. To minimize the uncontrolled influence of various minerals on hydrocarbon generation, the main minerals were selected for artificial preparation of the experimental samples. The study found that, except for the Junggar Basin, the main source rock series in each oil-bearing basin consists of shale, a small amount of marl, fine sandstone, and siliceous rocks. Generally, clay minerals, quartz, calcite, feldspar, and other minerals are mainly developed in shale, except for carbonate minerals in the shale from the lower 3rd member of the Shahejie Formation to the upper 4th member of the Shahejie Formation in the Bohai Bay Basin. The shale minerals in other basins are mainly clay minerals, with an average mineral content of 40–58.2%, and the quartz content is relatively high, with an average mineral content of 26.5–32.9%. In this experiment, four minerals (quartz, feldspar, carbonate, and clay minerals) were selected for preparation, with a clay content of 50%, quartz content of 25%, carbonate content of 15%, and feldspar content of 10%. The organic matter was kerogen prepared from immature samples of the Qingshankou Formation, and the prepared samples had a TOC of 5%. The formation water used in the experiment was collected from oilfield water, which belonged to the $CaCl_2$ type formation water with a salinity of 228 °g/L. The ion composition of the formation water was similar to the actual composition of the main Chinese lacustrine basins (Fig. 12.26).

The investigation and measured data indicate that actinides and transitional metal elements are significantly enriched in the main source rocks of each basin, with an abundance 2–5 times higher than that of sedimentary elements. The actinides, mainly Th and U, and the transition metal elements, primarily V, Zn, Cu, Co, Cd, Mo and Mn, are the most abundant. Furthermore, the metal elements in great basin source rocks exhibit a tendency of enrichment to crude oil, with higher abundances observed in crude oil. Therefore, for experimental study, a combination of radioactive elements represented by U and Th and transition metal elements represented by V, Zn, Mn, Cd, and Mo was selected, with the content of each metal element closer to that of crude oil in the basin (Table 12.6). The DYS-150 series experiments focus on analyzing the metal elements present in the interaction between source rocks and crude oil in both the Songliao and Ordos basins. Among these experiments, the concentration of Mn is

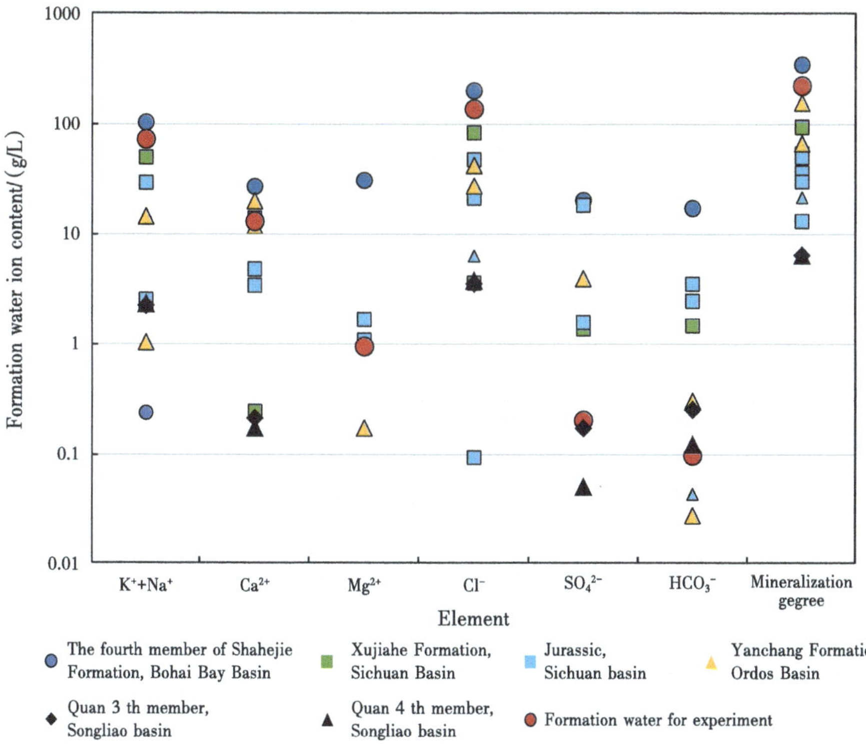

Fig. 12.26 Variations of ion content in formation water from different basins and in experiments

Table 12.6 Comparison between the experiment and the enrichment of metal elements in source rocks and crude oils in each basin (unit: μg/g)

Experimental sequence		Mo	Zn	Mn	Cd	Ni	V	U	Th
Bohai Bay Basin	Source rock		93.70	746.57	0.24	31.30	84.21	4.29	9.02
	Crude oil	96.00	2700		9.30		1000		
Songliao Basin	Source rock	3.42	109.46	504.72	0.18	26.05	101.23	6.24	13.18
	Crude oil	45.00	2500	6800	41.00	1700	330.00		
Sichuan Basin	Source rock		100	720.05		42.58	129.20	2.70	10.70
Ordos Basin	Source rock	48.72	102.83	341.10	0.83	31.18	125.47	17.91	17.89
	Crude oil	77.00	700		1.40	10,000			
Junggar Basin	Source rock	6.90	54.11	543.12		34.86	107.00	3.82	5.26
	Crude oil		1400	1000	1.40	3300	3000	100	
DYS-150		105.26		1368.42/ 2736.84			5263.16	21.05	
SL-200		210.53	2105.26	2105.26	210.53	2105.26			
LX-TM		31.58	2105.26	2105.26	94.74	968.42			
LX-U								31.58	10.53

Note Mn content of DYS-150 series experiments is 1368.42/36.84 μg/g, indicating that experiments are carried out with two concentrations of Mn

notably higher than that found in each respective basin, measuring at 52631.58 μg/g. Additionally, the metal element content observed in the SL-200 experiment closely aligns with that of crude oil discovered in the Songliao Basin.

12.4.1.2 Experimental Scheme

The experiment comprises two series of experiments, DYS-150 and SL-200, as shown in Table 12.7. The mineral composition of the two series of experiments is the same. The experiment temperature for DYS-150 is 150 °C, and the heating time is 30 days. The experimental variables involve individual metal elements, including Mn, V, U and Mo, to determine the effect of each metal element on oil and gas generation at low temperature and different concentrations. The experiment temperature for SL-200 is 200 °C, and the heating time is 90 days. The loaded metal elements are

Table 12.7 Experimental conditions of low temperature and longtime catalysis

Serial number	TEMP/ °C	Time/ day	Experimental number	Catalytic substance	Catalytic substance concentration/ (μg/g)	Formation pressure/ MPa	Lithostatic pressure/ MPa
1	150	30	DYS-150-0	–	–	25	62.5
2	150	30	DYS-150-5Mn	Mn solution	1368.42	25	62.5
3	150	30	DYS-150-10Mn	Mn solution	2736.84	25	62.5
4	150	30	DYS-150-Mn (Only)	Mn simple substance	52,631.58	25	62.5
5	150	30	DYS-150-10V	V solution	5263.16	25	62.5
6	150	30	DYS-150-10U	U solution	21.05	25	62.5
7	150	30	DYS-150-5Mo	Mo solution	105.26	25	62.5
8	200	90	SL-0-200	–	–	25	62.5
9	200	90	SL-CH-200	Mn, Mo, Zn, Cd, Ni	Mn, Zn, Ni (2105.26) Mo, Cd (210.53)	25	62.5

a combination of transition group and radioactive metal elements, with the aim of determining the effect of different metal elements on the transformation of organic matter into oil and gas.

12.4.1.3 Experimental Process

Metal Element Load

(1) According to the experimental scheme, an organic nitrate solution and smectite mineral powder are thoroughly and evenly mixed.
(2) Heat the mixture at 200 °C under vacuum for 6 h to remove excess water and activate the metal elements.
(3) After cooling to room temperature, use a grinder to grind the smectite minerals to 60 mesh to complete the loading of metal elements.

Sample Preparation of Source Rock

(1) Add 5 g of kerogen to a grinder and grind it thoroughly for 15 min.

(2) Add 45 g of clay minerals (smectite) loaded with metal elements in three portions, and grind for 5 min after each addition to ensure full contact between the organic matter and the clay minerals.
(3) Add 9 g of feldspar, 13.5 g of carbonate, and 22.5 g of quartz in sequence, and fully grind and mix the minerals after each addition.

Formation Water Treatment

(1) Remove components dissolved in organic solvents by chloroform extraction and fractionation.
(2) Add 50 mL of 30% mass fraction H_2O_2 solution into each liter of formation water, place an ultraviolet lamp in the formation water, and remove the organic components in the formation water by catalytic oxidation of organic substances.
(3) Heat the formation water solution to 100 °C and boil for 5 min to remove the residual H_2O_2.

Thermal Simulation Experiment of Heating and Pressurizing

(1) Leakage test: Place the source rock sample in the sample chamber and install the chamber in the hydrocarbon generation reactor. Start the oil pump to apply sealing pressure, vacuum the instrument pipeline and hydrocarbon generation reactor, and fill the system with 10 MPa inert gas. Let it stand for a 30 min test. If the system is well sealed, release the gas, vacuum it with a vacuum pump, then fill it with inert gas. Repeat this process 3–5 times and finally vacuum the system. If the system is not well sealed, it is necessary to identify the reason or reload the sample.
(2) Water injection: Fill the sample space of the source rock with 40 MPa high-pressure formation water using a high-pressure pump. This will completely fill the pore space of the source rock sample. During the water absorption process, the fluid pressure will continuously drop due to the pressed core sample. When the system fluid pressure no longer drops, it means that the pores have been filled with formation water. After water injection, let it sit for 12 h and try to detect any leakage. Excess formation water should be expelled through the vent valve before the temperature is raised, and the fluid pressure before the temperature rise should be maintained at 2–3 MPa.
(3) Increase pressure and temperature: Start the oil pump to apply lithostatic pressure to the core sample until it reaches the set value. Start the temperature control interface and set the temperature increase rate and heating time according to the experimental scheme. Turn on the heating and ventilation functions. At the same time, start the lithostatic pressure fine-tuning and set the pressure to increase synchronously with the formation temperature increase. This will simulate the process of sedimentary basin sinking and burial depth.

Collection and Quantification of Hydrocarbon Generation Products

(1) Gas collection and quantification: After the experiment, open the product valve to collect the generated oil–gas–water mixture, separate the oil–water mixture and gas through the gas–liquid separation tank cooled by water circulation. The oil–water mixture is frozen in the gas–liquid separation tank, and the gas is measured for volume in the metering tube, then collected for analysis of its composition using gas chromatography.
(2) Collection and quantification of light oil: This includes the oil in the gap between the inner wall of the hydrocarbon generating kettle and the sample chamber, the space in the connecting pipeline, and the oil products frozen in the gas–liquid separation tank during gas extraction.
(3) Collection and quantification of heavy oil and solid residual samples: After weighing the simulated residual samples of source rocks, chloroform is used to extract bitumen "A", which mainly consists of heavy oil. The total oil includes the sum of light oil and heavy oil, and the total hydrocarbon includes the sum of total oil and hydrocarbon gas.

12.4.2 Analysis of Simulation Experiment Results of Low Temperature Catalytic Oil Generation

12.4.2.1 Effects of Different Metal Catalysts on Gaseous Products

The gaseous products produced by low-temperature catalytic experiments consist mainly of C_1–C_5 hydrocarbons, as well as non-hydrocarbon products such as CO_2 and H_2. The addition of metal catalysts has a significant impact on the yields of different gaseous products. In the 30 day low-temperature experiment at 150 °C, the metal catalyst exhibited an inhibitory effect on the yield of hydrocarbon gas, resulting in varying degrees of reduction after the addition of metal elements such as Mn, Mo, and U (Fig. 12.27). The inhibitory effects of different elements on hydrocarbon gas are notably distinct, with U, V and Mo exhibiting stronger inhibitory effects than Mn. In particular, V can reduce the yield of hydrocarbon gas by 90%.

Different concentrations of the same solution can affect hydrocarbon gas yield differently. In the experiment, the hydrocarbon gas yield was higher for solution DYS-150-10Mn, which had twice the concentration of Mn compared to DYS-150-5Mn.

In the 90 day experiment at 200 °C, the addition of various transition metal elements such as Mn, Mo, U and Th resulted in a decrease of about 34% in the experimental hydrocarbon gas yield, indicating a certain inhibitory effect (Table 12.8).

The change in light hydrocarbon gas yield in the low-temperature experiment with metal elements was primarily reflected in the CH_4 yield, which showed an inhibitory

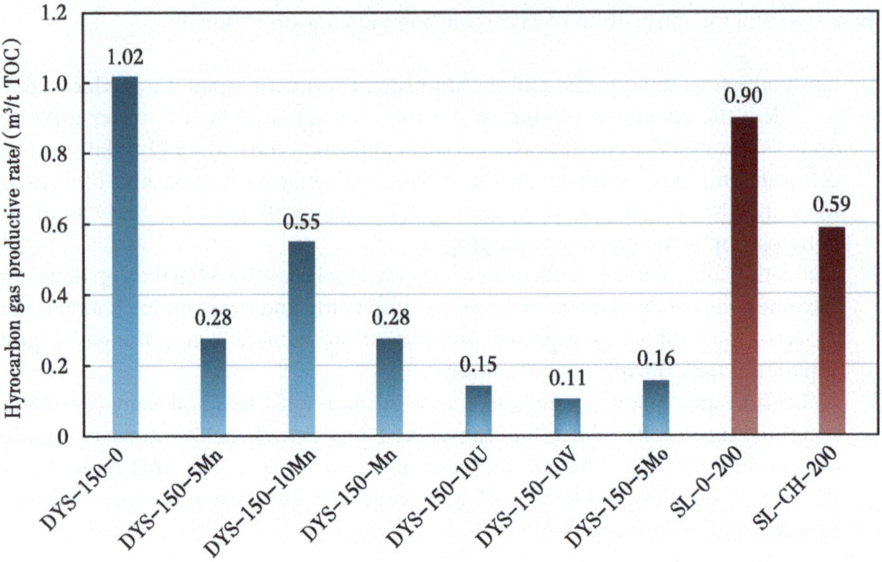

Fig. 12.27 Effect of different metal element catalysts on hydrocarbon gas generation rate

effect. The CH_4 yield decreased by 69–88%, with the V solution loading experiment showing the greatest reduction (Fig. 12.28).

The addition of metal elements has a significant impact on the yield of non-hydrocarbon gas, with notable differences in non-hydrocarbon gas yield after adding different metal elements. Excessive Mn has a strong promoting effect on H_2 generation, and the yield of H_2 is 2342 times higher than that of the experiment without adding metal elements. The U solution also has a strong stimulating effect on H_2 generation, while other metal elements and their combinations have little influence on H_2 generation (Fig. 12.29a). Whether the large amount of hydrogen produced during the experiment can provide a hydrogen source for oil generation on the scale of millions of years subsurface is worthy of in-depth consideration and study. The loading experiment of U solution and V solution has the most significant stimulating effect on CO_2 generation, and the CO_2 yield is 13 times and 7 times higher than that without metal elements (Fig. 12.29b). The yields of H_2 and CO_2 in the experimental products with the combination of transition metal elements decreased to varying degrees.

Comparing the low-temperature experiment with data from low-temperature catalytic gas experiments published in literature (Fig. 12.30), it is found that the yield of hydrocarbon gas and methane loaded with metal elements as catalysts are similar to those of previous experiments, and both are smaller than those of experiments without metal elements. This further confirms the inhibitory effect of metal elements on hydrocarbon gas generation at low temperature. The comparative experimental results also show that higher temperatures and longer heating times are beneficial to the generation of natural gas, but when the temperature is lower than 200 °C, the

Table 12.8 Yield of gaseous products in low temperature catalytic simulation experiments

Sample number	Temperature/°C	Lithology	CO_2 generation rate/ m^3/tTOC	H_2 generation rate/ m^3/tTOC	Hydrocarbon gas generation rate/m^3/tTOC	CH_4/m^3/tTOC	C_2H_6/m^3/tTOC	C_3H_8/m^3/tTOC
DYS-150-0	150	Mudstone	3.19	0.13	1.02	0.91	0.11	0
DYS-150-5Mn	150	Mudstone	1.59	0.09	0.28	0.19	0.09	0
DYS-150-10Mn	150	Mudstone	5.83	0	0.56	0.28	0.14	0.14
DYS-150-Mn	150	Mudstone	0.42	304.46	0.28	0.24	0.04	0
DYS-150-10U	150	Mudstone	8.74	72.13	0.15	0.15	0	0
DYS-150-10V	150	Mudstone	42.7	0.22	0.11	0.11	0	0
DYS-150-5Mo	150	Mudstone	20.83	0.31	0.16	0.16	0	0
SL-0-200	200	Mudstone	69.83	0.11	0.9	0.84	0.04	0.02
SL-CH-200	200	Mudstone	28.11	0.05	0.59	0.56	0.03	0

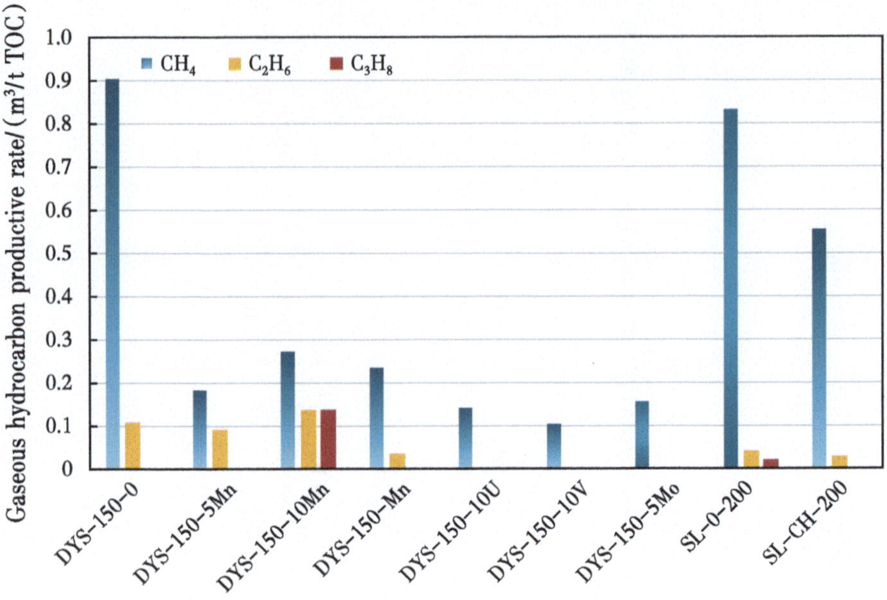

Fig. 12.28 Effects of different metal element catalysts on the yield of different gaseous hydrocarbons

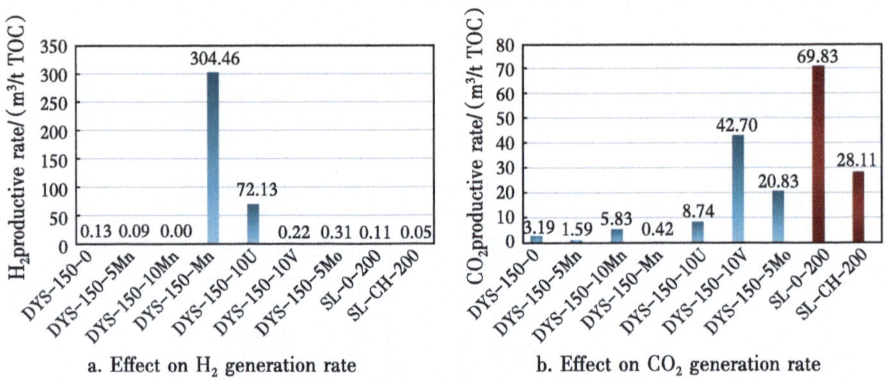

Fig. 12.29 Effect of different metal element catalysts on non-hydrocarbon generation rates

temperature has a weak effect on the yield of hydrocarbon gas. In the low-temperature catalytic experiments at 60, 80, 100, 120, and 150 °C, some hydrocarbon gas yields are still similar. The generation of hydrocarbon gas at low temperature is complicated, and there is no simple linear relationship. Even when subjected to similar simulation conditions, the yield of hydrocarbon gas exhibits fluctuations within a specific range. Through comparing the variations in CH_4 yield across different simulation experiments, it is observed that the influence of catalysis on hydrocarbon gas primarily

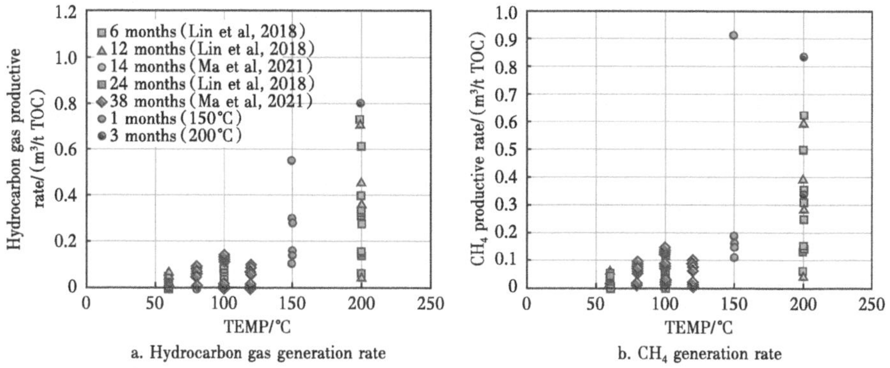

Fig. 12.30 Oil and gas generation rate in different simulation experiments (after Wei Lin 2018)

manifests through changes in CH_4 yield. These changes in CH_4 yield correspond closely with variations in the overall yield of hydrocarbon gas (Fig. 12.30b).

12.4.2.2 Influence of Different Metal Elements on the Formation of Liquid Hydrocarbons

Table 12.9 shows the results of the low-temperature simulation experiment on oil generation. The experiments loaded with metal element solutions all exhibit a stimulating effect on oil generation, with varying degrees of influence from different metal elements. During the 30 day experiment at 150 °C, the oil yield of the experiment without metal elements was 3.98 kg/tTOC, and the oil yield of the experiment with metal solutions showed improvement. Mo had the most notable stimulating effect on oil generation, with an oil yield of 25.15 kg/tTOC, which represents an approximately five-fold increase. Previous low-temperature experimental studies have shown that higher temperatures, longer heating times, and lower hydrostatic pressure are conducive to catalytic oil and gas generation, whereas low temperatures and shorter heating times are not. The oil yield of the experiment with Mo added at 150 °C for 30 days was higher than that of the experiment without metal added at 200 °C for 90 days. This indicates that the metal element catalyst partially compensates for the lower oil yield caused by temperature and time.

The catalytic effect of metal elements on oil generation is influenced by their state and concentration. For a given element, higher concentrations of Mn lead to a stronger promotion of oil generation. The experimental yield of oil generation for DSY-150-5Mn with lower Mn concentration is 18.19 kg/tTOC, which is higher than that of DSY-150-10Mn. The oil generation rate does not increase, but instead decreases when Mn is added, indicating an inhibitory effect on hydrocarbon generation. Experiments with various combinations of metal elements also showed a promoting effect on oil generation. In the 90 day experiment at 200 °C, the oil generation rate in experiments with added metal elements was significantly higher (Fig. 12.31).

Table 12.9 Oil generation in low temperature catalytic simulation experiment

Sample number	TEMP/°C	Lithology	Hydrocarbon gas generation rate/m^3/tTOC	Total oil weight/mg	Total oil yield/kg/tTOC	Total hydrocarbon generation rate/kg/tTOC
DYS-150-0	150	Mudstone	0.75	12.44	3.98	4.73
DYS-150-5Mn	150	Mudstone	0.26	56.86	18.19	18.45
DYS-150-10Mn	150	Mudstone	0.66	16.89	5.4	6.06
DYS-150-Mn	150	Mudstone	0.22	11.17	3.57	3.79
DYS-150-10U	150	Mudstone	0.1	19.27	6.17	6.27
DYS-150-10V	150	Mudstone	0.08	15.43	4.94	5.02
DYS-150-5Mo	150	Mudstone	0.11	78.61	25.15	25.26
SL-0-200	200	Mudstone	0.64	113.73	22.71	23.35
SL-CH-200	200	Mudstone	0.44	210.77	42.09	42.53

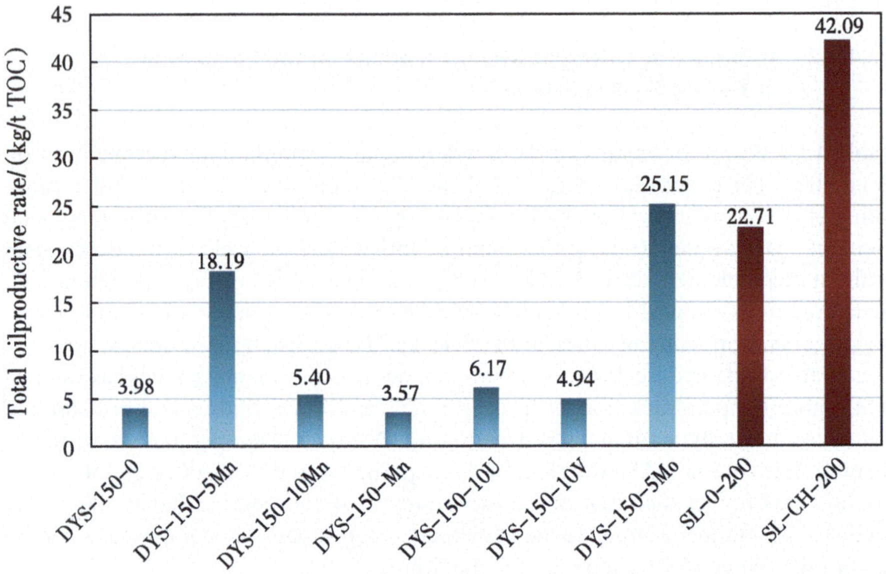

Fig. 12.31 Comparison of crude oil yield in low temperature catalytic simulation experiments

Saturated hydrocarbons are stable components in crude oil, and their composition characteristics are closely related to the properties of crude oil's original organic matter. They also show regular changes with the maturity and evolution of organic matter. Therefore, studying their compositions using gas chromatography can help us understand the nature of organic matter and the mature evolution characteristics of organic matter.

The Pr/Ph ratio increases with depth in both homogeneous sedimentary sequences and heterogeneous strata, indicating that maturity has an influence on the Pr/Ph ratio. Based on this, it is proposed that the higher the maturity, the higher the Pr/Ph ratio, while the Ph/n-C_{18} ratio decreases due to the generation and release of hydrocarbons. In this low-temperature catalytic experiment, the Pr/Ph values of the experiments loaded with metal elements all increased to some extent, and the Ph/n-C_{18} values all decreased. This indicates that the addition of metal elements led to a certain enhancement of the evolution of source rocks, that is, metal elements promoted hydrocarbon generation to some extent (Table 12.10 and Fig. 12.32).

Table 12.10 Organic geochemical characteristics of low temperature catalytic simulation experiment products

Sample number	Temperature/°C	Lithology	Main carbon peak	Odd–even predominance of n-alkane (OEP)	Pr/n-C_{17}	Ph/n-C_{18}	Pr/Ph
DYS-150-0	150	Mudstone	C_{27}	1.442	0.379	0.585	0.289
DYS-150-5Mn	150	Mudstone	C_{27}	1.096	0.276	0.27	0.473
DYS-150-10Mn	150	Mudstone	C_{27}	1.695	0.365	0.437	0.465
DYS-150-Mn	150	Mudstone	C_{27}	1.6	0.376	0.463	0.511
DYS-150-10U	150	Mudstone	C_{27}	1.317	0.327	0.333	0.517
DYS-150-10V	150	Mudstone	C_{27}	1.295	0.327	0.312	0.566

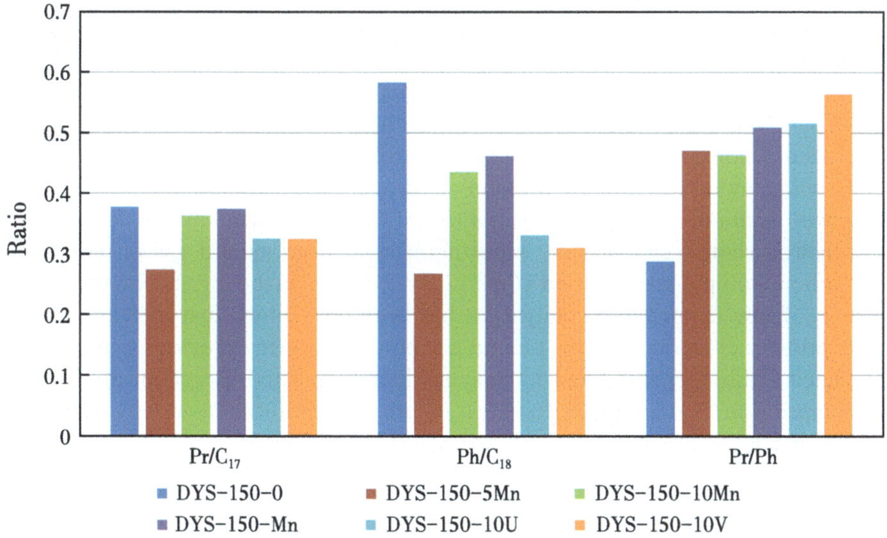

Fig. 12.32 Changes of geochemical parameters in low-temperature catalytic simulation experiment products

However, in high-temperature simulation experiments (>400 °C), the cracking of liquid hydrocarbons in the sample generates more low-carbon hydrocarbons, which leads to an increase in the Pr/n-C_{17} and Ph/n-C_{18} ratios. This is contrary to the low-temperature experiment (Liu et al. 2006). Their study also demonstrated that the reaction pathway differs under high-temperature conditions compared to low-temperature conditions, deviating to some extent from the actual process of oil and gas generation in source rocks.

12.4.2.3 Discussion on Low Temperature Catalytic Hydrocarbon Generation

The effect of temperature may be amplified in high-temperature experiments, which can obscure, amplify, or alter the effects of catalysts. However, previous low-temperature catalytic experiments had complicated conditions and lacked quantitative research on the catalytic gas generation effects of metal elements. This low-temperature catalytic experiment involves combining metal elements with clay minerals for quantitative study, which further confirms that metal elements have catalytic effects on hydrocarbon generation under geological low-temperature conditions. The most direct evidence is that the oil yield of the low-temperature catalytic experiment using transition metal elements and radioactive metal elements as catalysts has generally been greatly improved, and the catalytic effects of different metal elements vary, among which the catalytic effect of Mo solution is the most obvious.

Although the addition of catalysts significantly inhibits the generation of hydrocarbon gas, the composition characteristics of its products confirm the presence of catalysis. Research shows that methane is the dominant component of all forms of natural gas (50–100% in C_1–C_4; Fig. 12.33). Currently, laboratory pyrolysis of organic matter is unable to replicate the same methane concentration found in natural gas. In pyrolysis experiments, methane only accounts for 30–50% of C_1–C_4 products in kerogen decomposition and 10–40% in petroleum decomposition (Fig. 12.34a). Under mild conditions (190 °C), n-olefins, hydrogen, and carbonaceous sedimentary rocks with medium and high concentrations of transition metals will react, and the light hydrocarbons generated are indistinguishable from natural gas in molecular and carbon isotopic composition (Fig. 12.34b). Upon analyzing the experimental findings of low-temperature catalysis, it has been determined that methane serves as the primary component in the natural gas produced during previous experiments conducted within the project and documented literature. This observation aligns with the composition of natural gas found in nature and conforms to the hydrocarbon generation characteristics of the catalyst used (Fig. 12.34c–d).

The experimental results of low-temperature catalytic gas from shale demonstrate that methane, ethane, and propane in natural gas and low-temperature gas are in or near thermodynamic equilibrium. The products of natural gas from organic matter pyrolysis are primarily controlled by kinetics, and their products are out of thermodynamic equilibrium. The natural gas produced by thermal pyrolysis cannot generate these hydrocarbons in equilibrium, nor can it make them reach equilibrium within

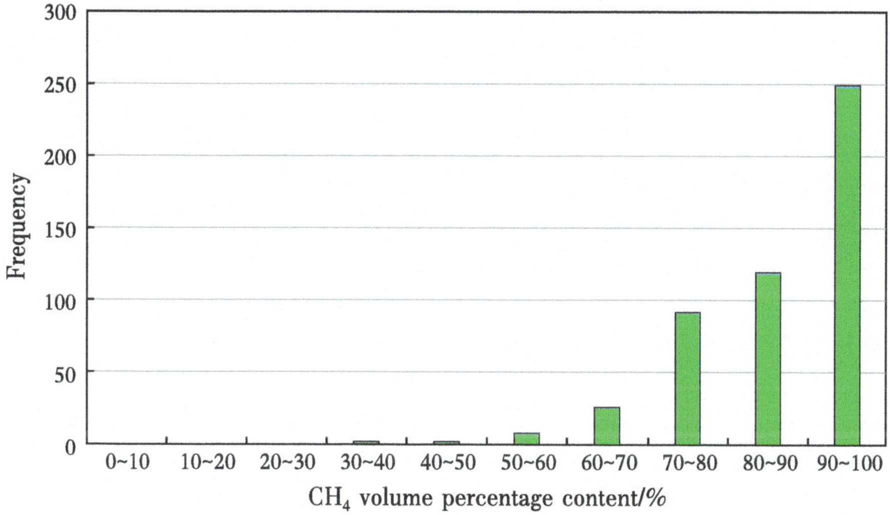

Fig. 12.33 Proportion of methane content in natural gas

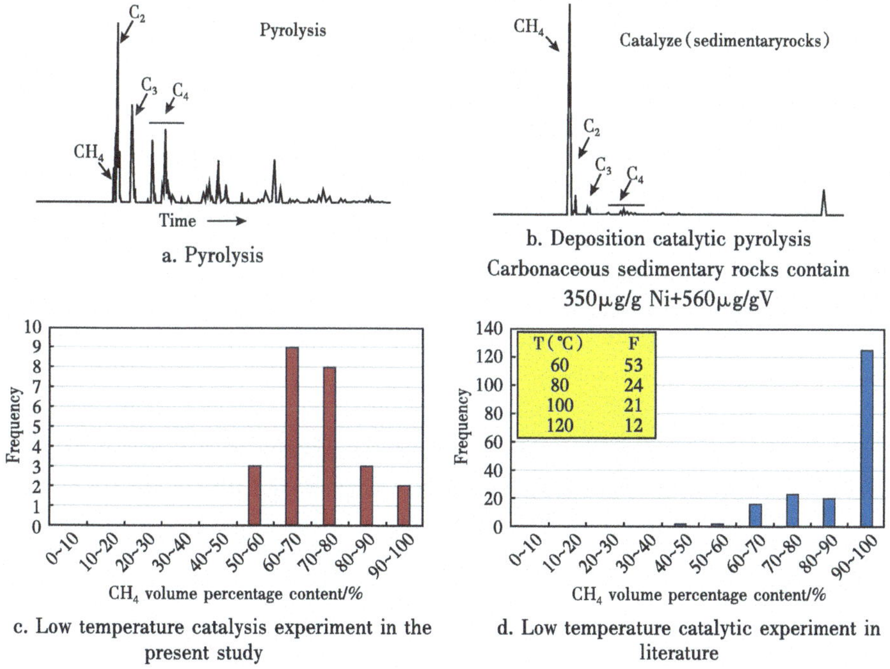

Fig. 12.34 Natural gas generation in different experiments

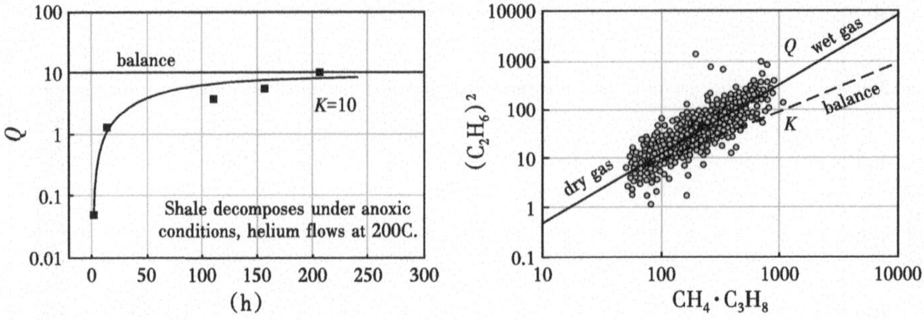

Fig. 12.35 Variation of gas composition with time and thermodynamic equilibrium

geological time. In contrast, catalytic reactions are mainly controlled by thermodynamics, and hydrocarbons can reach equilibrium through the double decomposition of hydrocarbon conversion, resulting in products that are close to thermodynamic equilibrium.

$$2C_nH_m \rightleftharpoons C_{n-1}H_{m-2} + C_{n+1}H_{m+2} \tag{12.10}$$

$$2C_2H_6 \rightleftharpoons CH_4 + C_3H_8 \tag{12.11}$$

$$\text{The equilibrium equation: } K(Q) = \left[(CH_4) \cdot \left(C_3H_g\right) / \left(C_2H_6\right)^2\right] \tag{12.12}$$

In the case of catalytic reactions, $(C_2)^2$–C_1–C_3 should exhibit a stable correlation, meaning that the equilibrium equation (12.12) will eventually tend towards stability (Fig. 12.35). The natural gas produced in this low-temperature catalytic experiment showed a clear positive correlation between $(C_2)^2$–C_1–C_3 and temperature, indicating that the generation of C_1–C_3 products was highly stable and mainly controlled by thermodynamics, which is consistent with the assumed catalytic reaction (Fig. 12.36).

12.4.3 Summary of Simulation Experiment of Low Temperature Catalytic Oil Generation

(1) The "clay mineral–metal element geological molecular sieve" promotes the evolution of oil and gas generation at low temperatures, and the Pr/Ph value representing maturity in the saturated hydrocarbon of the experimental crude oil loaded with metal elements increases to some extent. This directly affects the diagenesis and hydrocarbon generation reaction processes, as well as the properties of oil and gas products due to the differences in types, contents, and occurrence states of metal elements.

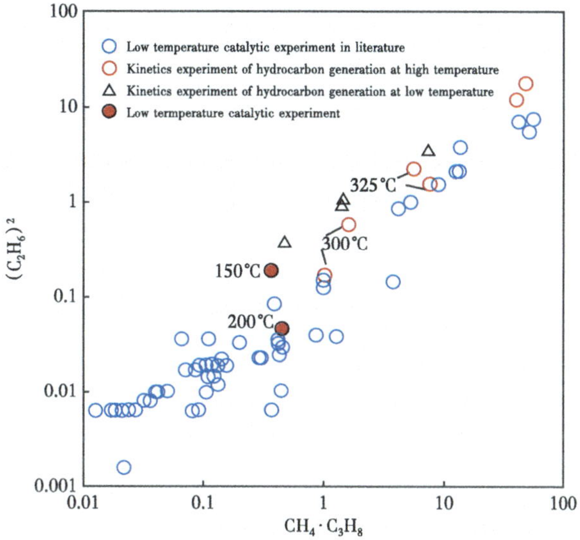

Fig. 12.36 The relationship of natural gas $(C_2)^2$–$C_1{\cdot}C_3$ generated in this experiment

(2) After adding a single catalytic substance loaded in a solution at 150 °C for 30 days, it can promote the oil generation of source rocks to some extent, and the Mo element has the most obvious stimulating effect on oil generation, increasing oil generation by about 5 times. However, the higher the concentration, the stronger the stimulating effect on oil generation. The experimental oil generation yield of DSY-150-5Mn with a lower Mn solution concentration is 18.19 kg/tTOC, which is higher than that of DSY-150-10Mn. The experimental oil generation rate with added Mn did not increase but decreased, showing an inhibitory effect.

(3) The "clay mineral–metal element geological molecular sieve" inhibits the generation of hydrocarbon gas at low temperatures, and the inhibition effects of different elements on hydrocarbon gas are obviously different. The experiment involving the addition of U, V, and Mo at 150 °C for 30 days demonstrated a more pronounced inhibitory effect on hydrocarbon gas compared to Mn. Specifically, the presence of element V resulted in a significant reduction in the yield of hydrocarbon gas by 90%. In the experiment at 200 °C and 90 days, the yield of experimental hydrocarbon gas decreased by about 34% when Mn, Mo, U, Th and other transition metal elements were added. Among them, the experiment of adding U solution and Mn simple substance has a strong promoting effect on H_2 generation, which may provide a hydrogen source for oil generation on the scale of millions of years subsurface.

(4) Although the addition of a catalyst inhibits the generation of hydrocarbon gas, the composition characteristics of its products confirm the existence of thermal catalysis. In the low-temperature catalytic experiment, methane dominates in all forms of natural gas (the mass fraction is 50–100% in C_1–C_4), which is close to the actual geological situation. In the previous high-temperature thermal

simulation experiment, methane only accounts for 30%–60% in all forms of natural gas.

12.4.4 Suggestions on Simulation Experiment of Low Temperature Catalytic Oil Generation

By employing the principles of organic geochemistry, an independent development of an oil generation simulation experiment instrument was conducted. This instrument facilitated a low-temperature (50–200 °C) catalytic oil generation simulation experiment under geological conditions. The experiment provided insights into the process and characteristics of pyrolysis oil generation in source rocks within geological settings. It also confirmed the potential influence of certain transition metal elements and radioactive metal elements in the oil generation process. However, it should be noted that the samples used in this study were prepared based on the current organic matter content, kerogen type, and overall rock composition of the primary source rocks found in various oil fields. As a result, the S2 in the samples could not undergo pyrolysis at low temperatures. Instead, a significant portion of the soluble organic matter that could be pyrolyzed had already transformed into soluble bitumen during geological history. Even samples with Ro < 0.5% had experienced substantial volatilization. Consequently, in the simulation experiment of low-temperature catalytic oil generation, the amount of soluble organic matter involved in the oil generation reaction was minimal. Thus, the simulation experiment primarily constituted the catalytic pyrolysis of a small quantity of soluble organic matter and the subsequent catalytic cracking of its pyrolysis products to produce hydrocarbon gas. This limitation resulted in the inability to accurately determine the true geological oil generation quantity, S1. Considering these limitations, it is recommended that future research endeavors focus on preparing samples rich in soluble organic matter either in the laboratory or by collecting samples from modern sediments abundant in original organic matter. These samples should be dried, compacted, and solidified under controlled conditions, utilizing the simulation experiment technology and instruments developed in this study. Conducting oil generation simulation experiments under geological conditions using these improved methods will provide a more accurate assessment of the actual oil generation potential of source rocks.

References

Defan G (2009) On basic research and theoretical innovation of petroleum geology. Sino-Global Energy 14(12):47–53

Defan G (2014) On marine oil generation and terrestrial oil generation. Sino-Global Energy 10:12

Defan G (2018) Development and experiment of oil generation simulation experimental instrument for source rocks. Sino-Global Energy 23(5):29–36

Defan G, Guoli W, Jingong Z et al (2004) Theoretical thinking of basin-forming hydrocarbon accumulation: from basin to oil and gas reservoir. Petroleum Industry Press, Beijing

Defan G, Xuhui X, Zhiming L et al (2008) Theoretical thinking of basin-forming hydrocarbon accumulation and hydrocarbon generation model in limited space. Oil Gas Geol 29(6):709–715

Defan G, Xuhui X, Zhiming L et al (2011) New progress in basic research on hydrocarbon generation and expulsion in limited space of source rocks. Pet Exp Geol 33(5):441–445

Feng JH, Fan TY, Jin ZY (1982) Analysis of deuterium in natural water by uranium metal method / / bulletin of the institute of mineral deposits. Chinese Acad Geol Sci 5(Geology Press):143–147

Frank D Mango (1992) Transition metal catalysis in the generation of petroleum: a genetic anomaly in Ordovician oils. Geochimica Et Cosmochimica Acta 56(10):3851–3854

Huimin L, Bingsong Y, Zhonghuai X et al (2018) Microfacies characteristics of organic-rich shale in continental lake basin and its indicative significance for shale oil enrichment-taking Jiyang Sub-basin in Bohai Bay Basin as an example. Acta Petrolei Sinica 39(12):1328–1343

Hunt JM (1979) Petroleum geochemistry and geology. W. H. Freeman and Company, San Francisco

Jianyi H, Difan H (1991) Theoretical basis of lacustrine petroleum geology in China. Petroleum Industry Press in Chinese

Kennedy MJ, Löhr SC, Fraser SA, Baruch ET (2014) Direct evidence for organic carbon preservation as clay-organic nanocomposites in a Devonian black shale; from deposition to diagenesis. Earth Planet Sci Lett 388:59–70

Li Y et al (2015) DRIFT spectroscopic study of diagenetic organic-clay interactions in argillaceous source rocks. Spectrochim Acta Part A Mol Biomol Spectrosc 148:138–145

Liu QY, Liu WH, Meng QX (2006) Organic geochemistry of N-alkanes from Tarim coal with differentmaterials in pyrolysis under closed system. Nat Gas Geosci 17(03):313–318

Lin W, Schimmelmann A, Maria M, et al (2018) Catalytic generation of methane at 60–100° C and 0.1–300 MPa from source rocks containing kerogen Types I, II, and III. Geochimica Et Cosmochimica Acta 231: 88–116

Mango FD, Jarvie D, Herriman E (2009) Natural gas at thermodynamic equilibrium Implications for the origin of natural gas. Geochem Transact 10(1):1–12

Mao GZ, Liu CY, Zhang DD, et al (2014) Experimental study on the role of uranium in the hydrocarbon generation and evolution of type III source rocks. Chinese Sci Earth Sci 44(08): 1740–1750.

Rahman HM, Kennedy M, Löhr S et al (2017) Clay-organic association as a control on hydrocarbon generation in shale. Organic Geochem 105:42–55

Seewald JS (2003) Organic–inorganic interactions in petroleum producing sedimentary basins. Nature 426(20):327–333

Taylor SR, Almond J, Arnott S, et al (2003) The brent field, Block 211/29, UK North Sea. Geol Soc London, Memoirs 20(1):233–250

Tissot BP, Welte DH (1982) Oil formation and distribution: A New Approach to Oil and Gas Exploratio. Hao S.H. et al. translated.Beijing: Petroleum Industry Press.

Wang DY (1986) Application and Protection of Uranium (238) in Catalysis. Modern Chem Ind 01):59–45.

Wang XX, Cai JG, Bao YJ (2006) Catalysis of Clay Mineral to Organic Matter in hydrocarbon Genesis. Mar Origin Petrol Geol 11(3):27–38

Zhiming L, Xiaolu Y, Ershe X et al (2010) Mineral composition characteristics and significance of effective source rocks in Dongying sag, Bohai Bay Basin. Pet Exp Geol 32(3):270–275

Zhiming L, Lunju Z, Zhongliang M et al (2011) Simulation of hydrocarbon generation and discharge in limited space of source rocks and its significance. Pet Exp Geol 33(5):447–451

Zhongliang M, Lunju Z, Zhiming L (2012) Simulation experimental study on hydrocarbon generation and expulsion controlled by temperature and pressure in limited space of source rocks. Acta Sedimenta Sinica 30(5):955–963

Zhu X, Cai J, Wang G et al (2018) Role of organo-clay composites in hydrocarbon generation of shale. Int J Coal Geol 192:83–90

Appendix
Theoretical Thinking on the Evaluation Method of Petroleum Resources in Lacustrine Oil-Bearing Basins

Oil reserves and resources play a crucial role in the economic development and national security of a country. As a result, both national governments and oil companies prioritize research efforts in this area. The objective is to identify and assess available resources ahead of time, providing a dependable foundation for future development planning. This research work holds significant value, as it allows for informed decision-making and ensures the sustainable utilization of oil resources for the benefit of the nation.

A. Three National Evaluation of Oil and Gas Resources in China

The evaluation of oil resources was pioneered by the United States, which conducted its first prediction and assessment in 1906. However, it wasn't until 1989 that a comprehensive evaluation and study of oil and gas resources took place in 13 oil-bearing areas (9 onshore and 4 offshore) over a span of two years. In contrast, China has undertaken three national evaluation studies on oil and natural gas resources since 1981. To gain a comprehensive overview of the advancements in this field, it would be advantageous to examine the progress made in all three research studies conducted in China. This review will provide valuable insights into China's exploration and assessment of its oil and natural gas resources, contributing to a broader understanding of the country's energy landscape.

A.1 The First National Oil and Gas Resources Evaluation

In 1981, the Science and Technology Department of China's Ministry of Petroleum Industry coordinated the participation of oilfields, universities, and research institutes in the country's inaugural national evaluation and research project on oil and natural

D. Guan et al., *Theory and Simulation Experimental Technology of Oil Generation From Nonmarine Source Rocks in China*, https://doi.org/10.1007/978-981-96-3145-2

gas resources. The primary objective of this research was to conduct a comprehensive study on petroleum geology, utilizing various methodologies such as statistical analysis, analogical reasoning, and bitumen "A" methods for evaluation. To illustrate the outcomes of this endeavor, this paper provides a concise overview of the research findings from Daqing Oilfield.

A.1.1 Research Ideas

The research begins by examining the formation and evolutionary processes of the Songliao Basin. Specifically, it focuses on identifying the "deep depression," which represents a long-term depressed area within the basin. The analysis aims to understand the correlation between the subsidence center and sedimentary center in this depression area, as well as the red-ox conditions of the depositional environment. Furthermore, investigations have been conducted to determine the formation conditions of lacustrine source rocks and to study the geochemical factors influencing the accumulation, preservation, and transformation of organic substances. As a result, the research enables the classification of the combination of source rocks, reservoirs, and cap rocks. Additionally, it explores the relationship between source rocks and oil and gas reservoirs and provides predictions regarding potential targets for oil and gas exploration in the area. By comprehensively studying these factors, a better understanding of the geological conditions and potential resources within the Songliao Basin is obtained.

A.1.2 Research Content

The research encompasses a wide range of topics related to the Songliao Basin. These include the formation, development, and structure of the basin itself. Furthermore, it explores the formation, development, and sedimentary characteristics specific to the depression area within the basin. The research also delves into the sedimentary environment characteristics of this depression area, including factors such as salinity, pH levels, reduction phase, and oxidation indexes of the ancient lake basin. Additionally, it investigates the paleoclimate of the lake basin and examines changes in the ancient water body within the basin. Moreover, the research focuses on understanding the geological and geochemical characteristics of source rocks. This includes studying their lithology, thickness, TOC content, as well as the geochemical aspects of organic matter transformation. The evaluation and division standards for source rocks are also addressed, as outlined in Table A.1 of the research. Furthermore, the research involves evaluating source rock segments and calculating oil generation. It also includes the division and evaluation of source-reservoir-cap assemblages, along with an analysis of the development characteristics of oil-generating areas. Finally, the study investigates the distribution of hydrocarbon reservoirs and offers predictions related to these reservoirs. By covering these diverse areas of investigation, the research provides a comprehensive understanding of the Songliao Basin, including its geological, sedimentary, and hydrocarbon-related aspects.

A.1.3 Evaluation of Source Rocks and Calculation of Oil Generation in Songliao Basin

By conducting a comprehensive analysis of various factors such as the organic matter richness in source rocks, the reducing degree of the sedimentary environment, the transformation conditions of organic matter, and the sedimentary conditions of source rocks, it has been determined that certain layers within different formations possess favorable conditions for oil generation.

The first member of the Qingshankou Formation is identified as having the best oil-generating conditions and the highest oil-generating potential, making it the most crucial layer for oil generation. Additionally, the second and third members of the Qingshankou Formation, along with the first member of the Nenjiang Formation, are also considered significant oil-generating layers. However, the second and third members of the Yaojia Formation exhibit limited oil-generating ability only within the middle portion of the formation, thus being categorized as favorable but less significant oil-generating layers. Recognizing that dispersed bitumen in source rocks represents a transitional product in the organic matter-to-petroleum conversion process, and that the oily component in bitumen "A" shares more similarities with petroleum found in asphalt, the following oil generation calculation formula is proposed:

$$Q = V\rho A_1 A_{\text{percentage of oil}} K \tag{A.1}$$

where: Q—oil generation, kg;

V— volume of source rock, m^3;

P—density of source rock, kg/m^3;

A_1—the content of bitumen "A" in the source rock;

$A_{\text{percentage of oil}}$—the content of oil components in bitumen "A";

K—the coefficient of oil-bearing components transforming into petroleum, and the lacustrine source rocks in the Songliao Basin are selected as 0.4. Calculation results of oil generation in each section:

The first member of Qingshankou Formation: 47.9×10^8t;
Member 2–3 of Qingshankou Formation: 16.3×10^8t;
Member 1 of Nenjiang Formation: 8.0×10^8t;
Member 2–3 of Yaojia Formation: 0.3×10^8t;
The total oil generation of source rocks in Songliao Basin is 72.5×10^8t.

Table A.1 Criteria for the classification of Lower Cretaceous source rocks in Songliao Basin (from internal report of the Songliao Basin, 1977)

Oil generation grade	Organic substance						Sedimentary conditions		
	Organic carbon	Fluorescent asphalt/dispersed asphalt					Reduced sulfur/%	K%	Geochemical facies
		Fluoresce bitumen/%	Percentage of oil/%	Resin and asphaltene/%	Bitumen "A"/%	Bitumen "A"/ alcohol benzene A			
Excellent source rock	>0.5	>0.15	>5.0	<85	>0.05	>0.8	>0.4	>0.4	Reducing-strong reducing
Good source rock	0.5–0.2	0.015–0.005	5.0–2.0	85–95	0.03–0.05	0.6–0.8	0.15–0.4	0.25–0.4	Weakly reducing subfacies
Poor source rock	<0.2	<0.005	<2.0	>95	<0.03	<0.6	<0.15	<0.25	Oxidation phase

Oil generation grade	Transformation condition			Geological sign		
	L1	L2	L3	Lithologic character	Sedimentary cycle	Sedimentary facies
Excellent source rock	>7	>20	>30	Dark gray mudstone, black shale	Mainly stable relative subsidence	Deep lake and deep lake facies
Good source rock	5–7	15–20	20–30	Dominated by ray mudstone, with some gray-green mudstone	The crust starts uplift.	Shallow lake facies
Poor source rock	<5	<15	<20	Purplish red, brownish red and grayish green mudstone	Mainly uplift or frequent oscillation	Lake beach, fluvial and partial shallow lake facies

Note The coefficient "K" represents the conversion factor for oil-based components into petroleum. "L1" refers to the initial bitumen degree, which is the ratio of carbon in the bitumen "A" within the rock to the residual organic carbon present in the rock. This ratio provides insight into the quantity and characteristics of organic matter that have undergone transformation into bitumen. "L2" represents the degree of bituminization for the secondary bitumen. It is calculated as the ratio of the carbon element within the total bitumen content in the rock to the residual organic carbon in the rock. This degree reflects the overall capacity of organic matter to be converted into bitumen. Similarly, "L3" denotes the degree of bituminization for the tertiary bitumen. It is determined by the ratio of the carbon content in bitumen "A" within the rocks to the total bitumen content. This degree indicates the abundance of neutral free bitumen within the overall bitumen composition

During the research phase, it has been established that lacustrine source rocks are primarily formed in continental basins that have developed over geological time. These basins have experienced prolonged and stable subsidence, resulting in the accumulation of extensive lacustrine strata. The presence of humid or semi-humid climates within these basins is conducive to the generation of oil and gas, as well as the formation of oil and gas fields. The accumulation of a substantial amount of organic matter in these lacustrine source rocks can be attributed to the prevalent humid climate. Additionally, the long-term depression of the Earth's crust in these areas further facilitates the generation of oil. The sedimentary environment found in freshwater lakes, characterized by reducing conditions, also contributes to the generation of oil. Based on the findings of this evaluation study, it has been determined that China's 143 sedimentary basins cover an area of approximately 469.63×10^4 km^2. Within these basins, there are significant oil resources estimated at approximately 744×10^8 metric tons, along with natural gas resources amounting to around 33×10^{12} cubic meters.

A.2 The Second National Evaluation of Oil and Gas Resources in Hina

In 1992, a collaborative effort between the China National Petroleum Corporation and China National Offshore Oil Corporation led to the second national evaluation and research on oil and natural gas resources. Basin simulation technology played a significant role in this evaluation and research process. The outcomes of this evaluation and research revealed that there were 150 sedimentary basins in China with a total sedimentary rock area estimated at approximately 430.47×10^4 km^2. The evaluation also indicated substantial oil resources, totaling around 940×108 metric tons, as well as significant natural gas resources amounting to approximately 38×10^{12} cubic meters.

A.3 A New Round of National Oil and Gas Resources Eva Uation in China

In 2003, a new round of national oil and gas evaluation and research was conducted through the collaborative efforts of the former Ministry of Land and Resources, the National Development and Reform Commission, and the Ministry of Finance. Various oil companies including CNPC, Sinopec, CNOOC, Yanchang Petroleum, as well as relevant universities and scientific research institutes, jointly undertook the responsibility for this evaluation and research project. Despite the national-level organization of this research endeavor, basin simulation technology remained the primary approach employed by each participating unit. The evaluation results indicated that the combined mineral rights area of 129 sedimentary basins in China was estimated to be approximately 330.28×10^4 km^2. It is important to note that this figure only considers the mineral rights area of 115 onshore sedimentary basins, excluding the 14 basins located in the southern part of the South China Sea. The prospective oil resources identified through the evaluation amounted to around 1287.11×10^8 metric tons, while prospective natural gas resources were estimated to be approximately 70.21×10^{12} cubic meters. Additionally, the petroleum geological resources were estimated at 895.10×10^8 metric tons, and the natural gas geological resources were estimated at around 43.87×10^{12} cubic meters.

A.4 Characteristics of Oil and Gas Resources Evaluation in China

China is adopting the American approach to oil and gas resource evaluation research, albeit with distinct research methods. In the American methodology, researchers primarily rely on existing exploration and production data to draw analogies or make extrapolations. They do not directly employ geochemical methods to calculate the potential quantities of oil generation, migration, accumulation, and preservation. This divergence in research methods can be attributed to several factors. Firstly, the underlying theory and mechanisms governing oil formation and subsurface migration are still not fully comprehended. As a result, certain parameters involved in the evaluation process, such as migration and accumulation coefficients, are susceptible to subjective elements, making accurate selection challenging. Secondly, the United States has drilled a considerable number of oil exploration and development wells, allowing them to accumulate a substantial volume of reliable data. This wealth of data serves as a foundation for their statistical extrapolations or analogies.

In contrast, China's approach to evaluation and research primarily revolves around the direct calculation methods of geochemistry, specifically employing the calculation method of oil generation in basin simulation technology. There are several reasons driving this methodology.

Firstly, the theory of oil generation through thermal degradation of kerogen, as introduced by Tissot and Welte during their lectures in China in 1978 and their monograph "Petroleum Formation and Occurrence: A New Approach to Oil and Gas Exploration," has been widely accepted and implemented in China. This theory has become the dominant approach, even within China's organic geochemistry community.

Secondly, the "mathematical model: a quantitative approach to the prospect evaluation of oil and natural gas" theory proposed by Tissot and Welte has paved the way for the development of basin simulation technology. This technology enables the quantitative calculation of oil generation from source rocks. Over time, this software has evolved into a three-dimensional visualization tool, enhancing its convenience for application in the evaluation and research of oil and natural gas resources.

Lastly, basin simulation technology and the oil generation calculation method have been mastered by all oil fields, as well as related teaching and scientific research institutions in China. This proficiency facilitates standardized management practices and establishes a unified data platform, allowing for the comparison of evaluation results across different basins.

The calculation software for basin simulation technology is based on the theory of thermal degradation of kerogen. This theory originated from the practical production of "man-made oil" by heating and retorting oil shale, which suggests that any rocks rich in kerogen can produce petroleum products through heating and retorting. As a result, the theory of kerogen thermal degradation oil generation became an important subject in organic chemistry research. Tissot and Welte conducted pyrolysis experiments on source rocks using a rock evaluation pyrolyzer and obtained similar pyrolysis parameters as those found in oil shale (namely S1, S2, S3, and Tmax). They concluded that "source rocks are essentially oil shale, and the kerogen contained within them is not significantly different. The process of pyrolysis, which generates (artificial) oil from oil shale, is very similar to the process of generating oil from deep-buried source rocks due to high temperatures. Therefore, any rock rich in type I or type II-1 organic matter can serve as a good source rock if it reaches sufficient burial depth, while shallowly buried rocks of this nature can be treated as oil shale." This understanding from the two scholars establishes a close connection between the concepts of organic geochemistry.

According to the principles of organic chemistry, pyrolysis of kerogen can generate (artificial) oil under controlled high-temperature conditions. This implies that all organic matter (S1 + S2) present in oil shale or source rocks can be pyrolyzed to produce oil (artificial oil). However, under subsurface geological conditions, S1 + S2 in oil shale cannot undergo pyrolysis to generate oil. This is evident from the fact that the Ro values (reflecting thermal maturity) of oil shale in various geological ages in China do not exceed 0.6%. Of the S1 + S2 in source rocks, only S1 can undergo pyrolysis and generate oil, while S2 remains as residual organic matter without undergoing pyrolysis. Experimental studies conducted by Tissot and Welte have confirmed that S2 can be completely pyrolyzed to produce oil only when the temperature exceeds 350 °C, reaching 500–550 °C. However, the subsurface source rocks in the oil-generating depression (or sag) do not experience temperatures above 50 °C. Therefore, it is clearly impossible for S2 to undergo pyrolysis and generate oil in this setting. This indicates that only S1 in the source rocks has the ability to pyrolyze and generate oil. Consequently, the oil generation calculation performed by basin simulation technology's oil generation software, considering both S1 and S2, does not represent actual oil generation in the subsurface oil-generating depression. After converting migration and accumulation coefficients, the calculated oil generation exhibits the peculiar phenomenon of "using non-existent subsurface oil generation to identify real oil resources."

In summary, the traditional method of evaluating source rocks based only on the content of organic carbon and the thinking of calculating S1 + S2 oil generation by basin simulation technology are incorrect. The real oil generation of source rocks is only the pyrolysis oil generation of S1. Thus, the core of petroleum resource evaluation is to study the oil generation of S1, the conditions of primary migration of source rocks, and the development characteristics of the sedimentary system of secondary migration.

B. Study on Oil Generation of Soluble Organic Matter S1 in Source Rocks

The lacustrine basins in China, particularly the deep depressions, are characterized by thick mudstone or shale deposits and a high abundance of organic matter. However, these basins are geographically distant from external provenances and contain few sandstone interlayers. Instead, turbidite sandstone is predominant. As a result, primary oil migration is not favorable, and the oil generated tends to become trapped in the mudstone or shale as droplets. To determine the original oil generation potential of source rocks, a method involves selecting cores from extensive mudstone or shale formations that lack sandstones or are located far away from sandstones. The volume parameters of these cores are determined, followed by crushing and extracting them using organic solvents like methanol or benzene. This process yields the S1 quantity, which represents the original oil generation of the source rocks.

In recent years, the Wuxi Institute of Petroleum Geology, part of the Research Institute of Petroleum Exploration and Development of SINOPEC, has employed a self-developed simulator for generation and expulsion of oil (hydrocarbon) from formation pores. They have systematically selected source rocks with a low maturity organic matter (Ro value approximately 0.5%) for pyrolysis experiments, obtaining S1 data. However, due to the challenges involved in sample selection, the obtained data do not entirely reflect the original oil generation potential of the main source rocks. Therefore, only qualitative evaluations of the source rocks are possible. For instance, the Wuxi Institute of Petroleum Geology has determined the S1 values of nine source rocks in North China and Northeast China, with S1 accounting for 6–14% of the total organic matter in the source rocks (refer to Table B.1).

The Quaternary argillaceous sediments contain a significant amount of soluble organic matter S1. These sediments, found in regions like the Yellow River Delta, the Yangtze River Delta, and various inland lake basins, are considered excellent samples for studying the potential original oil generation of source rocks such as mudstone or shale. It is crucial to conduct systematic and methodical basic research in this field. Through comprehensive basic research, we can acquire valuable insights into the evolution characteristics of the deposition and diagenetic processes of argillaceous sediments, with a particular focus on clay minerals and soluble organic matter. Additionally, we can explore the influence and catalytic properties of clay minerals on the pyrolysis process of organic matter. Ultimately, this research enables us to determine

Table B.1 Statistical results of S1 proportion of 9 source rock samples in North China and Northeast China

Serial number	Sample source	S1/(mg/g)	S2/(mg/g)	S1 + S2/(mg/g)	S1/(S1 + S2)/%
1	Huadian-3	6.47	96.68	103.15	6.27
2	Huadian-8	21.38	292.35	313.73	6.81
3	W161	11	136.96	147.96	7.43
4	Xinyang-1	8.95	110.59	119.54	7.49
5	Wang-24	20.13	210.89	231.02	8.71
6	Cha1	17.2	158.98	176.18	9.76
7	Well Bi215	9.12	81.03	90.15	10.12
8	Wei20	17.51	152.65	170.16	10.29
9	Pu1-154	4.3	27.58	31.88	13.49

the quantity of oil generated through the pyrolysis of soluble organic matter. These fundamental parameters are essential for accurately evaluating petroleum resources.

C. Study on the Conditions of Primary Migration and Secondary Migration of Source Rocks

The source rocks not only contain abundant S1 soluble organic matter, which serves as the material foundation for oil generation, but they also possess the necessary conditions for the primary and secondary migration of the generated oil. During the continuous subsidence period, lacustrine lake basins in China experienced significant development of external provenance systems. The gentle slopes of these basins predominantly developed river delta sedimentary systems, while the steep slopes were characterized by alluvial or alluvial fan deltas. In China's lacustrine oil-bearing basins, many oil fields are closely linked to the river delta sedimentary system. For example, the Daqing Oilfield is developed within a large-scale composite river delta sedimentary system. In the Bohai Bay Basin, sedimentary units such as the Jiyang Sub-basin, Liaohe Depression, and Huanghua Depression mainly exhibit medium-sized river delta sedimentary systems. Consequently, the major oil fields within the primary sedimentary depressions in the Bohai Bay Basin are associated with these systems. In the Ordos Basin, small river delta deposits are predominant, resulting in the development of small and medium-sized oil fields primarily in oil-generation depressions. As a result, several shale oil pools have formed in the basin.

The lacustrine lake basins in China have developed both large and small river delta sedimentary systems, which provide favorable conditions for the primary and secondary migration of oil from source rocks. The river delta can be divided into three parts: the delta plain belt, the delta front belt, and the pre-delta mud belt, progressing from the shore towards the center of the lake basin. Topographically, it transitions from high to low, and the sediment particles vary from coarse to fine. When the rate of sediment accumulation near the river mouth exceeds the settling speed (or the rising speed of the lake) in that area, the delta system gradually advances towards the center of the lake. In a vertical cross-section, it demonstrates the superposition of three layers, revealing the sedimentary characteristics of sand bodies during periods of water recession. Conversely, during periods of rising lake levels, there is the deposition of sand bodies during water invasion. These sand bodies are deposited along the river course and transformed into sheet-like formations by the flow of the lake, eventually being covered by mud from the lake environment. During the continuous subsidence stage of the entire lake basin, there are multiple instances of water invasion and water regression sedimentation, resulting in the vertical interbedding of mudstone and sandstone. Horizontally, the delta plain exhibits a higher proportion of sandstone compared to mudstone, with most of it being a mixture of sandstone and mudstone. As we move towards the delta front belt, the sandstone grains become finer and are thinly interbedded with mudstone. The pre-delta mud belt is dominated by dark mudstone mixed with sandstone, and it serves as the primary oil-generating area due to its transition towards the semi-deep lake region. These sedimentary characteristics create the material basis and conditions for oil generation and accumulation, contributing to the formation of source rocks, as well as the primary and secondary migration of oil after its generation. The specific performance is as follows:

(1) Within the pre-delta mud belt of the river delta sedimentary system, the presence of dark mudstone and thin sandstone formations contributes to the formation of source rocks and facilitates primary migration of oil.
(2) If these thin sandstones connect with the sand bodies in the delta front belt and the delta plain belt, oil entering these sandstones can migrate upwards, enabling secondary migration. Consequently, oil within the sandstone formations in the lake basin forms shale reservoirs, while oil within the sandstone outside the lake basin, along the shore, forms conventional reservoirs.
(3) In the case of lenticular thin sandstones, they can only form independent shale reservoirs within the lake basin.
(4) The river delta depositional systems of various sizes play a significant role in controlling the processes of source rock formation, oil generation, and reservoir development in the continental lake basins of China. The unusually well-developed sandstone within this depositional system allows for efficient accommodation of all the oil generated by the source rocks. Consequently, all the oil generated by the source rocks can be transported out, making it a valuable oil resource.
(5) Only the source rocks in the Da'anzhai Member of the Jurassic period in the Sichuan Basin, China, consist of dark mudstone with thin crust limestone. This indicates that the modes of oil generation and reservoir development in this area belong to independent lithologic shale reservoirs, without experiencing secondary migration of oil.
(6) The formation and development of lacustrine sedimentary basins in China are influenced by specific structural backgrounds and sedimentary conditions,

making them unique and incomparable. Each basin depression possesses its own characteristics. Therefore, accurate results can only be obtained through enhanced basic geological research, identification of petroleum geological characteristics in each exploration area, and collaboration with corresponding simulation experiments. Evaluation and research of petroleum resources in Chinese lacustrine sedimentary basins should be conducted by relevant oil fields engaged in exploration work, rather than through a national research approach with uniform norms and working methods.

The manufacturer's authorised representative in the EU is Springer Nature Customer Service Centre GmbH, Europaplatz 3, 69115 Heidelberg, Germany. If you have any concerns regarding our products, please contact ProductSafety@springernature.com

Printed and bound by CPI Group (UK) Ltd, Croydon, CR0 4YY
01/10/2026
02225936-0009